Power, Sex, Suicide
Mitochondria and the Meaning of Life

Power, Sex, Suicide

Mitochondria and the Meaning of Life

NICK LANE

OXFORD
UNIVERSITY PRESS

OXFORD
UNIVERSITY PRESS

Great Clarendon Street, Oxford OX2 6DP

Oxford University Press is a department of the University of Oxford.
It furthers the University's objective of excellence in research, scholarship,
and education by publishing worldwide in

Oxford New York

Auckland Cape Town Dar es Salaam Hong Kong Karachi
Kuala Lumpur Madrid Melbourne Mexico City Nairobi
New Delhi Shanghai Taipei Toronto

With offices in

Argentina Austria Brazil Chile Czech Republic France Greece
Guatemala Hungary Italy Japan Poland Portugal Singapore
South Korea Switzerland Thailand Turkey Ukraine Vietnam

Oxford is a registered trade mark of Oxford University Press
in the UK and in certain other countries

Published in the United States
by Oxford University Press Inc., New York

British Library Cataloguing in Publication Data

Data available

Library of Congress Cataloging in Publication Data

Data available

Typeset by Footnote Graphics Limited
Printed in Great Britain
on acid-free paper by
Clays Ltd., St. Ives plc

ISBN 0-19-280481-2   978-0-19-280481-5

1

30.00

For Ana
And for Eneko
Born, appropriately enough, in Part 6

Contents

List of Illustrations

Chapter heading illustrations © Ina Schuppe-Koistinen

The publishers apologize for any errors or omissions in the above list. If contacted they will be pleased to rectify these at the earliest opportunity.

Acknowledgements

Writing a book sometimes feels like a lonely journey into the infinite, but that is not for lack of support, at least not in my case. I am privileged to have received the help of numerous people, from academic specialists, whom I contacted out of the blue by email, to friends and family, who read chapters, or indeed the whole book, or helped sustain sanity at critical moments.

A number of specialists have read various chapters of the book and provided detailed comments and suggested revisions. Three in particular have read large parts of the manuscript, and their enthusiastic responses have kept me going through the more difficult times. Bill Martin, Professor of Botany at the Heinrich Heine University in Düsseldorf, has had some extraordinary insights into evolution that are matched only by his abounding enthusiasm. Talking with Bill is the scientific equivalent of being hit by a bus. I can only hope that I have done his ideas some justice. Frank Harold, emeritus Professor of Microbiology at Colorado State University, is a veteran of the Ox Phos wars. He was one of the first to grasp the full meaning and implications of Peter Mitchell's chemiosmotic hypothesis, and his own experimental and (beautifully) written contributions are well known in the field. I know of nobody who can match his insight into the spatial organization of the cell, and the limits of an overly genetic approach to biology. Last but not least, I want to thank John Hancock, Reader in Molecular Biology at the University of the West of England. John has a wonderfully wide-ranging, eclectic knowledge of biology, and his comments often took me by surprise. They made me rethink the workability of some of the ideas I put forward, and having done so to his satisfaction (I think) I am now more confident that mitochondria really do hold within them the meaning of life.

Other specialists have read chapters relating to their own field of expertise, and it is a pleasure to record my thanks. When ranging so widely over different fields, it is hard to be sure about one's grasp of significant detail, and without their generous response to my emails, nagging doubts would still beset me. As it is, I am hopeful that the looming questions reflect not just my own ignorance, but also that of whole fields, for they are the questions that drive a scientist's curiosity. In this regard, I want to thank: John Allen, Professor of Biochemistry, Queen Mary College, University of London; Gustavo Barja, Professor of Animal Physiology, Complutense University, Madrid; Albert Bennett, Professor of Evolutionary Physiology at the University of California, Irvine; Dr Neil Blackstone, Associate Professor of Evolutionary Biology at Northern Illinois University; Dr Martin Brand, MRC Dunn Human Nutrition Unit, Cambridge;

Dr Jim Cummins, Associate Professor of Anatomy, Murdoch University; Chris Leaver, Professor of Plant Sciences, Oxford University; Gottfried Schatz, Professor of Biochemistry, University of Basel; Aloysius Tielens, Professor of Biochemistry, University of Utrecht; Dr Jon Turney, Science Communication Group, Imperial College, London; Dr Tibor Vellai, Institute of Zoology, Fribourg University; and Alan Wright, Professor of Genetics, MRC Human Genetics Unit, Edinburgh University.

I am very grateful to Dr Michael Rodgers, formerly of OUP, who commissioned this book as one of his final acts before retiring. I am honoured that he retained an active interest in progress, and he cast his eagle eye over the first-draft manuscript, providing extremely helpful critical comments. The book is much improved as a result. In the same breath I must thank Latha Menon, Senior Commissioning Editor at OUP, who inherited the book from Michael, and invested it with her legendary enthusiasm and appreciation of detail as well as the larger picture. Many thanks too to Dr Mark Ridley at Oxford, author of *Mendel's Demon*, who read the entire manuscript and provided invaluable comments. I can't think of anyone better able to evaluate so many disparate aspects of evolutionary biology, with such a generous mind. I'm proud he found it a stimulating read.

A number of friends and family members have also read chapters and given me a good indication of what the general reader is prepared to tolerate. I want to thank in particular Allyson Jones, whose unfeigned enthusiasm and helpful comments have periodically sent my spirits soaring; Mike Carter, who has been friend enough to tell me frankly that some early drafts were too difficult (and that later ones were much better); Paul Asbury, who is full of thoughts and absorbing conversation, especially in wild corners of the country where talk is unconstrained; Ian Ambrose, always willing to listen and advise, especially over a pint; Dr John Emsley, full of guidance and inspiration; Professor Barry Fuller, best of colleagues, always ready to talk over ideas in the lab, the pub, or even the squash court; and my father, Tom Lane, who has read most of the book and been generous in his praise and gentle in pointing out my stylistic infelicities, while working to tight deadlines on his own books. My mother Jean and brother Max have been unstinting in their support, as indeed have my Spanish family, and I thank them all.

The frontispiece illustrations are by Dr Ina Schuppe-Koistinen, a researcher in biomedical sciences in Stockholm and noted watercolorist, who is making a name in scientific art. The series was specially commissioned for this book, and inspired by the themes of the chapters. I'm very grateful to her, as I think they bring to life the mystery of our microscopic universe, and give the book a unique flavour.

Special thanks to Ana, my wife, who has lived this book with me, through times best described as testing. She has been my constant sparring companion, bouncing ideas back and forth, contributing more than a few, and reading every word, well, more than once. She has been the ultimate arbiter of style, ideas, and meaning. My debt to her is beyond words.

Finally, a note to Eneko: he is antithetical to writing books, preferring to eat them, but is a bundle of joy, and an education in himself.

INTRODUCTION

Mitochondria
Clandestine Rulers of the World

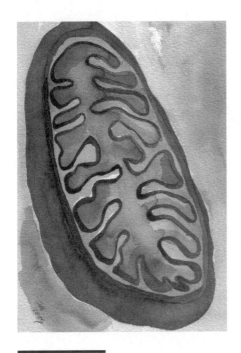

A mitochondrion—one of many tiny
power-houses within cells that control
our lives in surprising ways

Mitochondria are tiny organelles inside cells that generate almost all our energy in the form of ATP. On average there are 300–400 in every cell, giving ten million billion in the human body. Essentially all complex cells contain mitochondria. They look like bacteria, and appearances are not deceptive: they were once free-living bacteria, which adapted to life inside larger cells some two billion years ago. They retain a fragment of a genome as a badge of former independence. Their tortuous relations with their host cells have shaped the whole fabric of life, from energy, sex, and fertility, to cell suicide, ageing, and death.

Mitochondria are a badly kept secret. Many people have heard of them for one reason or another. In newspapers and some textbooks, they are summarily described as the 'powerhouses' of life—tiny power generators inside living cells that produce virtually all the energy we need to live. There are usually hundreds or thousands of them in a single cell, where they use oxygen to burn up food. They are so small that one billion of them would fit comfortably in a grain of sand. The evolution of mitochondria fitted life with a turbo-charged engine, revved up and ready for use at any time. All animals, the most slothful included, contain at least some mitochondria. Even sessile plants and algae use them to augment the quiet hum of solar energy in photosynthesis.

Some people are more familiar with the expression 'Mitochondrial Eve'—she was supposedly the most recent ancestor common to all the peoples living today, if we trace our genetic inheritance back up the maternal line, from child to mother, to maternal grandmother, and so on, back into the deep mists of time. Mitochondrial Eve, the mother of all mothers, is thought to have lived in Africa, perhaps 170 000 years ago, and is also known as 'African Eve'. We can trace our genetic ancestry in this way because all mitochondria have retained a small quota of their own genes, which are usually passed on to the next generation only in the egg cell, not in the sperm. This means that mitochondrial genes act like a female surname, which enables us to trace our ancestry down the female line in the same way that some families try to trace their descent down the male line from William the Conqueror, or Noah, or Mohammed. Recently, some of these tenets have been challenged, but by and large the theory stands. Of course, the technique not only gives an idea of our ancestry, but it also helps clarify who were *not* our ancestors. According to mitochondrial gene analysis, Neanderthal man *didn't* interbreed with modern *Homo sapiens*, but was driven to extinction at the margins of Europe.

Mitochondria have also made the headlines for their use in forensics, to establish the true identity of people or corpses, including several celebrated cases. Again, the technique draws on their small quota of genes. The identity of the last Russian Tzar, Nicholas II, was verified by comparing his mitochondrial genes with those of relatives. A 17-year-old girl rescued from a river in Berlin at the end of the First World War claimed to be the Tzar's lost daughter Anastasia, and was committed to a mental institution. After 70 years of dispute, her claim was finally disproved by mitochondrial analysis following her death in 1984.

More recently, the unrecognizable remains of many victims of the World Trade Center carnage were identified by means of their mitochondrial genes. Distinguishing the 'real' Saddam Hussein from one of his many doubles was achieved by the same technique. The reason that the mitochondrial genes are so useful relates partly to their abundance. Every mitochondrion contains 5 to 10 copies of its genes. Because there are usually hundreds of mitochondria in every cell, there are many thousands of copies of the same genes in each cell, whereas there are only two copies of the genes in the nucleus (the control centre of the cell). Accordingly, it is rare not to be able to extract any mitochondrial genes at all. Once extracted, the fact that all of us share the same mitochondrial genes with our mothers and maternal relatives means that it is usually possible to confirm or disprove postulated relationships.

Then there is the 'mitochondrial theory of ageing', which contends that ageing and many of the diseases that go with it are caused by reactive molecules called free radicals leaking from mitochondria during normal cellular respiration. The mitochondria are not completely 'spark-proof'. As they burn up food using oxygen, the free-radical sparks escape to damage adjacent structures, including the mitochondrial genes themselves, and more distant genes in the cell nucleus. The genes in our cells are attacked by free radicals as often as 10 000 to 100 000 times a day, practically an abuse every second. Much of the damage is put right without more ado, but occasional attacks cause irreversible mutations—enduring alterations in gene sequence—and these can build up over a lifetime. The more seriously compromised cells die, and the steady wastage underpins both ageing and degenerative diseases. Many cruel inherited conditions, too, are linked with mutations caused by free radicals attacking mitochondrial genes. These diseases often have bizarre inheritance patterns, and fluctuate in severity from generation to generation, but in general they all progress inexorably with age. Mitochondrial diseases typically affect metabolically active tissues such as the muscle and brain, producing seizures, some movement disorders, blindness, deafness, and muscular degeneration.

Mitochondria are familiar to others as a controversial fertility treatment, in which the mitochondria are taken from an egg cell (oocyte) of a healthy female donor, and transferred into the egg cell of an infertile woman—a technique known as 'ooplasmic transfer'. When it first hit the news, one British newspaper ran the story under the colourful heading 'Babies born with two mothers and one father'. This characteristically vivid product of the press is not totally wrong—while all the genes in the nucleus came from the 'real' mother, some of the mitochondrial genes came from the 'donor' mother, so the babies did indeed receive *some* genes from two different mothers. Despite the birth of more than 30 apparently healthy babies by this technique, both ethical and practical concerns later had it outlawed in Britain and the US.

Mitochondria even made it into a Star Wars movie, to the anger of some aficionados, as a spuriously scientific explanation of the famous force that may be with you. This was conceived as spiritual, if not religious, in the first films, but was explained as a product of 'midichlorians' in a later film. Midichlorians, said a helpful Jedi Knight, are 'microscopic life forms that reside in all living cells. We are symbionts with them, living together for mutual advantage. Without midichlorians, life could not exist and we would have no knowledge of the force.' The resemblance to mitochondria in both name and deed was unmistakeable, and intentional. Mitochondria, too, have a bacterial ancestry and live within our cells as symbionts (organisms that share a mutually beneficial association with other organisms). Like midichlorians, mitochondria have many mysterious properties, and can even form into branching networks, communicating among themselves. Lynn Margulis made this once-controversial thesis famous in the 1970s, and the bacterial ancestry of mitochondria is today accepted as fact by biologists.

All these aspects of mitochondria are familiar to many people through newspapers and popular culture. Other sides of mitochondria have become well known among scientists over the last decade or two, but are perhaps more esoteric for the wider public. One of the most important is apoptosis, or programmed cell death, in which individual cells commit suicide for the greater good—the body as a whole. From around the mid 1990s, researchers discovered that apoptosis is not governed by the genes in the nucleus, as had previously been assumed, but by the mitochondria. The implications are important in medical research, for the failure to commit apoptosis when called upon to do so is a root cause of cancer. Rather than targeting the genes in the nucleus, many researchers are now attempting to manipulate the mitochondria in some way. But the implications run deeper. In cancer, individual cells bid for freedom, casting off the shackles of responsibility to the organism as a whole. In terms of their early evolution, such shackles must have been hard to impose: why would potentially free-living cells accept a death penalty for the privilege of living in a larger community of cells, when they still retained the alternative of going off and living alone? Without programmed cell death, the bonds that bind cells in complex multicellular organisms might never have evolved. And because programmed cell death depends on mitochondria, it may be that multicellular organisms could not exist without mitochondria. Lest this sound fanciful, it is certainly true that all multicellular plants and animals *do* contain mitochondria.

Another field in which mitochondria figure very prominently today is the origin of the *eukaryotic* cell—those complex cells that have a nucleus, from which all plants, animals, algae, and fungi are constructed. The word *eukaryotic* derives from the Greek for 'true nucleus', which refers to the seat of the genes in the cell. But the name is frankly deficient. In fact, eukaryotic cells contain many other bits and pieces besides the nucleus, including, notably, the mitochondria. How these first complex cells evolved is a hot topic. Received wisdom says that they evolved step by step until one day a primitive eukaryotic cell engulfed a bacterium, which, after generations of being enslaved, finally became totally dependent and evolved into the mitochondria. The theory predicted that some of the obscure single-celled eukaryotes that *don't* possess mitochondria would turn out to be the ancestors of us all—they are relics from the days before the mitochondria had been 'captured' and put to use. But now, after a decade of careful genetic analysis, it looks as if all known eukaryotic cells either *have* or once *had* (and then lost) mitochondria. The implication is that the origin of complex cells is inseparable from the origin of the mitochondria: the two events were one and the same. If this is true, then not only did the evolution of multicellular organisms require mitochondria, but so too did the origin of their component eukaryotic cells. And if that's true, then life on earth would not have evolved beyond bacteria had it not been for the mitochondria.

Another more secretive aspect of mitochondria relates to the differences between the two sexes, indeed the requirement for two sexes at all. Sex is a well-known conundrum: reproduction by way of sex requires two parents to produce a single child, whereas clonal or parthenogenic reproduction requires just a mother; the father figure is not only redundant but a waste of space and resources. Worse, having two sexes means that we must seek our mate from just half the population, at least if we see sex as a means of procreation. Whether for procreation or not, it would be better if everybody was the same sex, or if there were an almost infinite number of sexes: two is the worst of all possible worlds. One answer to the riddle, put forward in the late 1970s and now broadly accepted by scientists, if relatively little known among the wider public, relates to the mitochondria. We need to have two sexes because one sex must specialize to pass on mitochondria in the egg cell, while the other must specialize *not* to pass on its mitochondria in the sperm. We'll see why in Chapter 6.

All these avenues of research place mitochondria back in a position they haven't enjoyed since their heyday in the 1950s, when it was first established that mitochondria are the seat of power in cells, generating almost all our energy. The top journal *Science* acknowledged as much in 1999, when it devoted its cover and a sizeable section of the journal to mitochondria under the heading 'Mitochondria Make A Comeback'. There had been two principal reasons for the neglect. One was that bioenergetics—the study of energy production in the mitochondria—was considered to be a difficult and obscure field, nicely summed up in the reassuring phrase once whispered around lecture theatres, 'Don't worry, nobody understands the mitochondriacs.' The second reason related to the ascendancy of molecular genetics in the second half of the twentieth century. As one noted mitochondriac, Immo Schaeffler, noted: 'Molecular biologists may have ignored mitochondria because they did not immediately recognize the far-reaching implications and applications of the discovery of the mitochondrial genes. It took time to accumulate a database of sufficient scope and content to address many challenging questions related to anthropology, biogenesis, disease, evolution, and more.'

I said that mitochondria are a badly kept secret. Despite their newfound celebrity, they remain an enigma. Many deep evolutionary questions are barely even posed, let alone discussed regularly in the journals; and the different fields that have grown up around mitochondria tend to be pragmatically isolated in their own expertise. For example, the mechanism by which mitochondria generate energy, by pumping protons across a membrane (chemiosmosis), is found in all forms of life, including the most primitive bacteria. It's a bizarre way of going about things. In the words of one commentator, 'Not since Darwin has biology come up with an idea as counterintuitive as those of, say, Einstein, Heisenberg or Schrödinger.' This idea, however, turned out to be true, and won Peter Mitchell a Nobel Prize in 1978. Yet the question is rarely posed: *Why* did such a peculiar means of generating energy become so central to so many different forms of life? The answer, we shall see, throws light on the origin of life itself.

Another fascinating question, rarely addressed, is the continued existence of mitochondrial genes. Learned articles trace our ancestry back to Mitochondrial Eve, and even use mitochondrial genes to piece together the relationships between different species, but seldom ask why they exist at all. They are just assumed to be a relic of bacterial ancestry. Perhaps. The trouble is that the mitochondrial genes can easily be transferred *en bloc* to the nucleus. Different species have transferred different genes to the nucleus, but *all* species with mitochondria have also retained exactly the same core contingent of mito-chondrial genes. What's so special about these genes? The best answer, we'll see, helps explain why bacteria never attained the complexity of the eukaryotes. It explains why life will probably get stuck in a bacterial rut elsewhere in the universe: why we might not be alone, but will almost certainly be lonely.

There are many other such questions, posed by perceptive thinkers in the specialist literature, but rarely troubling a wider audience. On the face of it, these questions seem almost laughably erudite—surely they would hardly exercise even the most pointy-headed boffins. Yet when posed together as a group, the answers impart a seamless account of the whole trajectory of evolution, from the origin of life itself, through the genesis of complex cells and multi-cellular organisms, to the attainment of larger size, sexes, warm-bloodedness, and into the decline of old age and death. The sweeping picture that emerges gives striking new insights into why we are here at all, whether we are alone in the universe, why we have our sense of individuality, why we should make love, where we trace our ancestral roots, why we must age and die—in short, into the meaning of life. The eloquent historian Felipe Fernández-Armesto wrote: 'Stories help explain themselves; if you know how something happened, you begin to see why it happened.' So too, the 'how' and the 'why' are intimately embraced when we reconstruct the story of life.

I have tried to write this book for a wide audience with little background in science or biology, but inevitably, in discussing the implications of very recent research, I have had to introduce a few technical terms, and assume a familiarity with basic cell biology. Even equipped with this vocabulary, some sections may still seem challenging. I believe it's worth the effort, for the fascination of science, and the thrill of dawning comprehension, comes from wrestling with the questions whose answers are unclear, yet touch upon the meaning of life. When dealing with events that happened in the remote past, perhaps billions of years ago, it is rarely possible to find definitive answers. Nonetheless, it is possible to use what we know, or think we know, to narrow down the list of possibilities. There are clues scattered throughout life, sometimes in the most unexpected places, and it is these clues that demand familiarity with modern molecular biology, hence the necessary intricacy of a few sections. The clues allow us to eliminate some possibilities, and focus on others, after the method of Sherlock Holmes. As Holmes put it: 'When you have eliminated the impossible, whatever remains, however improbable, must be the truth.' While it is dangerous to brandish terms like impossible at evolution, there is sleuthful satisfaction in reconstructing the most likely paths that life might have taken. I hope that something of my own excitement will transmit to you.

For quick reference I have given brief definitions of most technical terms in a glossary, but before continuing, it's perhaps valuable to give a flavour of cell biology for those who have no background in biology. The living cell is a minute universe, the simplest form of life capable of independent existence, and as such it is the basic unit of biology. Some organisms, like amoeba, or indeed bacteria, are simply single cells, or unicellular organisms. Other organisms are composed of numerous cells, in our own case millions of millions of them: we are multicellular organisms. The study of cells is known as *cytology*, from the Greek *cyto*, meaning cell (originally, hollow receptacle). Many terms incorporate the root cyto-, such as cytochromes (coloured proteins in the cell) and cytoplasm (the living matter of the cell, excluding the nucleus), or *cyte*, as in erythrocyte (red blood cell).

Not all cells are equal, and some are a lot more equal than others. The least equal are bacteria, the simplest of cells. Even when viewed down an electron microscope, bacteria yield few clues to their structure. They are tiny, rarely more than a few thousandths of a millimetre (microns) in diameter, and typically either spherical or rod-like in shape. They are sealed off from their external environment by a tough but permeable cell wall, and inside that, almost touching upon it, by a flimsy but relatively impermeable cell membrane, a few millionths of a millimetre (nanometres) thick. This membrane, so vanishingly thin, looms large in this book, for bacteria use it for generating their energy.

The inside of a bacterial cell, indeed any cell, is the cytoplasm, which is of gel-like consistency, and contains all kinds of biological molecules in solution or suspension. Some of these molecules can be made out, faintly, at the highest power magnification we can achieve, an amplification of a million-fold, giving the cytoplasm a coarse look, like a mole-infested field when viewed from the air. First among these molecules is the long, coiled wire of DNA, the stuff of genes, which tracks like the contorted earthworks of a delinquent mole. Its molecular structure, the famous double helix, was revealed by Watson and Crick more than half a century ago. Other ruggosities are large proteins, barely visible even at this magnification, and yet composed of millions of atoms, organized in such precise arrays that their exact molecular structure can be deciphered by the diffraction of X-rays. And that's it: there is little else to see, even though biochemical analysis shows that bacteria, the simplest of cells, are in fact so complex that we still have almost everything to learn about their invisible organization.

We ourselves are composed of a different type of cell, the most equal in our cellular farmyard. For a start they are much bigger, often a hundred thousand times the volume of a bacterium. You can see much more inside. There are great stacks of convoluted membranes, bristling with ruggosities; there are all kinds of vesicles, large and small, sealed off from the rest of the cytoplasm like freezer bags; and there is a dense, branching network of fibres that give structural support and elasticity to the cell, the cytoskeleton. Then there are the *organelles*—discrete organs within the cell that are dedicated to particular tasks, in the same way that a kidney is dedicated to filtration. But most of all, there is the nucleus, the brooding planet that dominates the little cellular universe. The planet of the nucleus is nearly as pockmarked with holes (in fact, tiny pores) as the moon. The possessors of such nuclei, the eukaryotes, are the most important cells in the world. Without them, our world would not exist, for all plants and animals, all algae and fungi, indeed essentially everything we can see with the naked eye, is composed of eukaryotic cells, each one harbouring its own nucleus.

The nucleus contains the DNA, forming the genes. This DNA is exactly the same in detailed molecular structure as that of bacteria, but it is very different in its large-scale organization. In bacteria, the DNA forms into a long and twisted loop. The contorted tracks of the delinquent mole finally close upon themselves to form a single circular chromosome. In eukaryotic cells, there are usually a number of different chromosomes, in humans 23, and these are linear, not circular. That is not to say that the chromosomes are stretched out in a straight line, but rather that each has two separate ends. Under normal working conditions, none of this can be made out down the microscope, but during cell division the chromosomes change their structure and condense into recognizable tubular shapes. Most eukaryotic cells keep two copies of each of their chromosomes—they are said to be diploid, giving humans a total of 46 chromo-. somes—and these pair up during cell division, remaining joined at the waist. This gives the chromosomes the simple star shapes that can be seen down the microscope. They are not composed only of DNA, but are coated in specialized proteins, the most important of which are called *histones*. This is an important difference with bacteria, for no bacteria coat their DNA with histones: their DNA is naked. The histones not only protect eukaryotic DNA from chemical attack, but also guard access to the genes.

When he discovered the structure of DNA, Francis Crick immediately understood how genetic inheritance works, announcing in the pub that evening that he understood the secret of life. DNA is a template, both for itself and for proteins. The two entwined strands of the double helix each act as a template for the other, so that when they are prized apart, during cell division, each strand provides the information necessary for reconstituting the full double helix, giving two identical copies. The information encoded in DNA spells out the molecular structure of proteins. This, said Crick, is the 'central dogma' of all biology: genes code for proteins. The long ticker tape of DNA is a seemingly endless sequence of just four molecular 'letters', just as all our words, all our books, are a sequence of only 26 letters. In DNA, the sequence of letters stipulates the structure of proteins. The *genome* is the full library of genes possessed by an organism, and may run to billions of letters. A gene is essentially the code for a single protein, which usually takes thousands of letters. Each protein is a string of subunits called *amino acids*, and the precise order of these dictates the functional properties of the protein. The sequence of letters in a gene specifies the sequence of amino acids in a protein. If the sequence of letters is changed —a 'mutation'—this may change the structure of the protein (but not always, as there is some redundancy, or technically degeneracy, in the code—several different combinations of letters can code for the same amino acid).

Proteins are the crowning glory of life. Their forms, and their functions, are almost endless, and the rich variety of life is almost entirely attributable to the rich variety of proteins. Proteins make possible all the physical attainments of life, from metabolism to movement, from flight to sight, from immunity to signalling. They fall into several broad groups, according to their function. Perhaps the most important group are the enzymes, which are biological catalysts that speed up the rate of biochemical reactions by many orders of magnitude, with an astonishing degree of selectivity for their raw materials.

Some enzymes can even distinguish between different forms of the same atom (isotopes). Other important groups of proteins include hormones and their receptors, immune proteins like antibodies, DNA-binding proteins like histones, and structural proteins, such as the fibres of the cytoskeleton.

The DNA code is inert, a vast repository of information housed out of the way in the nucleus, in the same way that valuable encyclopaedias are stored safely in libraries, rather than being consulted in factories. For daily use the cell relies on disposable photocopies. These are made of RNA, a molecule composed of similar building blocks to DNA, but spun-out in a single strand rather than the two strands of the double helix. There are several types of RNA, which fulfil distinct tasks. The first of these is messenger RNA, which equates in length, more or less, to a single gene. Like DNA, it, too, forms a string of letters, and their sequence is an exact replica of the gene sequence in the DNA. The gene sequence is *transcribed* into the slightly different calligraphy of messenger RNA, converted from one font into another without losing any meaning. This RNA is a winged messenger, and passes physically from the DNA in the nucleus, through the pores that pockmark its surface like the moon, and out into the cytoplasm. There it docks onto one of the many thousands of protein-building factories in the cytoplasm, the *ribosomes*. As molecular structures these are enormous; as visible entities they are miniscule. They can be seen studding some of the cell's internal membranes, giving them a rough impression on the electron microscope, and dotting through the cytoplasm. They are composed of a mixture of other types of RNA, and protein, and their job is to *translate* the message encoded in messenger RNA into the different language of proteins— the sequence of amino acids. The whole process of transcription and translation is controlled and regulated by numerous specialized proteins, the most important of which are called *transcription factors*. These regulate the expression of genes. When a gene is expressed, it is converted from the somnolent code into an active protein, with business about the cell or elsewhere.

Armed with this basic cell biology, let's now return to the mitochondria. They are organelles in the cell—one of the tiny organs dedicated to a specific task, in this case energy production. I mentioned that mitochondria were once bacteria, and in appearance they still look a bit like bacteria (Figure 1). Typically depicted as sausages or worms, they're able to take many twisted and contorted shapes, including corkscrews. They're usually of bacterial size, a few thousandths of a millimetre in length (1 to 4 microns), and perhaps half a micron in diameter. The cells that make up our bodies typically contain numerous mitochondria, the exact number depending on the metabolic demand of that particular cell. Metabolically active cells, such as those of the liver, kidneys, muscles, and brain, have hundreds or thousands of mitochondria, making up some 40 per cent of the cytoplasm. The egg cell, or oocyte, is exceptional: it

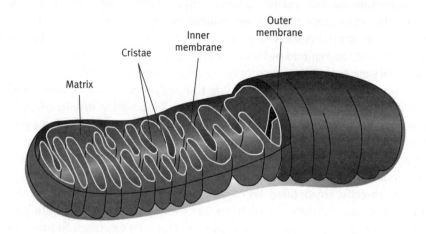

1 Schematic representation of a single mitochondrion, showing the outer and inner membranes; the inner membrane is convoluted into numerous folds known as cristae, which are the seat of respiration in the cell.

passes on around 100 000 mitochondria to the next generation. In contrast, blood cells and skin cells have very few, or none at all; sperm usually have fewer than 100. All in all, there are said to be 10 million billion mitochondria in an adult human, which together constitute about 10 per cent of our body weight.

Mitochondria are separated from the rest of the cell by two membranes, the outer being smooth and continuous, and the inner convoluted into extravagant folds or tubules, called *cristae*. Mitochondria don't lie still, but frequently move around the cell to the places they are needed, often quite vigorously. They divide in two like bacteria, with apparent independence, and even fuse together into great branching networks. Mitochondria were first detected using light microscopy, as granules, rods, and filaments in the cell, but their provenance was debated from the beginning. Among the first to recognize their importance was the German Richard Altmann, who argued that the tiny granules were in truth the fundamental particles of life, and accordingly named them *bioblasts* in 1886. For Altmann, the bioblasts were the only living components of the cell, which he held to be little more than a fortified community of bioblasts living together for mutual protection, like the people of an iron-age fortification. Other structures, such as the cell membrane and the nucleus, were constructed by the community of bioblasts for their own ends, while the cytosol (the watery part of cytoplasm), was just that: a reservoir of nutrients enclosed in the microscopic fortress.

Altmann's ideas never caught on, and he was ridiculed by some. Others claimed that bioblasts were a figment of his imagination—merely artefacts of his elaborate microscopic preparation. These disputes were aggravated by the fact that cytologists had become entranced by the stately dance of the chromosomes during cell division. To visualize this dance, the transparent components of the cell had to be coloured using a stain. As it happened, the stains that were best able to colour the chromosomes were acidic. Unfortunately, these stains tended to dissolve the mitochondria; their obsession with the nucleus meant that cytologists were simply dissolving the evidence. Other stains were ambivalent, colouring mitochondria only transiently, for the mitochondria themselves rendered the stain colourless. Their rather ghostly appearance and disappearance was scarcely conducive to firm belief. Finally Carl Benda demonstrated, in 1897, that mitochondria do have a corporeal existence in cells. He defined them as 'granules, rods, or filaments in the cytoplasm of nearly all cells . . . which are destroyed by acids or fat solvents.' His term, *mitochondria* (pronounced 'my-toe-con-dree-uh'), was derived from the Greek *mitos*, meaning thread, and *chondrin*, meaning small grain. Although his name alone stood the test of time, it was then but one among many. Mitochondria have revelled in more than thirty magnificently obscure names, including chondriosomes, chromidia, chondriokonts, eclectosomes, histomeres, microsomes, plastosomes, polioplasma, and vibrioden.

While the real existence of mitochondria was at last ceded, their function remained unknown. Few ascribed to them the elementary life-building properties claimed by Altmann; a more circumscribed role was sought. Some considered mitochondria to be the centre of protein or fat synthesis; others thought they were the residence of genes. In fact, the ghostly disappearance of mitochondrial stains finally gave the game away: the stains were rendered colourless because they had been *oxidized* by the mitochondria—a process analogous to the oxidation of food in cell respiration. Accordingly, in 1912, B. F. Kingbury proposed that mitochondria might be the respiratory centres of the cell. His suggestion was demonstrated to be correct only in 1949, when Eugene Kennedy and Albert Lehninger showed that the respiratory enzymes were indeed located in the mitochondria.

Though Altmann's ideas about bioblasts fell into disrepute, a number of other researchers also argued that mitochondria were independent entities related to bacteria, *symbionts* that lived in the cell for mutual advantage. A symbiont is a partner in a symbiosis, a relationship in which both partners benefit in some way from the presence of the other. The classic example is the Egyptian plover, which picks the teeth of Nile crocodiles, providing dental hygiene for the crocodile while gaining an easy lunch for itself. Similar mutual relationships can exist among cells such as bacteria, which sometimes live inside larger cells as *endosymbionts*. In the first decades of the twentieth century, virtually all parts of the cell were considered as possible endosymbionts, perhaps modified by their mutual coexistence, including the nucleus, the mitochondria, the chloroplasts (responsible for photosynthesis in plants), and the centrioles (the cell bodies that organize the cytoskeleton). All these theories were based on appearance and behaviour, like movement and apparently autonomous division, and so could never be more than suggestive. What's more, their protagonists were all too often divided by struggles over priority, by war and language, and rarely agreed among themselves. As the science historian Jan Sapp put it, in his fine book *Evolution by Association*: 'Thus unfolds an ironic tale of the fierce individualism of many personalities who pointed to the creative power of associations in evolutionary change.'

Matters came to a head after 1918, when the French scientist Paul Portier published his rhetorical masterpiece *Les Symbiotes*. He was nothing if not bold, claiming that: 'All living beings, all animals from Amoeba to Man, all plants from Cryptogams to Dicotyledons are constituted by an association, the *emboîtement* of two different beings. Each living cell contains in its protoplasm formations, which histologists designate by the name of mitochondria. These organelles are, for me, nothing other than symbiotic bacteria, which I call symbiotes.'

Portier's work attracted high praise and harsh criticism in France, though it was largely ignored in the English-speaking world. For the first time, however, the case did not stand on the morphological similarities between mitochondria and bacteria, but turned on attempts to cultivate mitochondria as a cell culture. Portier claimed to have done so, at least with 'proto-mitochondria', which he argued had not yet become fully adapted to their life inside cells. His findings were publicly contested by a panel of bacteriologists at the Pasteur Institute, who were unable to replicate them. And sadly, once he had secured his chair at the Sorbonne, Portier abandoned the field, and his work was quietly forgotten.

A few years later, in 1925, the American Ivan Wallin independently put forward his own ideas on the bacterial nature of mitochondria, claiming that such intimate symbioses were the driving force behind the origin of new species. His arguments again turned on culturing mitochondria, and he, too, believed that he had succeeded. But for a second time interest waned with the failure to replicate his work. This time symbiosis was not ruled out with quite the same venom, but the American cell biologist E. B. Wilson summed up the prevailing attitude in his famous remark: 'To many, no doubt, such speculations may appear too fantastic for present mention in polite biological society; nevertheless it is within the range of possibility that they may some day call for some serious consideration.'

That day turned out to be half a century later: aptly enough for the tale of an intimate symbiotic union, in the summer of love. In June 1967, Lynn Margulis submitted her famous paper to the *Journal of Theoretical Biology*, in which she resurrected the 'entertaining fantasies' of previous generations and cloaked them in newly scientific apparel. By then the case was much stronger: the existence of DNA and RNA in mitochondria had been proved, and examples of 'cytoplasmic heredity' catalogued (in which inherited traits were shown to be independent of the nuclear genes). Margulis was then married to the cosmologist Carl Sagan, and she took a similarly cosmic view of the evolution of life, considering not just the biology, but also the geological evidence of atmospheric evolution, and fossils of bacteria and early eukaryotes. She brought to the task a consummate discernment of microbial anatomy and chemistry, and applied systematic criteria to determine the likelihood of symbiosis. Even so, her work was rejected. Her seminal paper was turned down by 15 different journals before James Danielli, the far-seeing editor of the *Journal of Theoretical Biology*, finally accepted it. Once published, there were an unprecedented 800 reprint requests for the paper within a year. Her book, *The Origin of Eukaryotic Cells*, was rejected by Academic Press, despite having been written to contract, and was eventually published by Yale University Press in 1970. It was to become one of the most influential biological texts of the century. Margulis marshalled the evidence so convincingly that biologists now accept her once-heterodox view as fact, at least when applied to mitochondria and chloroplasts.

Bitter arguments persisted for well over a decade, and were arcane but vital. Without them, the final agreement would have been less secure. Everyone accepted that there are indeed parallels between mitochondria and bacteria, but not everyone agreed about what these really meant. Certainly the mitochondrial genes are bacterial in nature: they sit on a single circular chromosome (unlike the linear chromosomes of the nucleus) and are 'naked'—they're not wrapped up in histone proteins. Likewise, the transcription and translation of DNA into proteins is similar in bacteria and mitochondria. The physical assembly of proteins is also managed along similar lines, and differs in many details from standard eukaryotic practice. Mitochondria even have their own ribosomes, the protein-building factories, which are bacterial in appearance. Various antibiotics work by blocking protein assembly in bacteria, and also block protein synthesis in the mitochondria, but not from the nuclear genes in eukaryotes.

Taken together, these parallels might sound compelling, but in fact there are possible alternative interpretations, and it was these that underpinned the long dispute. In essence, the bacterial properties of mitochondria could be explained if the speed of evolution was slower in the mitochondria than in the nucleus. If so, then the mitochondria would have more in common with bacteria simply because they had not evolved as fast, and so as far. They would retain more atavistic traits. Because the mitochondrial genes are not recombined by sex, this position was sustainable, if somewhat unsatisfying. It could only be refuted when the actual rate of evolution was known, which in turn required the direct sequencing of mitochondrial genes, and the comparison of sequences. Only after Fred Sanger's group in Cambridge had sequenced the human mitochondrial genome in 1981 did it transpire that the evolution rate of mitochondrial genes was *faster* than that of the nuclear genes. Their atavistic properties could only be explained by a direct relationship; and this relationship was ultimately shown to be with a very specific group of bacteria, the α-proteobacteria.

Even the visionary Margulis was not correct about everything, luckily for the rest of us. Aligning herself with the earlier advocates of symbiosis, Margulis had argued that it would one day prove possible to grow mitochondria in culture—it was only a matter of finding the right growth factors. Today, we know that this is not possible. The reason was also made clear by the detailed sequence of the mitochondrial genome: the mitochondrial genes only encode a handful of proteins (13 to be exact), along with all the genetic machinery needed to make them. The great majority of mitochondrial proteins (some 800) are encoded by the genes in the nucleus, of which there are 30 000 to 40 000 in total. The apparent independence of mitochondria is therefore truly apparent, and not genuine. Their reliance on two genomes, the mitochondrial and the nuclear, is evident even at the level of a few proteins that are composed of multiple subunits, some of which are encoded by the mitochondrial genes, and others by the nuclear genes. Because they rely on both genomes, mitochondria can only be cultured within their host cells, and are correctly designated 'organelles', rather than symbionts. Nonetheless, the word 'organelle' gives no hint of their extraordinary past, and affords no insight into their profound influence on evolution.

There is another sense in which many biologists today still disagree with Lynn Margulis, and that relates to the evolutionary power of symbiosis in general. For Margulis, the eukaryotic cell is the product of multiple symbiotic mergers, in which the component cells have been subsumed into the greater whole to varying degrees. Her theory has been dubbed the 'serial endosymbiosis theory', meaning that eukaryotic cells were formed by a succession of such mergers between cells, giving rise to a community of cells living within one another. Besides chloroplasts and mitochondria, Margulis cites the cell skeleton with its organizing centre, the centriole, as the contribution of another type of bacteria, the *Spirochaetes*. In fact, according to Margulis the whole organic world is an elaboration of collaborative bacteria—the microcosm. The idea goes back to Darwin himself, who wrote in a celebrated passage: 'Each living being is a microcosm—a little universe formed of self-propagating organisms inconceivably minute and numerous as the stars in the heavens.'

The idea of a microcosm is beautiful and inspiring, but raises a number of difficulties. Cooperation is not an alternative to competition. A collaboration between different bacteria to form new cells and organisms merely raises the bar for competition, which is now between the more complex organisms rather than their collaborative subunits—many of which, including the mitochondria, turn out to have retained plenty of selfish interests of their own. But the biggest difficulty with an all-embracing view of symbiosis is the mitochondria themselves, which wag a cautionary finger at the power of microscopic collaboration. It seems that all eukaryotic cells either have, or once had (and then lost), mitochondria. In other words, possession of mitochondria is a *sine qua non* of the eukaryotic condition.

Why on earth should this be? If collaboration between bacteria were so commonplace, we might expect to find all sorts of distinct 'eukaryotic' cells, each composed of a different set of collaborative microorganisms. Of course, we do—there is a great range of eukaryotic collaboration, especially in the more obscure microscopic communities living in inaccessible places, such as the mud of the sea floor. But the astonishing finding is that all these far-flung eukaryotes share the same ancestry—and they *all* either have or once had mitochondria. This is not true of any other collaboration between microorganisms in eukaryotes. In other words, the collaborations that attained fulfilment in eukaryotic organisms are contingent on the existence of mitochondria. If the original merger had not taken place, then neither would any of the others. We can say this with near certainty, because the bacteria have been collaborating and competing among themselves for nearly four billion years, and yet only came up with the eukaryotic cell once. The acquisition of mitochondria was the pivotal moment in the history of life.

We are discovering new habitats and relationships all the time. They are a fabulously rich testing ground of ideas. To give just a single example, one of the more surprising discoveries at the turn of the millennium was the abundance of tiny, so-called *pico-eukaryotes*, which live among the micro-plankton in extreme environments, such as the bottom of the Antarctic oceans, and in acidic, iron-rich rivers, like the Rio Tinto in southern Spain (known by the ancient Phoenicians as the 'river of fire' because of its deep red colour). In general, such environments were considered to be the domain of hardy, 'extremophile' bacteria, and the last place one might expect to find fragile eukaryotes. The pico-eukaryotes are about the same size as bacteria and favour similar environments, and so generated a lot of interest as possible intermediates between bacteria and eukaryotes. Yet despite their small size and unusual predilection for extreme conditions, all turned out to fit into known groups of eukaryotes: genetic analysis showed they don't challenge the existing classification system at all. Astonishingly, this new bubbling fountain of variations on a eukaryotic theme adds up to no more than *subgroups* to existing groups, all of which we have known about for many years.

In these unsuspected environments, the very places we would expect to find a tapestry of unique collaborations, we do not. Instead, we find more of the same. Take the smallest known eukaryotic cell, for example, *Ostreococcus tauri*. It is less than a thousandth of a millimetre (1 micron) in diameter, rather smaller than most bacteria, yet it is a perfectly formed eukaryote. It has a nucleus with 14 linear chromosomes, one chloroplast—and, most remarkably of all, several tiny mitochondria. It is not alone. The unexpected fountain of eukaryotic variation in extreme conditions has thrown up perhaps 20 or 30 new subgroups of eukaryotes. It seems that all of them have, or once had, mitochondria, despite their small size, unusual lifestyles, and hostile surroundings.

What does all this mean? It means that mitochondria are not just another collaborative player: they hold the key to the evolution of complexity. This book is about what the mitochondria did for us. I ignore many of the technical aspects that are discussed in textbooks—incidental details like porphyrin synthesis and even the Krebs cycle, which could in principle take place anywhere else in the cell, and merely found a convenient location in the mitochondria. Instead, we'll see why mitochondria made such a difference to life, and to our own lives. We'll see why mitochondria are the clandestine rulers of our world, masters of power, sex, and suicide.

PART 1

Hopeful Monster
The Origin of the Eukaryotic Cell

All true multicellular life on earth is made up of eukaryotic cells—cells with a nucleus. The evolution of these complex cells is shrouded in mystery, and may have been one of the most unlikely events in the entire history of life. The critical moment was not the formation of a nucleus, but rather the union of two cells, in which one cell physically engulfed another, giving rise to a chimeric cell containing mitochondria. Yet one cell engulfing another is commonplace; what was so special about the eukaryotic merger that it happened only once?

The first eukaryote—one cell engulfed another to form an extraordinary chimera two billion years ago

Are we alone in the universe? Ever since Copernicus showed that the earth and planets orbit the sun, science has marched us away from a deeply held anthropocentric view of the universe to a humbling and insignificant outpost. From a statistical point of view, the existence of life elsewhere in the universe seems to be overwhelmingly probable, but on the same basis it must be so distant as to be meaningless to us. The chances of meeting it would be infinitesimal.

In recent decades, the tide has begun to turn. The shift coincides with the mounting scientific respectability afforded to studies on the origin of life. Once a taboo subject, dismissed as ungodly and unscientific speculation, the origin of life is now seen as a solvable scientific conundrum, and is being inched in upon from both the past and the future. Starting at the beginning of time and moving forwards, cosmologists and geologists are trying to infer the likely conditions on the early earth that might have given rise to life, from the vaporizing impacts of asteroids and the hell-fire forces of vulcanism, to the chemistry of inorganic molecules and the self-organizing properties of matter. Starting in the present and moving backwards in time, molecular biologists are comparing the detailed genetic sequences of microbes in an attempt to construct a universal tree of life, right down to its roots. Despite continuing controversies about exactly how and when life began on earth, it no longer seems as improbable as we once imagined, and probably happened much faster than we thought. The estimates of 'molecular clocks' push back the origin of life to a time uncomfortably close to the period of heavy bombardment that cratered the moon and earth 4000 million years ago. If it really did happen so quickly in our boiling and battered cauldron, why not everywhere else?

This picture of life evolving amidst the fire and brimstone of primordial earth gains credence from the remarkable capacity of bacteria to thrive, or at least survive, in excessively hostile conditions today. The discovery, in the late 1970s, of vibrant bacterial colonies in the high pressures and searing temperatures of sulphurous hydrothermal vents at the bottom of the oceans (known as 'black smokers') came as a shock. The complacent belief that all life on earth ultimately depended on the energy of the sun, channelled through the photosynthesis of organic compounds by bacteria, algae, and plants, was overturned at a stroke. Since then, a series of shocking discoveries has revolutionized our perception of life's orbit. Self-sufficient (autotrophic) bacteria live in countless numbers in the 'deep-hot biosphere', buried up to several miles deep in the

rocks of the earth's crust. There they scrape a living from the minerals themselves, growing so slowly that a single generation may take a million years to
reproduce—but they are undoubtedly alive (rather than dead or latent). Their
total biomass is calculated to be similar to the total bacterial biomass of the
entire sunlit surface world. Other bacteria survive radiation at the genetically
crippling doses found in outer space, and thrive in nuclear power stations or
sterilized tins of meat. Still others flourish in the dry valleys of Antarctica, or
freeze for millions of years in the Siberian permafrost, or tolerate acid baths and
alkaline lakes strong enough to dissolve rubber boots. It is hard to imagine that
such tough bacteria would fail to survive on Mars if seeded there, or could not
hitch a lift on comets blasted across deep space. And if they could survive there,
why should they not evolve there? When handled with the adept publicity of
NASA, ever eager to scrutinize Mars and the deepest reaches of space for signs
of life, the remarkable feats of bacteria have fostered the rise and rise of the
nascent science of astrobiology.

The success of life in hostile conditions has tempted some astrobiologists to
view living organisms as an emergent property of the universal laws of physics.
These laws seem to favour the evolution of life in the universe that we see
around us: had the constants of nature been ever-so-slightly different, the stars
could not have formed, or would have burnt out long ago, or never generated
the nurturing warmth of the sun's rays. Perhaps we live in a multiverse, in
which each universe is subject to different constants and we, inevitably, live
in what Astronomer Royal Martin Rees calls a *biophilic* universe, one of a small
set in which the fundamental constants favour life. Or perhaps, by an unknown
quirk of particle physics, or a breathtaking freak of chance, or by the hand of a
benevolent Creator, who put in place the biophilic laws, we are lucky enough to
live in a true universe that does favour life. Either way, our universe apparently
kindles life. Some thinkers go even further, and see the eventual evolution of
humanity, and in particular of human consciousness, as an inevitable outcome
of the universal laws, which is to say the precise weightings of the fundamental
constants of physics. This amounts to a modern version of the clockwork universe of Leibniz and Newton, parodied by Voltaire as 'All is for the best in the
best of all possible worlds.' Some physicists and cosmologists with a leaning
towards biology find a spiritual grandeur in this view of the universe as the midwife of intelligence. Such insights into the innermost workings of nature are
celebrated as a 'window' into the mind of God.

Most biologists are more cautious, or less religious. Evolutionary biology
holds more cautionary tales than just about any other science, and the erratic
meanderings of life, throwing up weird and improbable successes, and demol-
ishing whole phyla by turns, seems to owe more to the contingencies of history
than to the laws of physics. In his famous book *Wonderful Life*, Stephen Jay

Gould wondered what might happen if the film of life were to be replayed over and over again from the beginning: would history repeat itself, leading inexorably each time to the evolutionary pinnacle of mankind, or would we be faced with a new, strange, and exotic world each time? In the latter case, of course, 'we' would not have evolved to see it. Gould has been criticized for not paying due respect to the power of convergent evolution, which is the tendency of organisms to develop similarities in physical appearance and performance, regardless of their ancestry, so that anything which flies will develop similar-looking wings; anything that sees will develop similar-looking eyes. This criticism was propounded most passionately and persuasively by Simon Conway Morris, in his book *Life's Solution*. Conway Morris, ironically, was one of the heroes of Gould's book, *Wonderful Life*, but he opposes that book's sweeping conclusion. Play back the film of life, says Conway Morris, and life will flow down the same channels time and time again. It will do so because there are only so many possible engineering solutions to the same problems, and natural selection means that life will always tend to find the same solutions, whatever they may be. All of this boils down to a tension between contingency and convergence. To what extent is evolution ruled by the chance of contingency, versus the necessity of convergence? For Gould all is contingent; for Conway Morris, the question is, would an intelligent biped still have four fingers and a thumb?

Conway Morris's point about convergent evolution is important in terms of the evolution of intelligence here or anywhere else in the universe. It would be disappointing to discover that no form of higher intelligence had ever managed to evolve elsewhere in the universe. Why? Because very different organisms should converge on intelligence as a good solution to a common problem. Intelligence is a valuable evolutionary commodity, opening new niches for those clever enough to occupy them. We should not think only of ourselves in this sense: some degree of intelligence, and in my view conscious self-awareness, is widespread among animals, from dolphins to bears to gorillas. Humanity evolved quickly to fill the 'highest' niche, and a number of contingent factors no doubt facilitated this rise; but who is to say that, given a vacated niche and a few tens of millions of years, the kind of foraging bears that break into cars and dustbins could not evolve to fill it? Or why not the majestic and intelligent giant squid? Perhaps it was little more than chance and contingency that led to the rise of *Homo sapiens*, rather than any of the other extinct lines of *Homo*, but the power of convergence always favours the niche. While we are the proud possessors of uniquely well-developed minds, there is nothing particularly improbable about the evolution of intelligence itself. Higher intelligence could evolve here again, and by the same token anywhere else in the universe. Life will keep converging on the best solutions.

The power of convergence is illustrated by the evolution of 'good tricks' like flight and sight. Life has converged on the same solutions repeatedly. While repeated evolution does not imply inevitability, it does change our perception of probability. Despite the obviously difficult engineering challenges involved, flight evolved independently no less than four times, in the insects, the pterosaurs (such as pterodactyls), the birds, and the bats. In each case, regardless of their different ancestries, flying creatures developed rather similar-looking wings, which act as aerofoils—and we too have paralleled this design feature in aeroplanes. Similarly, eyes have evolved independently as many as forty times, each time following a limited set of design specifications: the familiar 'camera eye' of mammals and (independently) the squid; and the compound eyes of insects and extinct groups such as trilobites. Again, we too have invented cameras that work along similar principles. Dolphins and bats developed sonar navigation systems independently, and we invented our own sonar system before we knew that dolphins and bats took soundings in this way. All these systems are exquisitely complex and beautifully adapted to needs, but the fact that each has evolved independently on several occasions implies that the odds against their evolution were not so very great.

If so, then convergence outweighs contingency, or necessity overcomes chance. As Richard Dawkins concluded, in *The Ancestors Tale*: 'I am tempted by Conway Morris's belief that we should stop thinking of convergent evolution as a colourful rarity to be remarked and marvelled at when we find it. Perhaps we should come to see it as the norm, exceptions to which are occasions for surprise.' So if the film of life is played back over and over again, we may not be here to see it ourselves, but intelligent bipeds ought to be able to gaze up at flying creatures, and ponder the meaning of the heavens.

If the origin of life amidst the fire and brimstone of early Earth was not as improbable as we once thought (more on this in Part 2), and most of the major innovations of life on Earth all evolved repeatedly, then it is reasonable to believe that enlightened intelligent beings will evolve elsewhere in our universe. This sounds reasonable enough, but there is a nagging doubt. On Earth, all of this engineering flamboyance evolved in the last 600 million years, barely a sixth of the time in which life has existed. Before that, stretching back for perhaps more than 3000 million years, there was little to see but bacteria and a few primitive eukaryotic organisms like algae. Was there some other brake on evolution, some other contingency that needed to be overcome before life could really get going?

The most obvious brake, in a world dominated by simple single-celled organisms, is the evolution of large multicellular creatures, in which lots of cells collaborate together to form a single body. But if we apply the same yardstick of repeatability, then the odds against multicellularity do not seem particularly high. Multicellular organisms probably evolved independently quite a few times. Animals and plants certainly evolved large size independently; so too (probably) did the fungi. Similarly, multicellular colonies may have evolved more than once among the algae—the red, brown, and green algae are ancient lineages, which diverged more than a billion years ago, at a time when single-celled forms were predominant. There is nothing about their organization or genetic ancestry to suggest that multicellularity arose only once among the algae. Indeed, many are so simple that they are better viewed as large colonies of similar cells, rather than true multicellular organisms.

At its most basic level a multicellular colony is simply a group of cells that divided but failed to separate properly. The difference between a colony and a true multicellular organism is the degree of specialization (differentiation) among genetically identical cells. In ourselves, for example, brain cells and kidney cells share the same genes but are specialized for different tasks, switching on and off whichever genes are necessary. At a simpler level, there are numerous examples of colonies, even bacterial colonies, in which some differentiation between cells is normal. Such a hazy boundary between a colony and a multicellular organism can confound our interpretation of bacterial colonies, which some specialists argue are better interpreted as multicellular organisms, even if most ordinary people would view them as little more than slime. But the important point is that the evolution of multicellular organisms does not appear to have presented a serious obstacle to the inventive flow of life. If life got stuck in a rut, it wasn't because it was so hard to get cells to cooperate together.

In Part 1, I shall argue that there was one event in the history of life that was genuinely unlikely, which was responsible for the long delay before life took off in all its extravagance. If the film of life were played back over and over again, it seems to me likely that it would get stuck in the same rut virtually every time: we would be faced with a planet full of bacteria and little else. The event that made all the difference here was the evolution of the *eukaryotic cell,* the first complex cells that harbour a nucleus. An esoteric term like 'eukaryotic cell' might seem a quibbling exception, but the fact is that all true multicellular organisms on earth, including ourselves, are built only from eukaryotic cells: all plants, animals, fungi, and algae are eukaryotes. Most specialists agree the eukaryotic cell evolved only once. Certainly, all known eukaryotes are related—all of us share exactly the same genetic ancestry. If we apply the same rules of probability, then the origin of the eukaryotic cell looks far more improbable than the evolution of multicellular organisms, or flight, sight, and intelligence. It looks like genuine contingency, as unpredictable as an asteroid impact.

What has all this to do with mitochondria, you may be wondering? The answer stems from the surprising finding that all eukaryotes either have, or once had, mitochondria. Until quite recently, mitochondria had seemed almost incidental to the evolution of eukaryotes, a nicety rather than a necessity. The really important development, after which eukaryotes are named, was the evolution of the nucleus. But now this is perceived differently. Recent research suggests that the acquisition of mitochondria was far more important than simply plugging an efficient power-supply into an already complicated cell, with a nucleus brimming with genes—it was the single event that made the evolution of complex eukaryotic cells possible at all. If the mitochondrial merger had not happened then we would not be here today, nor would any other form of intelligent or genuinely multicellular life. So the question of contingency boils down to a practical matter: how did mitochondria evolve?

1

The Deepest Evolutionary Chasm

The void between bacterial and eukaryotic cells is greater than any other in biology. Even if we begrudgingly accept bacterial colonies as true multicellular organisms, they never got beyond a very basic level of organization. This is hardly for lack of time or opportunity—bacteria dominated the world for two billion years, have colonized all thinkable environments and more than a few unthinkable ones, and in terms of biomass still outweigh all multicellular life put together. Yet for some reason, bacteria never evolved into the kind of multi-cellular organism that a man on the street might recognize. In contrast, the eukaryotic cell appeared much later (according to the mainstream view) and in the space of just a few hundred million years—a fraction of the time available to bacteria—gave rise to the great fountain of life we see all around us.

The Nobel laureate Christian de Duve has long been interested in the origin and history of life. He suggests in a wise final testament, *Life Evolving*, that the origin of the eukaryotes may have been a bottleneck rather than an improbable event—in other words, their evolution was an almost inevitable consequence of a relatively sudden change in the environmental conditions, such as a rise in the amount of oxygen in the atmosphere and oceans. Of all the populations of proto-eukaryotes living at the time, one form simply happened to be better adapted and expanded rapidly through the bottleneck to take advantage of the changing circumstances: it prospered, while less well-adapted competing forms died out, giving a misleading impression of chance. This possibility depends on the actual sequence of events and selection pressures involved, and can't be ruled out until these are known with certainty. And of course, when we are talking of selection pressures exerted two billion years ago, it is unlikely that we can ever be certain; nonetheless, as I mentioned in the Introduction, it is possible to exclude some of the possibilities by considering modern molecular biology, and to narrow down a list of the most likely possi-bilities.

Despite my enormous respect for de Duve, I don't find his bottleneck thesis very convincing. It is too monolithic, and the sheer variety of life weighs against it—there seems to be a place for almost everything. The whole world did not change at once, and many varied niche environments persisted. Perhaps most importantly, environments lacking in oxygen (anoxic or hypoxic environments) persisted on a large scale, and do so to this day. To survive in such environments calls for a very different set of biochemical skills from those needed to survive in the new oxygenated surroundings. The fact that some eukaryotes already existed should not have precluded the evolution of a variety of different 'eukaryotes' in different environments, such as the stagnant sludge at the bottom of the oceans. Yet this is not what happened. The astonishing fact is that all the single-celled eukaryotes that live there are related to the oxygen-breathing organisms living in fresh air. I find it highly improbable that the first eukaryotes were so competitive that they annihilated all competition from every environment, even those most unsuited to their own character. Certainly the eukaryotes are not so competitive that they annihilated the competition of bacteria: they took their place alongside them, and opened up new niches for themselves. Nor can I think of any parallels elsewhere in life, on any scale. The fact that the eukaryotes became the masters of oxygen respiration did not lead to its disappearance among the bacteria. And more generally, many types of bacteria persisted for billions of years despite unceasing and unforgiving competition for the same resources.

Let's consider a single example, the methanogens. These bacteria (more technically, Archaea) scratch a living by generating methane gas from hydrogen and carbon dioxide. We'll consider this briefly as the methanogens are important to our story later on. The problem for methanogens is that, though carbon dioxide is plentiful, hydrogen is not: it quickly reacts with oxygen to form water, and so is not found in oxygenated environments for any long period. The methanogens can therefore only survive in environments where they have access to hydrogen gas—usually environments totally lacking in oxygen, or with constant volcanic activity, replenishing the source of hydrogen faster than it is used up. But the methanogens are not the only type of bacteria that use hydrogen—and they are not particularly efficient at extracting hydrogen from the environment. Another type of bacteria, so-called *sulphate-reducing bacteria*, makes a living by converting (or reducing) sulphate into hydrogen sulphide—the gas that stinks of rotten eggs (in fact rotten eggs reek of hydrogen sulphide). To do so, they too can use hydrogen gas, and they usually outcompete the methanogens for this scarce resource. Even so, the methanogens have survived for three billion years in niche environments, where the sulphate-reducing bacteria are penalized in some other respect—usually for the lack of sulphate. For example, because freshwater lakes are impoverished in sulphate, the sulphate-reducing bacteria can't establish themselves; and in the sludge at the bottom of such lakes, or in stagnant marshes, the methanogens live on. The methane gas they emit is known as swamp gas, and at times it sets alight with a mysterious blue flame that plays over the marshes, a phenomenon known as 'will-o-the-wisp', which explains many 'sightings' of ghosts and UFOs. But the productions of the methanogens are far from insubstantial. Anyone who advocates switching from exploiting oil reserves to natural gas can thank the methanogens—they are responsible for essentially our entire supply. Methanogens are also found in the guts of cattle and even people, as the hindgut is exceedingly low in oxygen. The methanogens thrive in vegetarians because grass, and vegetation in general, is low in sulphur compounds. Meat is much richer in sulphur; so sulphate-reducing bacteria usually displace methanogens in carnivores. Change your diet, and you will notice the difference in polite company.

The point I want to make about methanogens is that they were the losers in the race through a bottleneck, yet nonetheless survived in niche environments. Similarly, on a larger scale, it is rare for the loser to disappear completely, or for the latecomers never to gain at least a precarious foothold. The fact that flight had already evolved among birds did not preclude its later evolution in bats, which became the most numerous mammalian species. The evolution of plants did not lead to the disappearance of algae, or indeed the evolution of vascular plants to the disappearance of mosses. Even mass extinctions rarely lead to the disappearance of whole classes. If the dinosaurs disappeared, the reptiles are nevertheless still among us, despite stiff competition from birds and mammals. It seems to me that the only bottleneck in evolution comparable to that which de Duve postulates for the eukaryotic cell is the origin of life itself, which may have happened once, or perhaps numerous times with only one form ultimately surviving—in which case this, too, was a bottleneck. Perhaps, but this is not a good example, for we simply don't know. All we can say for sure is that all life living today ultimately shares the same ancestor, and so sprang from the same progenitor. Incidentally, this rules out the view, expressed by some, that our planet has been populated by successive waves of invasions from outer space—such a view is not compatible with the deep biochemical relatedness of all known life on earth.

If the origin of the eukaryotic cell was not a bottleneck, then it was probably a genuinely unlikely sequence of events, for it happened only once. Speaking as a multicellular eukaryote, I might be biased, but I do not believe that bacteria will ever ascend the smooth ramp to sentience, or anywhere much beyond slime, here or anywhere else in the universe. No, the secret of complex life lies in the chimeric nature of the eukaryotic cell—a hopeful monster, born in an improbable merger 2000 million years ago, an event still frozen in our innermost constitution and dominating our lives today.

Richard Goldschmidt first advanced the concept of a hopeful monster in 1940—the year that Oswald Avery showed that the genes are composed of DNA. Goldschmidt's name has since been derided by some writers, and held up as an anti-Darwinian hero by others. He deserves neither epitaph as his theory is neither impossible nor anti-Darwinian. Goldschmidt argued that the gradual accumulation of small genetic changes, *mutations*, was important, but could only account for the variation within a species: it wasn't a sufficiently powerful source of evolutionary novelty to explain the origin of a new species. Goldschmidt believed that the big genetic differences between species could not be mounted by a succession of tiny mutations, but required more profound 'macro-mutations'—monster leaps across 'genetic space', which is to say the gulf between two different genetic sequences (the number of changes required to get from one to another). He appreciated that random macro-mutations, sudden large changes in gene sequence, were far more likely to produce an unworkable mutant, and so he christened his one-in-a-million success a 'hopeful monster'. For Goldschmidt, a hopeful monster was the lucky outcome of a large and sudden genetic change, rather than a succession of tiny mutations— the kind of thing an archetypal mad scientist might produce in the laboratory after a lifetime of dedicated and deranging failure. With our modern understanding of genetics, we now know that macro-mutations don't account for speciation, at least not in multicellular creatures (though they may in bacteria, as argued by Lynn Margulis). However, it seems to me that the fusion of two whole genomes to create the first eukaryotic cell is better viewed as a macro-mutation to create a 'hopeful monster' than purely as a succession of small genetic changes.

So what kind of a monster was the first eukaryote, and why was its origin so improbable? To understand the answers, we need to think first about the nature of eukaryotic cells, and the many striking ways in which they differ from bacteria. We have already touched on this in the Introduction; here, we need to focus on the magnitude of the differences, the wide yawn of the chasm.

Differences between bacteria and eukaryotes

Compared with bacteria, most eukaryotic cells are enormous. Bacteria are rarely larger than a few thousandths of a millimetre (a few microns) or so in length. In contrast, although some eukaryotes, known as the *pico-eukaryotes*, are of bacterial size, the majority are ten to a hundred times those dimensions, giving them a cell volume about 10 000 to 100 000 times that of bacteria.

Size is not the only thing that matters. The cardinal feature of eukaryotes, from which their Greek name derives, is the possession of a 'true' nucleus. This nucleus is typically a spherical, dense mass of DNA (the genetic matter) wrapped up in proteins and enveloped in a double membrane. Here, already, are three big differences with bacteria. First, the bacteria lack a nucleus at all, or else have a primitive version that is not enclosed by a membrane. For this reason bacteria are also termed 'prokaryotes', from the Greek 'before the nucleus'. While this is potentially a prejudgement—some researchers argue that cells with a nucleus are just as ancient as those without—most specialists agree that prokaryotes are well named: they really did evolve before cells with a nucleus (the eukaryotes).

The second big difference between bacteria and eukaryotes is the size of their genomes as a whole—the total number of genes. Bacteria generally have far less DNA than even simple single-celled eukaryotes such as yeast. This difference can be measured either in terms of the total number of genes—usually adding up to hundreds or thousands—or the total DNA content. This latter value is known as the C-value, and is measured in 'letters' of DNA. It includes not only the genes, but also the stretches of so-called *non-coding* DNA—DNA that does not code for proteins, and so can't really be called 'genes'. The differences in both the number of genes and the C-value are revealing. Single-celled eukaryotes like yeasts have several times as many genes as most bacteria, whereas humans have perhaps twenty times as many. The difference in the C-value, or total DNA content, is even more striking, as eukaryotes contain far more non-coding DNA than bacteria. The total DNA content of eukaryotes spans an extraordinary five orders of magnitude. The genome of a large amoeba, *Amoeba dubia*, is more than 200 000 times larger than that of the tiny eukaryotic cell, *Encephalitozoon cuniculi*. This enormous range is unrelated to complexity, or the total number of genes. *Amoebae dubia* actually has 200 times more DNA than do humans, even though it has far fewer genes and is obviously less complex. This odd discrepancy is known as the C-value paradox. Whether all this non-coding DNA has any evolutionary purpose is debated. Some of it certainly does, but a large part remains puzzling, and it is hard to see why an amoeba should need so much (we will return to this in Part 4). Nonetheless, it is a fact, requiring an explanation, that eukaryotes generally have orders of magnitude more DNA than prokaryotes. This is not without a cost. The energy required to copy all this extra DNA, and to ensure it is copied faithfully, affects the rate and circumstances of cell division, with implications that we will explore later.

The third big difference lies in the packing and organization of DNA. As we noted in the Introduction, most bacteria possess a single circular chromosome. This is anchored to the cell wall, but otherwise floats freely around the cell, ready for quick replication. Bacteria also carry genetic 'loose change' in the form of tiny rings of DNA called plasmids, which replicate independently and can be passed from one bacterium to another. The daily exchange of loose plasmids in this way is equivalent to shopping with loose change, and explains how the genes for drug resistance spread so quickly in a population of bacteria— just as a coin may find itself in twenty different pockets in a day. Returning to their main gene bank, few bacteria wrap their main chromosome in proteins—

rather, their genes are 'naked', making them easily accessible—a current account rather than a savings account. Bacterial genes tend to be ordered in groups that serve a similar purpose, and act as a functional unit, which are known as *operons*. In contrast, eukaryotic genes give no semblance of order. Eukaryotic cells possess quite a number of disparate, straight chromosomes, which are usually doubled up to give pairs of equivalent chromosomes, such as the 23 pairs of chromosomes found in humans. In eukaryotes, the genes are strung along these chromosomes in virtually a random order, and to make matters worse they are often fragmented into short sections with long stretches of non-coding DNA breaking up the flow. To build a protein, a great tract of DNA often needs to be read off, before it is spliced up and melded together to form a coherent transcript that codes for the protein.

Eukaryotic genes are not just randomized and fragmented, they are also tricky to get at. The chromosomes are tightly wrapped in proteins called histones, which block access to the genes. When the genes are being replicated during cell division, or copied to make transcripts for building proteins, the configuration of the histones must be altered to allow access to the DNA itself. This in turn has to be controlled by proteins called transcription factors.

Altogether, the organization of the eukaryotic genome is a complicated business that fills library after library with footnotes. We'll come to another aspect of this complicated set-up in Part 5 (sex, which is not found in bacteria). For now, though, the most important take-home point is that there is an energetic cost to all of this complexity. Where bacteria are almost always ruthlessly streamlined and efficient, most eukaryotes are lumbering and labyrinthine.

A skeleton and many closets

Outside the nucleus, eukaryotic cells are also very different to bacteria. Eukaryotic cells have been described as cells with 'things inside' (Figure 2). Most of the things inside are membrane structures, made of a vanishingly thin sandwich of fatty molecules called lipids. The membranes form into vesicles, tubes, cisterns, and stacks, enclosing spaces—closets—that are physically separated, by the lipid barrier, from the watery cytosol. Different membrane systems are specialized for various tasks, such as building the components of the cell, or breaking down food to generate energy, or for transport, storage, and degradation. Interestingly, for all their variety of size and shape, most of the eukaryotic closets are variations on the simple vesicle: some are elongated and flattened, others are tubular, and some are simply bubbles. The most unexpected is the nuclear membrane, which looks like a continuous double membrane enclosing the nucleus, but is in fact a series of large flattened vesicles welded together, and rather surprisingly, continuous with other membrane compartments in

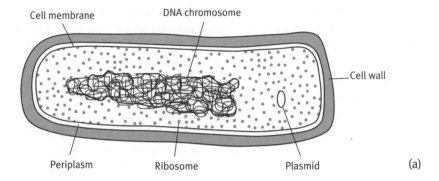

Cell membrane   DNA chromosome

Cell wall

Periplasm   Ribosome   Plasmid   (a)

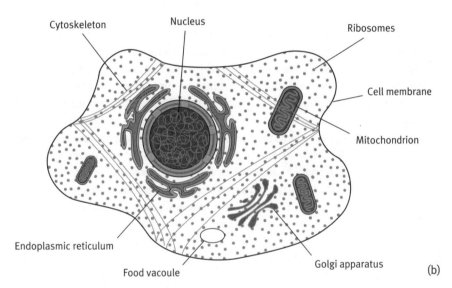

Cytoskeleton   Nucleus

Ribosomes

Cell membrane

Mitochondrion

Endoplasmic reticulum

Food vacuole

Golgi apparatus   (b)

2 Schematic illustrations of a bacterial cell (a) compared with a eukaryotic cell (b). The illustrations are not drawn to scale; bacteria are about the same size as the mitochondria in (b). The membrane structures are in fact sparsely depicted in the eukaryotic cell, for clarity, and in reality the differences in internal structure are even more marked. Bacteria are remarkable for their inscrutability, even by electron microscopy.

the rest of the cell. The nuclear membrane is therefore distinct in structure from the external membranes of any cell, which are always a continuous single (or double) layer.

Then there are the tiny organs within cells, the so-called organelles, such as the mitochondria and the chloroplasts in plants and algae. The chloroplasts are worth a special mention. They are responsible for photosynthesis, the process by which solar energy is converted into the currency of biological molecules, which possess their own chemical energy. Like the mitochondria, the chloroplasts derive from bacteria, in this case the *cyanobacteria*, the only group of bacteria capable of true photosynthesis (to generate oxygen). It is notable that both mitochondria and chloroplasts were once free-living bacteria, and still retain a number of partially independent traits, including a contingent of their own genes. Both are involved in energy generation for their host cells. Both these organelles are tangibly different from the other membrane systems of eukaryotic cells, and these differences set them apart. Like the nucleus, the mitochondria and chloroplasts are enclosed in a double membrane, but unlike the nucleus their membranes form a true continuous barrier. Along with their own DNA, their own ribosomes and protein assembly, and their semi-autonomous manner of division, the double membranes of the mitochondria and chloroplasts are among the features that point an incriminating finger at their bacterial ancestry.

If eukaryotic cells have things inside, bacteria are inscrutable. They have little of the eukaryotic riot of internal membrane systems, apart from their single external cell membrane, which is occasionally folded in upon itself to give some texture to the cell. Even so, the flourishing eukaryotic membranes share the same basic composition with the sparse membranes of bacteria. Both are composed of a water-soluble 'head' of glycerol phosphate, to which are bound several long fatty chains, which are soluble only in oils. Just as detergents form naturally into tiny droplets, so too the chemical structure of lipids enables them to coalesce naturally into membranes, in which the fatty chains are buried inside the membrane, while the water-soluble heads protrude from either side. This kind of consistency in the bacteria and eukaryotes helps to convince biochemists that both ultimately share a common inheritance.

Before we move on to consider the meaning of all these similarities and differences, let's just complete our whistlestop tour of the eukaryotic cell. There are two remaining differences with bacteria that I'd like to touch on. First, besides their membrane structures and organelles, eukaryotic cells contain a dense internal scaffolding of protein fibres known as the cytoskeleton. Second, unlike bacteria, eukaryotes do not have a cell wall, or at least not a bacterial-style cell wall (plant cells, and some algae and fungi, do actually possess cell walls, but these are very different from bacterial walls, and evolved much later).

The internal cytoskeleton and the external cell wall are utterly different conceptions, but nonetheless have equivalent functions—both provide structural support, in the same way that the external cuticle of an insect and own internal skeleton both provide structural support. Bacterial cell walls vary in structure and composition, but in general they provide a rigid skeleton that maintains the shape of the bacterium, preventing it from swelling to bursting point, or collapsing, if its environment suddenly changes. In addition, the bacterial cell wall provides a solid surface for anchoring the chromosome (containing its genes), along with various locomotive devices, such as whip-like threads, or flagellae. In contrast, eukaryotic cells usually have a flexible outer membrane, which is stabilized by the internal cytoskeleton. This is not at all a fixed structure but is constantly being remodelled—a highly energetic process—giving the cytoskeleton a dynamism unattainable by a cell wall. This means that eukaryotic cells (or protozoa at least) are not as robust as bacteria, but they have the immeasurable advantage that they can change shape, often quite vigorously. The classic example is the amoeba, which crawls around and engulfs its food by *phagocytosis*: temporary cellular projections, known as pseudopodia (literally, false feet) flow around the prey and meld together again, forming a food vacuole inside the cell. The pseudopodia are stabilized by dynamic changes in the cytoskeleton. They meld together again so easily because the lipid membranes are as fluid as soap bubbles, and can easily bud off into vesicles, then meld back together. Their ability to change shape and engulf food by phagocytosis enables single-celled eukaryotic organisms to become true predators, setting them apart from bacteria.

The road less travelled—from bacteria to eukaryotes

Eukaryotic cells and bacteria are constructed from essentially the same building materials (nucleic acids, proteins, lipids, and carbohydrates). They have exactly the same genetic codes, and very similar membrane lipids. Clearly they share a common inheritance. On the other hand, the eukaryotes are different from bacteria in virtually every aspect of their structure. Eukaryotic cells are, on average, 10 000 to 100 000 times the volume of bacteria, and contain a nucleus and many membranes and organelles. They generally carry orders of magnitude more genetic material and fragment their genes into short sections, in no particular order. Their chromosomes are straight rather than circular, and are wrapped in histone proteins. Most reproduce by sex, at least occasionally. They are supported internally by a dynamic cytoskeleton and may lack an external cell wall, which enables them to scavenge food and ingest whole bacteria.

The mitochondria are only one element in this catalogue of differences, and might seem to be just another added extra. They are not, as we shall see. But we are left with the question: why did the eukaryotes make such a complicated evolutionary pilgrimage while bacteria barely changed in nearly four billion years?

The origin of the eukaryotic cell is one of the hottest topics in biology, what Richard Dawkins has termed the 'Great Historic Rendezvous'. It furnishes exactly the right balance of science and speculation to generate violent passions among supposedly dispassionate scientists. Indeed, it sometimes feels as if each new piece of evidence throws up a new hypothesis to explain the evolutionary roots of the eukaryotic cell. Such hypotheses have traditionally fallen into two groups, those which try to explain the eukaryotes on the basis of mergers between a variety of bacterial cells, and those which try to derive most eukaryotic features from within the group, without recourse to so many mergers. As we saw in the Introduction, Lynn Margulis argued that both mitochondria and chloroplasts are derived from free-living bacteria. She also argued that several other features of eukaryotic cells, including the cytoskeleton, along with its organizing centres, the centrioles, are derived from bacterial mergers, but she has been less successful at drawing the field with her. The problem is that resemblances in cellular structures may derive from a direct evolutionary relationship, in which the endosymbiont has degenerated to the point that its ancestry can only just be made out. Alternatively, similarities in structure may be the result of convergent evolution, in which similar selection pressures inevitably generate similar structures, as there are only a few possible engineering solutions to a particular problem, as discussed earlier.

In the case of cellular objects like the cytoskeleton, which, unlike the chloroplasts and the mitochondria, do not have a genome of their own, it's difficult to establish provenance. If genealogy can't be traced directly, it is not easy to prove whether an organelle is symbiotic or an invention of the eukaryotes. Most biologists lean towards the simplest view, that most eukaryotic traits, including the nucleus and the organelles, except for the mitochondria and chloroplasts, are purely eukaryotic inventions.

To trace a path through this maze of contradictions we'll consider just two of the competing theories on the origin of the eukaryotic cell, which seem to me to be the most likely possibilities—the 'mainstream' view and the 'hydrogen hypothesis'. The mainstream view has superseded Lynn Margulis' original ideas in many details, and in its present form is largely attributable to Oxford biologist Tom Cavalier-Smith. Few researchers have quite as detailed an understanding of the molecular structures of cells and their evolutionary relationships as Cavalier-Smith, and he has put forward numerous important and contentious theories on cellular evolution. The hydrogen hypothesis is an utterly different theory, argued forcefully by Bill Martin, an American biochemist at Heinrich-Heine University in Düsseldorf, Germany. Martin is a geneticist by background, and tends to prefer biochemical, rather than structural, insights into the origins of the eukaryotes. His ideas are counter-intuitive, and have generated a heated, even vitriolic, response in some quarters, but they are underpinned by a crisp ecological logic that cannot be ignored. The pair often clash at conferences, and their views seem to hang over such meetings with an almost Victorian sense of melodrama, reminiscent of Conan Doyle's Professor

Challenger. At a splendid discussion meeting on the origin of eukaryotic cells at the Royal Society of London in 2002, Cavalier-Smith and Martin contested each other's views throughout the meeting, and I was impressed to find them still embroiled in debate hours afterwards in the local pub.

2

Quest for a Progenitor

How did the eukaryotic cell evolve from bacteria? The mainstream view assumes that it was by way of a sequence of tiny steps, through which a bacterium was gradually transformed into a primitive eukaryotic cell, possessing everything that characterises the modern eukaryotes, except for mitochondria. But what were these steps? And how did they get started down a path that in the end found a way across the deep chasm separating the eukaryotes from bacteria?

Tom Cavalier-Smith has argued that the key step forcing the evolution of the eukaryotes was the catastrophic loss of the cell wall. According to the Oxford English Dictionary, the word 'catastrophe' means 'a calamitous fate' or 'an event producing a subversion of the order of things'. For any bacteria that lose their cell wall, either definition may easily come true. Most wall-less bacteria are extremely fragile, and unlikely to survive long outside the cosy laboratory environment. This does not mean that such calamities are rare events, though. In the wild, bacterial cell walls might be lost quite often, either by mutation or active sabotage. For example, some antibiotics (such as penicillin) work by blocking the formation of the cell wall. Bacteria engaged in chemical warfare may well have produced such antibiotics. This is not at all improbable—most new antibiotics are isolated from bacteria and fungi engaged in exactly this kind of struggle. So, the first step, the calamitous loss of the cell wall, might not have posed any problem. What of the second step: survival and subversion of the order of things?

As we noted in the previous chapter, there are potentially big advantages to getting rid of the unwieldy cell wall, not least being able to change shape and engulf food whole by phagocytosis. According to Cavalier-Smith, phagocytosis is the defining feature that set the eukaryotes apart from bacteria. Any bacterium that solved the problem of structural support and movement could certainly subvert the established order of things. Yet, for a long time, it looked as if surviving without a cell wall was a magic trick equivalent to pulling a rabbit out of a hat. Bacteria were believed to lack an internal cytoskeleton, and if that was the case, the eukaryotes must have evolved their complex skeleton in a single generation, or faced extinction. In fact this assumption turns out to be groundless. In two seminal papers, published in the journals *Cell* and *Nature* in 2001,

Laura Jones and her colleagues at Oxford, and Fusinita van den Ent and her colleagues in Cambridge, showed that some bacteria do indeed have a cytoskeleton as well as a cell wall—they wear a belt and braces, as Henry Fonda put it in *Once Upon a Time in the West* ('never trust a man who can't even trust his own trousers'). Unlike Fonda's risk-averse cowboy, however, bacteria do need both to maintain their shape.

Many bacteria are spherical (*cocci*) while others are rod-shaped (*bacilli*), filamentous, or helical. Some oddballs have been found that even have triangular or square shapes. Quite what advantages these different shapes might confer is an interesting question, but it seems that the default bacterial shape is spherical, and any other shape requires internal support. Non-spherical bacteria possess protein filaments very similar in microscopic structure to those found in eukaryotes like yeast, as well as in humans and plants. In each case, the cytoskeleton filaments are composed of a protein akin to actin, best known for its role in muscle contraction. In non-spherical bacteria, these filaments form into a helical swirl underneath the cell membrane, which apparently provides structural support. What is clear is that if the genes encoding the filaments are deleted then bacteria that are normally rod-like in shape (bacilli) develop as spherical cocci instead. Impressions resembling bacilli have been found in rocks 3500 million years old, so it is conceivable that the cytoskeleton evolved not long after the appearance of the earliest cells. This reverses the problem. If a cytoskeleton was there all along, then why do so few bacteria survive the loss of the cell wall? We'll return to this theme in Part 3. For now, let's satisfy ourselves with the possible consequences.

'Discovery' of the archaea—a missing link?

Only two groups of cells have thrived in the absence of a cell wall—the eukaryotes themselves, and the *Archaea*, a remarkable group of prokaryotes (cells that lack a nucleus, like bacteria). The *Archaea* were discovered by Carl Woese and George Fox at the University of Illinois in 1977, and named from the Greek for 'ancient'. Most archaea do, in fact, have a cell wall, but their walls are rather different in chemical composition from those of bacteria, and some groups (such as the boiling-acid loving *Thermoplasma*) do not have a cell wall at all. Curiously, antibiotics like penicillin don't affect the synthesis of archaeal cell walls, lending support to the idea that cell walls might have been the target of bacterial chemical warfare. Like bacteria, archaea are tiny, typically measuring a few thousandths of a millimetre (microns) across, and they do not have a nucleus. Like bacteria, they have a single circular chromosome. Again, like bacteria, the archaea take on many shapes and forms, and so presumably have some sort of cytoskeleton. One reason why they were discovered so recently is that archaea are mostly 'extremophiles', that is, they thrive in the most extreme and arcane of environments, from boiling acid-baths beloved of *Thermoplasma*, to putrid marshes (inhabited by marsh-gas producing methanogens) and even buried oilfields. In the latter case, the archaea responsible have attracted commercial interest, or rather annoyance, as they 'sour' the wells— they raise the sulphur content of oil, which corrodes the well-casings and metal pipelines. Greenpeace could hardly conceive a more wily sabotage.

The 'discovery' of the archaea is a relative term, as some of them had been known about for decades (particularly the oil-souring archaea and swamp-gas producing methanogens), but their small size and lack of nucleus meant that they were invariably mistaken for bacteria. In other words, they were not so much discovered as reclassified; and even now, some researchers prefer to classify them with the bacteria, as just another diverse group of inventive prokaryotes. But the painstaking genetic studies of Woese and others have convinced most impartial observers that the archaea really do differ in profound ways from bacteria, ways that go well beyond the construction of their cell walls. We now know that about 30 per cent of archaeal genes are unique to the group. These unique genes code for forms of energy metabolism (such as the generation of methane gas) and cell structures (such as membrane lipids) that are not found in any other bacteria. The differences are important enough for most scientists to regard the archaea as a separate 'domain' of life. This means that we now classify all living things into three great domains—the bacteria, the archaea, and the eukaryotes (which, as we have seen, includes all multicellular plants, animals, and fungi). The bacteria and the archaea are both prokaryotic (lacking a cell nucleus) while the eukaryotes all do have a nucleus.

Despite their love of extreme environments and unique characteristics, the archaea also share a mosaic of traits with both bacteria and eukaryotes. I say 'mosaic' advisedly, as many of these traits are self-contained modules, encoded by groups of genes that work together as a unit (such as the genes for protein synthesis, or for energy metabolism). These individual modules fit together like the pieces of a mosaic, to construct the overall pattern of an organism. In the case of the archaea, some pieces are similar to those used by eukaryotes, while others are more reminiscent of bacteria. It is almost as if they were selected at random from a lucky dip of cell characteristics. So, for example, even though the archaea are prokaryotes, easily mistaken for bacteria when viewed down the microscope, some of them nonetheless wrap their chromosome in histone proteins, in a very similar manner to eukaryotes.

The parallels between archaea and eukaryotes go further. The presence of histones means that archaeal DNA is not easily accessible, so, like the eukaryotes, archaea need complicated transcription factors to copy or to transcribe their DNA (reading off the genetic code to construct a protein). The detailed mechanism of genetic transcription in the archaea parallels that in eukaryotes, albeit in a simpler fashion. There are also similarities in the way that the two groups construct their proteins. As we saw in the Introduction, all cells assemble their proteins using the tiny molecular factories called ribosomes. The ribosomes are broadly similar in all three domains of life, implying that they share a common ancestry, but they differ in many details. Interestingly, there are more differences between the bacterial and archaeal ribosomes than there are between archaeal and eukaryotic ribosomes. For example, toxins like diphtheria toxin block protein assembly on ribosomes in both the archaea and eukaryotes, but not in bacteria. Antibiotics like chloramphenicol, strepto-mycin, and kanamycin block protein synthesis in the bacteria, but not in the archaea or eukaryotes. These patterns are explained by differences in the way that protein synthesis is initiated, and in the detailed structure of the ribosome factories themselves. The ribosomes of eukaryotes and archaea have more in common with each other than either do with bacteria.

All this means the archaea are about as close to a missing link between the bacteria and the eukaryotes as we are ever likely to find. The archaea and the eukaryotes probably share a relatively recent common ancestor, and are best seen as 'sister' groups. This seems to back up Cavalier-Smith's view that the loss of the cell wall, possibly in the common ancestor of the archaea and the eukaryotes, was the catastrophic step that later propelled the evolution of eukaryotes. The earliest eukaryotes may have looked a little like modern archaea. Intriguingly, though, no archaea ever learnt to change shape to scav-enge a living by engulfing food in the eukaryotic fashion. On the contrary, instead of developing a flexible cytoskeleton as the eukaryotes did, the archaea developed quite a stiff membrane system, and remained nearly as rigid as bacterial cells. So there is more to being 'eukaryotic' than just lacking a cell wall; but might it be no more complex than lifestyle? Were the ancestral eukaryotes simply wall-less archaea, which modified their existing cytoskeleton into a more dynamic scaffolding that enabled them to change shape and eat food in lumps, by phagocytosis? Might this alone account for how they came by their mitochondria—they simply ate them? And if so, might there still be a few living fossils from the age before mitochondria lurking in hidden corners, relics of those primitive eukaryotes that shared more traits with the archaea?

The archezoa—eukaryotes without mitochondria

According to the theory put forward by Cavalier-Smith as long ago as 1983, some of the simple single-celled eukaryotes living today *do* still resemble the earliest eukaryotes. More than a thousand species of primitive eukaryotes do not possess mitochondria. While many of these probably lost their mitochondria later, simply because they didn't need them (evolution is always quick to jettison unnecessary traits), Cavalier-Smith argued that at least a few of these species were probably 'primitively amitochondriate'—in other words, they never did have any mitochondria, but were instead primitive relics of the age before the eukaryotic merger. To generate their energy, most of these cells depend on fermentations in the same way as yeast. While a few of them tolerate the presence of oxygen, most grow best at very low levels or even in the complete absence of the gas, and thrive today in low-oxygen environments. Cavalier-Smith named this hypothetical group the 'archezoa' in deference to their ancient roots and their animal-like, scavenging mode of living, as well as their similarities to the archaea. The name 'archezoa' is unfortunate, in that it is confusingly similar to 'archaea'. I can only apologize for this confusion. The archaea are prokaryotes (without a nucleus), one of the three domains of life, while the archezoa are eukaryotes (with a nucleus) that never had any mito-chondria.

Like any good hypothesis, Cavalier-Smith's was eminently testable by the genetic sequencing technologies then reaching fruition—the capacity to work out the precise sequence of letters in the code of genes. By comparing the gene sequences of different eukaryotes, it is possible to determine how closely related different species are to each other—or conversely, how remote the archezoa are from more 'modern' eukaryotes. The reasoning is simple. Gene sequences consist of thousands of 'letters'. For any gene, the sequence of these letters drifts slowly over time as a result of mutations, in which particular letters are lost or gained, or substituted one for another. Thus, if two different species have copies of the same gene, then the exact sequence of letters is likely to be slightly different in the two different species. These changes accumulate very slowly over millions of years. Other factors need to be considered, but to a point the number of changes in the sequence of letters gives an indication of the time elapsed since the two versions diverged from a common ancestor. These data can be used to build a branching tree of evolutionary relationships—the universal tree of life.

If the archezoa really could be shown to be among the oldest of eukaryotes, then Cavalier-Smith would have found his missing link—a primitive eukaryotic cell, that had never possessed any mitochondria, but which did have a nucleus and a dynamic cytoskeleton, enabling it to change shape and feed by phago-cytosis. The first answers became available within a few years of Cavalier-Smith's hypothesis, and apparently satisfied his predictions in full. Four groups of primitive-looking eukaryotes, which not only lacked mitochondria but also most other organelles, were confirmed by genetic analysis to be amongst the oldest of the eukaryotes.

The first genes to be sequenced, by Woese's group in 1987, belonged to a tiny parasite, no larger than a bacterium, which lives inside other cells—indeed, can *only* live inside other cells. This was the microsporidium *V. necatrix.* As a group, the microsporidia are named after their infective spores, all of which come replete with a projecting coiled tube, through which spores extrude their contents into a host cell, then multiply to begin their life cycle afresh, ultimately producing more infective spores. Perhaps the best-known representative of the microsporidia is *Nosema,* which is notorious for causing epidemics in honeybees and silkworms. When feeding inside the host cell, *Nosema* behaves like a minute amoeba, moving around and engulfing food by phagocytosis. It has a nucleus, a cytoskeleton and small bacterial-style ribosomes, but has no mitochondria or any other organelles. As a group, the microsporidia infect a wide variety of cells from many branches of the eukaryote tree-of-life, including vertebrates, insects, worms, and even single-celled ciliates (cells named after their tiny hair-like 'cilia', used for feeding and locomotion). As all microsporidia are parasites that can survive only inside other eukaryotic cells, they can't truly represent the first eukaryotes (because they would have had nothing to infect) but the diverse range of organisms that they do infect suggests that they have ancient origins, going back to the roots of the eukaryotic tree. This assumption seemed to be confirmed by genetic analysis, but there was a catch, as we shall see in a moment.

Over the next few years, the ancient status of the three other groups of primitive eukaryotes was confirmed by genetic analyses—the archamoebae, the metamonads, and the parabasalia. All three groups are best known as parasites, but free-living forms do also exist, perhaps fitting them better than the microsporidia as the earliest eukaryotes. As parasites, these three groups occasion much misery, illness, and death; how ironic that these repellent and life-threatening cells should be singled out as our own early ancestors. The archamoeba are best represented by *Entamoeba histolytica,* which causes amoebic dysentery, with symptoms ranging from diarrhoea to intestinal bleeding and peritonitis. The parasites burrow through the wall of the intestine to gain access to the bloodstream, from where they infect other organs, including the liver, lungs, and brain. In the long term, they may form enormous cysts on these organs, especially the liver, causing up to 100 000 deaths worldwide each year. The other two groups are less deadly but no less smelly. The best-known metamonad is *Giardia lamblia,* another intestinal parasite. *Giardia* does not invade the intestinal walls or enter the bloodstream, but the infection is still thoroughly unpleasant, as any travellers who have incautiously drunk water from infected streams know to their cost. Watery diarrhoea and 'eggy' flatulence may persist for weeks or months. Turning to the third group, the parabasalia, the best known is *Trichomonas vaginalis,* which is among the most prevalent, albeit least menacing, of the microbes that cause sexually transmitted diseases (though the inflammation it produces may increase the risk of contracting other diseases such as AIDS). *T. vaginalis* is transmitted mainly by vaginal intercourse but can also infect the urethra in men. In women, it causes vaginal inflammation and the discharge of a malodorous yellowish-green fluid. All in all, this portfolio of foul ancestors just goes to prove that we can choose our friends but not our relatives.

The eukaryote's progress

For all their unpleasantness, the archezoa nonetheless fitted the bill as primitive eukaryotes, survivors from the earliest days before the acquisition of mitochondria. Genetic analysis confirmed that they did branch away from more modern eukaryotes at an early stage of evolution, some two thousand million years ago, while their uncluttered morphology was compatible with a simple early lifestyle as scavengers that engulfed their food whole by phagocytosis. Presumably, one fine morning, two thousand million years ago, a cousin of these simple cells engulfed a bacterium, and for some reason failed to digest it. The bacterium lived on and divided inside the archezoon. Whatever the original benefit might have been to either party the intimate association was eventually so successful that the chimeric cell gave rise to all modern eukaryotes with mitochondria—all the familiar plants, animals, and fungi.

According to this reconstruction, the original benefit of the merger was probably related to oxygen. Presumably it was not a coincidence that the merger took place at a time when oxygen levels were rising in the air and the oceans. A great surge in atmospheric oxygen levels certainly occurred around two billion years ago, probably in the wake of a global glaciation, or 'snowball earth'. This timing corresponds closely to that of the eukaryotic merger. Modern mitochondria make use of oxygen to burn sugars and fats in cell respiration, so it is not surprising that mitochondria should have become established at a time when oxygen levels were rising. As a form of energy-generation, oxygen respiration is much more efficient than other forms of respiration, which generate energy in the absence of oxygen (anaerobic respiration). All that said, it is unlikely that superior energy generation could have been the original advantage. There is no reason why a bacterium living inside another cell should pass on its energy to the host. Modern bacteria keep all their energy for themselves, and the last thing they do is export it benevolently to their neighbouring cells. Thus while there is a clear advantage for the ancestors of the mitochondria, which had intimate access to any of the host's nutrients, there is no apparent advantage to the host cell itself.

Perhaps the initial relationship was actually parasitic—a possibility first suggested by Lynn Margulis. Important work from Siv Andersson's laboratory at the University of Uppsala in Sweden, published in *Nature* in 1998, showed that the genes of the parasitic bacterium *Rickettsia prowazekii*, the cause of typhus, correspond closely with those of human mitochondria, raising the possibility that the original bacterium might have been a parasite not unlike *Rickettsia*. Even if the original invading bacterium was a parasite, the unbalanced 'partnership' may have survived, as long as its unwelcome guest did not fatally weaken the host cell. Many infections today become less virulent over time, as parasites also benefit from keeping their host alive—they do not have to search for a new home every time their host dies. Diseases like syphilis have become much less virulent over the centuries, and there are hints that a similar attenuation is already underway with AIDS. Interestingly, such attenuation over generations also takes place in amoebae such as proteus. In this case, the infecting bacteria initially often kill the host amoebae, but eventually become necessary for their survival. The nuclei of infected amoebae become incompatible with the original amoebae, and ultimately lethal to them, effectively forcing the origin of a new species.

In the case of the eukaryotic cell, the host is good at 'eating' and through its predatory lifestyle provides its guest with a continuous supply of food. We are told that there is no such thing as a free lunch, but the parasite might simply burn up the metabolic waste-products of the host without weakening it much at all, which is not far short of a free lunch. Over time the host learned to tap into the energy-generating capacity of its guest, by inserting membrane channels, or 'taps'. The relationship reversed. The guest had been the parasite of the host, but now it became the slave, its energy drained off to serve the host.

This scenario is only one of several possibilities, and perhaps the timing holds the key. Even if energy was not the basis of the relationship, the rise in oxygen levels might still explain the initial benefits. Oxygen is toxic to anaerobic (oxygen-hating) organisms—it 'corrodes' unprotected cells in the same way that it rusts iron nails. If the guest was an aerobic bacterium, using oxygen to generate its energy, while the host was an anaerobic cell (generating energy by fermentation), then the aerobic bacterium may have protected its host against toxic oxygen—it could have worked as an internally fitted 'catalytic converter', guzzling up oxygen from the surroundings and converting it into harmless water. Siv Andersson calls this the 'Ox-Tox' hypothesis.

Let's recapitulate the argument. A bacterium loses its cell wall but survives because it has an internal cytoskeleton, which it had made use of before to keep in shape. It now resembles a modern archaeon. With a few modifications to its cytoskeleton, the wall-less archaeon learns to eat food by phagocytosis. As it grows larger it wraps its genes in a membrane and develops a nucleus. It has now turned into an archezoon, perhaps resembling cells like *Giardia*. One such hungry archezoon happens to engulf a smaller aerobic bacterium but fails to digest it, let's say because the bacterium is a parasite like the modern *Rickettsia*, and has learned to evade the defences of its host. The two get along together in a benign parasitic relationship, but as atmospheric oxygen levels rise, the relationship begins to pay dividends to both the host and parasite: the parasite still gets its free lunch, but the host is now getting a better deal—it's protected from toxic oxygen from within by its catalytic converter. Then, finally, in an act of breathtaking ingratitude, the host plugs a 'tap' into the membrane of its guest and drains off its energy. The modern eukaryotic cell is born, and never looks back.

This long chain of reasoning is a good example of how science can piece together a plausible story and back it up with evidence at almost every point. To me there is a feeling of inevitability about the whole process: it could happen here and it could happen anywhere else in the universe—no single step is particularly improbable. There is simply a bottleneck, as postulated by Christian de Duve, in which the evolution of the eukaryotes is unlikely when there is not much oxygen around, but almost inevitable as soon as the oxygen levels rise. While everybody agrees this story is broadly speculative, it was widely believed to be plausible, and made use of most of the known facts. Nothing prepared the field for the reversal that was to follow in the late 1990s. As sometimes happens to the 'good' stories in science, virtually the entire edifice collapsed in the space of just five years. Nearly every point has now been contradicted. But perhaps the writing was on the wall. If the eukaryotes only evolved once, then a plausible story may be exactly the wrong kind of story.

Reversal of a paradigm

The first stone to crumble was the 'primitively amitochondriate' status of the archezoa. This term, if you recall, means that the archezoa never did have any mitochondria. But when more genes from different archezoa were sequenced, it began to look as if postulated progenitors of eukaryotic cells, such as *Entamoeba histolytica* (the cause of amoebic dysentery), were not the earliest representatives of their group after all. Other types of cell in the same group appeared to be even older—but did have mitochondria. Unfortunately, the genetic dating techniques were approximate and liable to error, and so the results were controversial. But if the estimated dates were correct, then the results could only mean that *Entamoeba histolytica* did have ancestors that had once possessed mitochondria, and so must have lost its own, rather than never having had any at all. If the archezoa are defined as a group of primitive eukaryotes that never had mitochondria, then *E. histolytica* could not be an archezoon.

In 1995, Graham Clark at the National Institutes of Health in the United States, and Andrew Roger at Dalhousie University in Canada, went back to look more closely at *E. histolytica* to see if there were any traces that it had formerly possessed mitochondria. There were. Hidden away in the nuclear genome were two genes that, from their DNA sequences, almost certainly derived from the original mitochondrial merger. They were presumably transferred from the early mitochondria to the host cell nucleus, and the cell later lost all physical traces that it had ever had any mitochondria. We should note that the transfer of genes from the mitochondria to the host is quite normal, for reasons that we'll consider in Part 3. Modern mitochondria have retained only a handful of genes, and the rest were either lost altogether or transferred across to the nucleus. The proteins encoded by these nuclear genes are often targeted back to the mitochondria. Interestingly *E. histolytica* does actually possess some oval organelles that might be the corrupt remains of mitochondria; they resemble mitochondria in their size and shape, and several of the proteins that have been isolated from them have also been found in the mitochondria in other organisms.

Not surprisingly, the burning question transferred to the other supposedly primitively amitochondriate groups. Had they, too, once possessed mitochondria? Similar studies were carried out, and so far all the 'archezoa' that have been tested turn out to have once possessed mitochondria, and lost them later on. For example, not only did *Giardia* apparently once have mitochondria, but it, too, may still preserve relics, in the form of tiny organelles called mitosomes, which continue to carry out some of the functions of mitochondria (if not the best known, aerobic respiration). Perhaps the most surprising results concerned the microsporidia. This supposedly ancient group not only *did* possess mitochondria in the past, but now turns out *not* to be an ancient group at all— they are most closely related to the higher fungi, a relatively recent group of eukaryotes. The apparent antiquity of the microsporidia is merely an artefact of their parasitic lifestyle inside other cells. And the fact that they infect so many different groups is but a testament to their success.

While it remains possible that the real archezoa are still out there, just waiting to be found, the consensus view today is that the entire group is a mirage— every single eukaryote that has ever been examined either has, or once had, mitochondria. If we believe the evidence, then there never were any primitive archezoa. And if this is true, then the mitochondrial merger took place at the very beginning of the eukaryotic line, and was perhaps inseparable from it: the merger *was* the unique event that gave rise to the eukaryotes.

If the prototype eukaryote was not an archezoon—in other words, not a simple cell that made its living by engulfing its food by phagocytosis—then what *did* it look like? The answer might possibly lie in the detailed DNA sequences of eukaryotes living today. We have seen it is possible to identify ex-mitochondrial genes by comparing their gene sequences; perhaps we can do the same with those genes inherited from the original host. The idea is simple. Because we know that mitochondria are related to a particular group of bacteria, the α-proteobacteria, we can exclude any genes that seem to derive from this source, and look to see where the rest come from. Of the rest, we can assume that some are unique to eukaryotes—they evolved in the last two thousand million years since the merger—while some might have been transferred from elsewhere. Even so, at least a few ought to line up with the original host. These genes should have been inherited by all the descendents of the original merger, and gradually accumulated modifications ever since; but they should still bear *some* resemblance to the original host cell.

This was the approach employed by Maria Rivera and her colleagues at the University of California, Los Angeles, published in 1998 and in more detail in *Nature* in 2004. This team compared complete genome sequences from representatives of each of the three domains of life, and found that eukaryotes possess two distinct classes of genes, which they referred to as *informational* and *operational* genes. The *informational* genes encoded all the fundamental inheritance machinery of the cell, enabling it to copy and transcribe DNA, to replicate itself, and to build proteins. The *operational* genes encoded the workaday proteins involved in cellular metabolism—in other words, the proteins responsible for generating energy and manufacturing the basic building blocks of life, such as lipids and amino acids. Interestingly, almost all the operational genes came from the α-proteobacteria, presumably by way of the mitochondria, and the only real surprise was how many more of these genes there were than expected—it seems the genetic contribution of the ancestor of the mitochondria was greater than anticipated. But the biggest surprise was the allegiance of the *informational* genes. These genes lined up with the archaea, as anticipated, but they bore a strong resemblance to the genes in a completely unexpected group of archaea: they were most similar to *methanogens*, those swamp lovers that shun oxygen and produce the marsh gas methane.

This is not the only piece of evidence to point a suspicious finger at the methanogens. John Reeve and his colleagues at Ohio State University, Columbus, have shown that the structure of eukaryotic histones (the proteins that wrap DNA) is closely related to methanogen histones. This similarity is surely no coincidence. Not only are the structures of the histones themselves closely related, but also the three-dimensional conformation of the whole DNA-protein package is amazingly similar. The chances of finding exactly the same structure in two organisms that are supposedly unrelated, like the methanogens and the eukaryotes, is equivalent to finding the same jet engine in two aeroplanes produced independently by two competing companies. Of course, we might well find the same engine, but we'd be incredulous if told that it had been 'invented' twice, without any knowledge of the rival company's version, or of the prototype: we would assume that the engine had been bought or stolen from another company. In the same way, the packaging of DNA with histones is so similar in the methanogens and the eukaryotes that the most likely explanation is that they derived the full package from a common ancestor—both were developed from the same prototype.

All this adds up to quite a package. Two tell-tale wisps of smoke curl out of the same smoking gun. If these wisps are believable, it seems we inherited both our informational genes and our histone proteins from the methanogens. Suddenly our most venerable ancestor is no longer the vile parasite that we suspected, but an even more alien entity, which survives today in stagnant swamps and the intestinal tract of animals. The original host in the eukaryotic merger was a methanogen.

We are now in a position to see what kind of a hopeful monster the first eukaryotic cell might have been—the product of a merger between a methanogen (which gained its energy by generating methane gas) and an α-proteobacterium, for example a parasite like *Rickettsia*. This is a startling paradox. Few organisms hate oxygen more than the methanogens do—they can only be found living in the stagnant, oxygen-free pits of the world. Conversely, few organisms depend more on oxygen than *Rickettsia*—they are tiny parasites living inside other cells, and have streamlined themselves to their specialist niche by throwing away redundant genes, leaving them with only the genes needed to reproduce themselves—and the genes needed for oxygen respiration. Everything else has gone. So the paradox is this: if the eukaryotic cell was supposedly born of a symbiosis between an oxygen-hating methanogen and an oxygen-loving bacterium, how could the methanogen possibly benefit from having α-proteobacteria inside it? For that matter, how did the α-proteobacteria benefit from being inside? Indeed, if the host was incapable of phagocytosis—and methanogens are certainly not able to change shape and eat other cells—how on earth did it get inside?

It is possible that Siv Andersson's Ox-Tox hypothesis still applies—in other words, the oxygen-guzzling bacterium protected its host from toxic oxygen, enabling the methanogen to venture into pastures new. But there is a big difficulty with this scenario now. Such a relationship makes sense for a primitive archezoon that lives by fermenting organic remains. This will prosper if it is able to migrate to any environment where such remains can be found. Such scavenging cells are the single-celled equivalent of jackals prowling Africa, covering vast distances in the search for a fresh carcass. But this roving existence would kill a methanogen. A methanogen is as tied to a low-oxygen environment as a hippo is to waterholes. The methanogens can *tolerate* the presence of oxygen, but they can't generate any energy in its presence, because they depend on hydrogen for fuel, and this is very rarely found in the same environment as oxygen. So if a methanogen does leave its watering hole, it must starve until it gets back: festering organic remains mean nothing to a methanogen—it would do better never to leave. Thus there is a deep tension between the interests of the methanogen, which gains nothing from venturing to pastures new, and those of an oxygen-guzzling parasite, which can't generate any energy at all in the anoxic environment favoured by methanogens.

This paradox is heightened because, as we have seen, their relationship could not have depended on energy in the form of exchangeable ATP—bacteria do not have ATP exporters, and never benevolently 'feed' each other. The tryst could still have been a parasitic relationship, in which the bacteria consumed the organic products of the methanogen from within—but again, there are problems with this, as an oxygen-dependent bacterium could not generate any energy from the innards of a methanogen unless it could persuade the methanogen to leave its waterhole, those comfortable oxygen-free surroundings. One might picture the α-proteobacteria herding the methanogens and driving them like cattle to an oxygen-rich slaughter field, but for bacteria this is nonsense. In short, the methanogens would starve if they left their waterhole; the oxygen-dependent bacteria would starve if they lived in the waterhole, and the middle ground, a little oxygen, must have been equally disadvantageous to both parties. Such a relationship seems to be mutually insufferable—is this really how the stable symbiotic relationship of the eukaryotic cell began? It is not just improbable, but downright preposterous. Luckily there is another possibility, which until recently seemed fanciful, but is now looking far more persuasive.

3

The Hydrogen Hypothesis

The quest to find the progenitor of the eukaryotic cell has run into dire straits. The idea that there might have been a primitive intermediate, a missing link with a nucleus but no mitochondria, has not been rigorously disproved, but looks more and more unlikely. Every promising example has turned out *not* to be a missing link at all, but rather to have adapted to a simpler lifestyle at a later date. The ancestors of all these apparently primitive groups *did* possess mitochondria, and their descendents eventually lost them while adapting to new niches, often as parasites. It seems possible to *be* a eukaryote without having mitochondria—there are a thousand such species among the protozoa—but it does not seem possible to be a eukaryote without once having *had* mitochondria, deep in the past. If the only way to be a eukaryotic cell is via the possession of mitochondria, then it might be that the eukaryotic cell itself was originally crafted from a symbiosis between the bacterial ancestors of the mitochondria and their host cells.

If the eukaryotic cell was born of a merger between two types of cell, the question becomes more pressing—what types of cell? According to the textbook view, the host cell was a primitive eukaryotic cell, without mitochondria, but this obviously can't be true if there never was a primitive eukaryotic cell that lacked mitochondria. In her endosymbiosis theory, Lynn Margulis had in fact proposed a union between two different types of bacteria, and her hypothesis looked set for a return to prominence after the demise of the missing link. Even so, Margulis and everyone else were thinking along the same lines—the host, they imagined, must have relied on fermentation to produce its energy, in the same way that yeasts do today, and the advantage that the mitochondria brought with them was an ability to deal with oxygen, giving their hosts a more efficient way of generating energy. The exact identity of the host could potentially be traced by comparing the gene sequences of modern eukaryotes with various groups of bacteria and archaea—and modern sequencing technology was just beginning to make that possible. But, as we have just seen, the apparent answer came as another shock: the genes of eukaryotic cells seem to be related most closely to *methanogens*, those obscure methane-producing archaea that live in swamps and intestines.

Methanogens! This answer is an enigma. In Chapter 1, we noted that the methanogens live by reacting hydrogen gas with carbon dioxide, and evanescing methane gas as a waste product. Free hydrogen gas only exists in the absence of oxygen, so the methanogens are restricted to *anoxic* environments—any marginal places where oxygen is excluded. It's actually worse than that. Methanogens can tolerate some oxygen in their surroundings, just as we can survive underwater for a short time by holding our breath. The trouble is that methanogens can't generate any energy in these circumstances—they have to 'hold their breath' until they get back to their preferred anoxic surroundings, because the processes by which they generate their energy can *only* work in the strict absence of oxygen. So if the host cell really was a methanogen, this raises a serious question about the nature of the symbiosis—why on earth would a methanogen form a relationship with any kind of bacteria that relied upon oxygen to live? Today, modern mitochondria certainly depend on oxygen, and if it was ever thus, neither party could make a living in the land of the other. This is a serious paradox and did not seem possible to reconcile in conventional terms.

Then in 1998, Bill Martin, whom we met in Chapter 1, stepped into the frame, presenting a radical hypothesis in *Nature* with his long-term collaborator Miklós Müller, from the Rockefeller University in New York. They called their theory the 'hydrogen hypothesis', and as the name implies it has little to do with oxygen and much to do with hydrogen. The key, said Martin and Müller, is that hydrogen gas can be generated as a waste product by some strange mitochondria-like organelles called *hydrogenosomes*. These are found mostly among primitive single-celled eukaryotes, including parasites such as *Trichomonas vaginalis*, one of the discredited 'archezoa'. Like mitochondria, hydrogenosomes are responsible for energy generation, but they do this in bizarre fashion by releasing hydrogen gas into their surroundings.

For a long time the evolutionary origin of hydrogenosomes was shrouded in mystery, but a number of structural similarities prompted Müller and others, notably Martin Embley and colleagues at the Natural History Museum in London, to propose that hydrogenosomes are actually related to mitochondria—they share a common ancestor. This was difficult to prove as most hydrogenosomes have lost their entire genome, but it is now established with some certainty.[1] In other words, whatever bacteria entered into a symbiotic

---

[1] In 1998, Johannes Hackstein and his colleagues at the University of Nijmegen in Holland discovered a hydrogenosome that had retained its genome, albeit a small one. The isolation of this genome deserved a medal: the hydrogenosome belonged to a parasite that could not be grown in culture and so had to be 'micro-manipulated' from its comfortable home in the hind-gut of cockroaches. Having achieved the unthinkable, Hackstein's group published the complete gene sequence in *Nature* in 2005, and confirmed that hydrogenosomes and mitochondria do have a common α-proteobacterial ancestor.

relationship in the first eukaryotic cell, its descendents numbered among them both mitochondria *and* hydrogenosomes. Presumably, said Martin—and this is the crux of the dilemma faced today—the original bacterial ancestor of the mitochondria *and* hydrogenosomes was able to carry out the metabolic functions of both. If so, then it must have been a versatile bacterium, capable of oxygen respiration as well as hydrogen production. We'll return to this question in a moment. For now, lets simply note that the 'hydrogen hypothesis' of Martin and Müller argues that it was the hydrogen metabolism of this common ancestor, not its oxygen metabolism, which gave the first eukaryote its evolutionary edge.

Martin and Müller were struck by the fact that eukaryotes containing hydrogenosomes sometimes play host to a number of tiny methanogens, which have gained entry to the cell and live happily inside. The methanogens align themselves with the hydrogenosomes, almost as if feeding (Figure 3). Martin and Müller realized that this was exactly what they were doing—the two entities live together in a kind of metabolic wedlock. Methanogens are unique in that they can generate all the organic compounds they need, as well as all their energy, from nothing more than carbon dioxide and hydrogen. They do this by attaching hydrogen atoms (H) onto carbon dioxide (CO_2) to produce the basic building blocks needed to make carbohydrates like glucose ($C_6H_{12}O_6$), and from these they can construct the entire repertoire of nucleic acids, proteins, and lipids. They also use hydrogen and carbon dioxide to generate energy, releasing methane in the process.

While methanogens are uniquely resourceful in their metabolic powers, they nonetheless face a serious obstacle, and we have already noted the reason in Chapter 1. The trouble is that, while carbon dioxide is plentiful, hydrogen is hard to come by in any environment containing oxygen, as hydrogen and oxygen react together to form water. From the point of view of a methanogen, then, anything that provides a little hydrogen is a blessing. Hydrogenosomes are a double boon, because they release both hydrogen gas and carbon dioxide, the very substances that methanogens crave, in the process of generating their own energy. Even more importantly, they don't need oxygen to do this—quite the contrary, they prefer to avoid oxygen—and so they function in the very low-oxygen conditions required by methanogens. No wonder the methanogens suckle up to hydrogenosomes like greedy piglets! The insight of Martin and Müller was to appreciate that this kind of intimate metabolic union might have been the basis of the original eukaryotic merger.

Bill Martin argues that the hydrogenosomes and the mitochondria stand at opposite ends of a little-known spectrum. Rather surprisingly, to anyone who is most familiar with textbook mitochondria, many simple single-celled eukaryotes have mitochondria that operate in the absence of oxygen. Instead of using

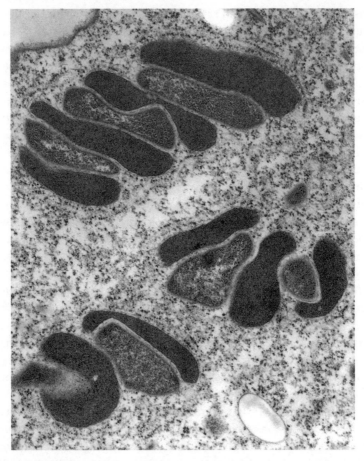

3 The image shows methanogens (light grey) and hydrogenosomes (dark grey). All are living inside the cytoplasm of a much larger eukaryotic cell, specifically the marine ciliate *Plagiopyla frontata*. According to the hydrogen hypothesis, such a close metabolic relationship between methanogens (which need hydrogen to live) and hydrogen-producing bacteria (the ancestor of the mitochondria as well as hydrogenosomes) may have ultimately given rise to the eukaryotic cell itself: the methanogens became larger, to physically engulf the hydrogen-producing bacteria.

oxygen to burn up food, these 'anaerobic' mitochondria use other simple compounds like nitrate or nitrite. In most other respects, they operate in a very similar fashion to our own mitochondria, and are unquestionably related. So the spectrum stretches from aerobic mitochondria like our own, which are dependent on oxygen, through 'anaerobic' mitochondria, which prefer to use other molecules like nitrates, to the hydrogenosomes, which work rather differently but are still related. The existence of such a spectrum focuses attention on the identity of the ancestor that eventually gave rise to the entire spectrum. What, asks Martin, might this common ancestor have looked like?

This question has profound significance for the origin of the eukaryotes, and so for all complex life on earth or anywhere else in the universe. The common ancestor could have taken one of two forms. It could have been a sophisticated bacterium with a large bag of metabolic tricks, which were later distributed to its descendents, as they adapted to their own particular niches. If that were the case, then the descendents could be said to have 'devolved', rather than 'evolved', for they became simpler and more streamlined as they grew specialized. The second possibility is that the common ancestor was a simple oxygen-respiring bacterium, perhaps the free-living ancestor of *Rickettsia* we discussed in the previous chapter. If that were the case, then its descendents must have become more diverse over evolution—they 'evolved' rather than 'devolved'. The two possibilities generate specific predictions. In the first case, if the ancestral bacterium was metabolically sophisticated, then it was in a position to hand down specialized genes directly to its ancestors, such as those for hydrogen production. Any eukaryotes adapting to hydrogen production could have inherited its genes from this common ancestor, regardless of how diverse they were to become later. Hydrogenosomes are found in diverse groups of eukaryotes. If they inherited their hydrogen-producing genes from the same ancestor, then these genes should be closely related to each other, regardless of how diverse their host cells became later. On the other hand, if all the diverse groups had originally inherited simple, oxygen-respiring mitochondria, they had to invent all the different forms of anaerobic metabolism independently, whenever they happened to adapt to a low-oxygen environment. In the case of the hydrogenosomes, the hydrogen-producing genes would necessarily have evolved independently in each case (or transferred randomly by lateral gene transfer), and so their evolutionary history would be just as varied as that of their host cells.

These possibilities give a plain choice. If the ancestor was metabolically sophisticated, then all the hydrogen-producing genes should be related, or at least *could* be related. On the other hand, if it was metabolically simple, then all these genes should be unrelated. So which is it? The answer is as yet unproved, but with a few exceptions, most evidence seems to favour the former proposition. Several studies published in the first years of the millennium attest to a single origin for at least a few genes in the anaerobic mitochondria and hydrogenosomes, as predicted by the hydrogen hypothesis. For example, the enzyme used by hydrogenosomes to generate hydrogen gas (the pyruvate: ferredoxin oxidoreductase, or PFOR), was almost certainly inherited from a common ancestor. Likewise the membrane pump that transports ATP out of both mitochondria and hydrogenosomes seems to share a similar ancestry; and an enzyme required for the synthesis of a respiratory iron-sulphur protein also appears to derive from a common ancestor. These studies imply that the common ancestor was indeed metabolically versatile and could respire using oxygen or other molecules, or generate hydrogen gas, as the circumstances dictated. Critically, such versatility (which might otherwise sound somewhat hypothetical) does exist today in some groups of α-proteobacteria such as *Rhodobacter*, which might therefore resemble the ancestral mitochondria better than does *Rickettsia*.

If so, why is the *Rickettsia* genome so similar to modern mitochondria? Martin and Müller argue that the parallels between *Rickettsia* and mitochondria derive from two factors. First, *Rickettsia* are α-proteobacteria, so their genes for *aerobic* (oxygen-dependent) respiration should indeed be related to the genes of *aerobic* mitochondria, as well as to those of other free-living oxygen-dependent α-proteobacteria. In other words, the mitochondrial genes are similar to those of *Rickettsia* not because they are necessarily *derived* from *Rickettsia*, but because *Rickettsia* and mitochondria both derived their genes for aerobic respiration from a common ancestor that might have been very different to *Rickettsia*. If that is the case, it begs the question: why, if they derived from a very different ancestor, did they eventually become so similar? This brings us to the second point postulated by Martin and Müller—they did so by *convergent evolution*, as discussed at the beginning of Part 1. Both *Rickettsia* and mitochondria share a similar lifestyle and environment: both generate energy by aerobic respiration inside other cells. Their genes are subject to similar selection pressures, which might easily bring about convergent changes in both the spectrum of surviving genes, and in their detailed DNA sequence. If convergence is responsible for the similarities, then the genes of *Rickettsia* should only be similar to the mammalian *oxygen-dependent* mitochondria, and not to the other types of *anaerobic* mitochondria that we have been discussing in the last few pages. If the common ancestor was very different to *Rickettsia*—if it was actually a versatile bacterium like *Rhodobacter*, with a bag of metabolic tricks—we wouldn't expect to find parallels between *Rickettsia* and these anaerobic mitochondria; and for the most part we do not.

From addict to world-beater

At present, evidence suggests that the two protagonists of the eukaryotic merger were a methanogen and a metabolically versatile α-proteobacterium, like *Rhodobacter*. The hydrogen hypothesis reconciles the apparently discordant ecological requirements of these protagonists by arguing that the deal revolved around the methanogen's metabolic addiction to hydrogen, and the bacterium's ability to provide it. But for many people this simple solution raises as many questions as it answers. How did a merger that could *only* work under anaerobic conditions (or low oxygen levels) bring about the glorious flowering of the eukaryotes, and especially multicellular eukaryotes, virtually all of which are now totally *dependent* on oxygen? Why did it happen at a time when oxygen levels were rising in the atmosphere and oceans—are we to believe that this was merely coincidence? If the first eukaryote lived in strictly anaerobic conditions, why did it not lose all its oxygen-respiring genes by evolutionary attrition, as anaerobic eukaryotes living today have done? And if the host was not a primitive eukaryotic cell, capable of changing shape and engulfing whole bacteria, how did the α-proteobacteria gain entrance at all?

Along with more recent evidence, the hydrogen hypothesis can explain each of these difficult questions, and remarkably, without even calling for a single evolutionary innovation (the evolution of new traits). Having struggled with these ideas myself, from a position, I should confess, of initial hostility, I believe that something of the sort almost certainly did happen. The chain of events proposed by Martin and Müller has an inexorable evolutionary logic about it, but critically, is dependent on the environment—on evolutionary selection pressures provided by a set of contingent circumstances, which we know happened on Earth at this time. The question is, if the film of life were to be replayed over and over again, as advocated by Stephen Jay Gould, would the same chain of events be repeated? I doubt it, for it seems to me unlikely that the particular train of events proposed by Martin and Müller would repeat itself in a hurry, if at all. My doubts are stronger still for an alien planet, with a different set of contingent circumstances. This is why I suspect that the evolution of the eukaryotic cell was fundamentally a chance event, and happened but once on Earth. Let's consider what might have happened. I'll narrate it as a 'just-so' story, for clarity, and omit the many 'may haves' that clutter the essential meaning (see also Figure 4).

Once upon a time, a methanogen and an α-proteobacterium lived side-by-side, deep in the ocean where oxygen was scarce. The α-proteobacterium was a scavenger, which plied its living in many ways, but often generated energy by way of fermenting food (the remains of other bacteria), excreting hydrogen and carbon dioxide as waste. The methanogen lived happily on these waste products, for it could use them to build everything it needed. The arrangement was so cosy and convenient that the two partners grew closer every day, and the methanogen gradually changed its shape (it has a cytoskeleton and shape can be selected for) to embrace its benefactor. Such shape changes can be seen in Figure 3.

As time passed, the embrace became quite suffocating, and the poor α-proteobacterium didn't have much surface left to absorb its food. It would

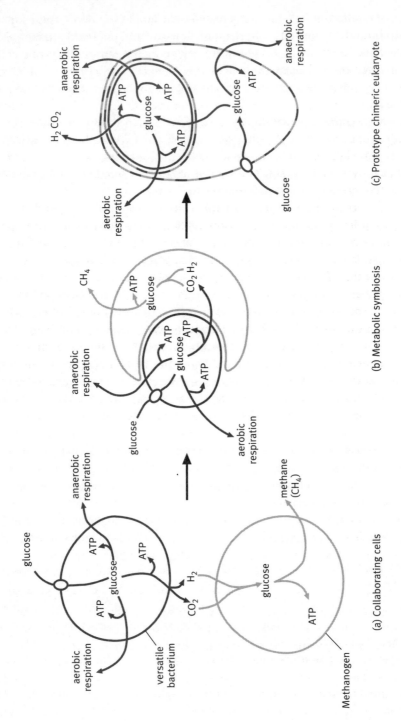

(a) Collaborating cells

(b) Metabolic symbiosis

(c) Prototype chimeric eukaryote

die of starvation unless a compromise could be found, but by now it was tightly bound to the methanogen and couldn't just leave. One possibility would have been for it to physically move inside the methanogen. The methanogen could then use its own surface to absorb all the food needed, and the two could continue their cosy arrangement. So the α-proteobacterium moved in.

Before we continue with our just-so story, let's just note that there are several examples of bacteria living inside other bacteria; it's not necessary to be phagocytosed. The best-known example is *Bdellovibrio*, a fearsome bacterial predator that moves quickly, at about 100 cell-lengths per second, until it collides with a host bacterium. Just before colliding it spins rapidly and penetrates the cell wall. Once inside, it breaks down the host's cellular constituents and multiplies, completing its life cycle within 1–3 hours. How many non-predatory bacteria gain access to other bacteria or archaea is a moot question, but the basic postulate of hydrogen hypothesis, that phagocytosis is not necessary to breech another cell, does not sound unreasonable. Indeed, a discovery in 2001 makes it seem more reasonable: mealybugs, the small, white, cotton ball-like insects found living on many house plants, contain β-proteobacteria living within some of their cells as endosymbionts (collaborative bacteria living inside other cells). Incredibly, these endosymbiotic bacteria contain even smaller γ-proteobacteria living inside them. Thus one bacterium lives in another, which in turn lives inside an insect cell, showing that bacteria can indeed live peacefully inside one another. The discovery smacks of the old verse: 'big fleas have little fleas upon their backs to bite 'em; and little fleas have smaller fleas, and so ad infinitum.'

Let's resume our story. The α-proteobacterium has now found itself inside a methanogen; so far so good. But there was a new problem. The methanogen wasn't practiced at absorbing its food—it normally made its own from hydrogen and carbon dioxide, and so it couldn't feed its benefactor after all. Luckily, the α-proteobacterium came to the rescue. It had all the genes necessary for absorbing food, so it could hand them over to the methanogen and all would

4 Hydrogen hypothesis. Simplified schematic showing the relationship between a versatile bacterium and a methanogen. (a) The bacterium is capable of different forms of aerobic and anaerobic respiration, as well as fermentation to generate hydrogen; under anaerobic conditions the methanogen makes use of the hydrogen and carbon dioxide given off by the bacterium. (b) The symbiosis becomes closer as the methanogen is now dependent on hydrogen produced by the bacterium, which is gradually engulfed. (c) The bacterium is now completely engulfed. Gene transfer from the bacterium to the host enables the host to import and ferment organics in the same way as the bacterium, freeing it from its commitment to methanogenesis. The dashed line indicates that the cell is chimeric.

be well. The methanogen could now absorb food from outside, and this should have enabled the α-proteobacterium to continue supplying it with hydrogen and carbon dioxide. But the problem was not so easily resolved. The methanogen was now absorbing food, and converting this into glucose on behalf of the α-proteobacterium. The trouble was that methanogens normally use glucose to build up complex organic molecules, whereas the α-proteobacteria break it down for energy. The glucose was subject to a tug of war: instead of passing it on to the greedy bacteria living inside it, which in turn would feed it, the methanogen was inadvertently diverting the food supply to construction projects. If this persisted, both would starve. The α-proteobacterium could solve the problem by handing over more of its genes, which would enable the methanogen to ferment some glucose into breakdown products that the α-proteobacterium could then use. So it handed over the genes.

How, you may be wondering, do unthinking bacteria come to hand over all the genes needed to make such a deal work? This kind of question troubles any discussion of natural selection, but it is answered, as most of them are, simply by thinking about the problem in terms of a population. In this case, we are thinking about a population of cells, some of which thrive, some die, and some continue as they are. Consider a population of methanogens, all of which have many small α-proteobacteria living with them in close proximity. Some individual relationships are relatively 'distant', in that the α-proteobacteria are not physically enveloped by the methanogen; they get along fine, but quite a lot of hydrogen is lost to the surroundings, or to other methanogens. These 'loose' relationships may lose out to closer relationships, in which the α-proteobacteria are being enveloped, and less hydrogen is lost. Of course, each methanogen is likely to harbour a number of α-proteobacteria, a few of which are probably more enveloped than others. So while the overall assemblage might function happily, a few particular α-proteobacteria might be suffocating from the closeness of the embrace. What happens if they die of suffocation? Assuming there are others to take their place, the overall union of the symbiosis might not be affected at all, but the dying α-proteobacterium spills its genes into the environment. Some of these will be taken up by the methanogen in the usual manner by lateral gene transfer, and some will become incorporated into the methanogen's chromosome. Let's assume that this process is happening simultaneously in hordes of many millions, perhaps many billions, of symbiotic methanogens. By the law of averages, some at least will happen to transfer all the right genes (which are in any case grouped together as a single functional unit, or an *operon*). If so, then the methanogen will be able to absorb organic compounds from the surroundings. Exactly the same process would account for the transfer of genes for fermentation to the methanogen; and indeed there is no reason why both sets of genes should not be transferred simultaneously. It's all down to population dynamics: if the beneficiary happens to be more successful than its brethren, then the power of natural selection will soon amplify the fruit of the successful union.

But there is a startling ending to this story. Having acquired two sets of genes by lateral gene transfer, the methanogen could now do everything. It could absorb food from the surroundings, and ferment it to produce energy. Like an ugly duckling transforming into a swan it suddenly didn't need to be a methanogen any more. It was free to roam and no longer needed to avoid oxygenated surroundings, which, once upon a time, would have blocked its only source of energy—methane production. What's more, when roaming in aerobic conditions, the internalized α-proteobacteria could use the oxygen to generate energy much more efficiently, so they too benefited. All that the host (we can't reasonably call it a methanogen any more) needed was a tap, an ATP pump, which it could plug into the membrane of its α-proteobacterial guest to drain off its ATP, and the entire world would be its stage. The ATP pumps are indeed a eukaryotic invention, and if we are to believe the gene sequences of different groups of eukaryotes, they evolved very early in the history of the eukaryotic union.

So the answer to the question of life, the universe, and everything, or the origin of the eukaryotic cell, was simply gene transfer. Through a series of small and realistic steps, the hydrogen hypothesis explains how a chemical dependency between two cells evolved to become a single chimeric cell containing organelles that function as mitochondria. This cell is able to import organic molecules, like sugars, across its external membrane and to ferment them in its cytoplasm, in the same fashion as yeast. It is able to pass the fermentation products onwards to the mitochondria, which can then oxidize them using oxygen, or for that matter other molecules such as nitrate. This chimeric cell does not yet have a nucleus. It may or may not have lost its cell wall. It does have a cytoskeleton, but has probably not adapted it for changing shape like an amoeba; it merely provides rigid structural support. In short, we have derived a 'prototype' eukaryote without a nucleus. We'll return in Part 3 to how this prototype might have gone on to become a fully-fledged eukaryote. To end this chapter, though, let's consider the play of chance in the hydrogen hypothesis.

Chance and necessity

Each step of the hydrogen hypothesis depends on selection pressures that may or may not have been strong enough to force that particular adaptation, and each step depends utterly on the last—hence the deep uncertainty about whether exactly the same sequence of steps would be repeated if the film of life were to be played again. For opponents of the theory, the greatest problem lies in the last few steps, the transition from a chemical dependency that only works in the absence of oxygen, to the flowering of the eukaryotic cell as an oxygen-dependent cell which thrives in aerobic conditions. For this to happen, all the genes needed for oxygen respiration must have survived intact throughout the early chimera years, despite falling into prolonged disuse. If the theory is correct then obviously they did; but if the transition had taken just a little longer, then the genes for oxygen respiration may easily have been lost by mutation, and so the oxygen-dependent multicellular eukaryotes would never have been born; and neither would we, nor anything else beyond bacterial slime.

The fact that these genes were not lost sounds like an outrageous fluke, and perhaps this alone accounts for why the eukaryotes only evolved once. But perhaps there was also something about the environment that gave our ancestor a nudge in the right direction. In *Science*, in 2002, Ariel Anbar at the University of Rochester, and Andrew Knoll at Harvard, suggested that the changing chemistry of the oceans might explain why the eukaryotes evolved when they did, at a time of rising oxygen levels, despite their strictly anaerobic lifestyle. As atmospheric oxygen levels rose, so too did the sulphate concentration of the oceans (because the formation of sulphate, SO_4^{2-}, requires oxygen). This in turn led to a massive rise in the population of another type of bacteria, the sulphate-reducing bacteria, which we met briefly in Chapter 1. There, we noted that the sulphate-reducing bacteria almost invariably out-compete the methanogens for hydrogen in today's ecosystems, so the two species are rarely found living together in the oceans.

When we think of a rise in oxygen levels, we tend to think of more fresh air, but the effects can actually be startlingly counterintuitive. As I discussed in an earlier book, *Oxygen: The Molecule that Made the World*, what actually happens is this. The foul sulphurous fumes emanating from volcanoes contain sulphur in forms such as elemental sulphur and hydrogen sulphide. When this sulphur reacts with oxygen, it is oxidized to produce sulphates. This is the same problem we face today with acid rain—the sulphur compounds released into the atmosphere from factories become oxidized by oxygen to form sulphuric acid, H_2SO_4. The 'SO_4' is the sulphate group, and it is this group that the sulphate-reducing bacteria need to oxidize hydrogen—which in chemical terms is exactly the same thing as reducing the sulphate, hence the name of the bacteria. Here is the rub. When oxygen levels rise, sulphur is oxidized to form sulphates, which accumulate in the oceans—the more oxygen, the more sulphate. This is the raw material needed by the sulphate-reducing bacteria, which convert sulphate into hydrogen sulphide. Although a gas, hydrogen sulphide is actually heavier than water, and so it sinks down towards the bottom of the oceans. What happens next depends on the dynamic balance in the concentrations of sulphate, oxygen, and so on. However, if hydrogen sulphide is formed more rapidly than oxygen in the deep oceans (where photosynthesis is less active because sunlight does not permeate down) then the outcome is a 'stratified' ocean. The best example today is the Black Sea. In general, in stratified oceans the depths become stagnant, reeking of hydrogen sulphide (or technically, 'euxinic'), whereas the sunlit surface waters fill up with oxygen. Geological evidence shows that this is exactly what happened in the oceans throughout the world two billion years ago, and the stagnant conditions apparently persisted for at least a billion years, and probably longer.

Now to my point. When the oxygen levels rose, so too did the population of sulphate-reducing bacteria. If, like today, the methanogens couldn't compete with these voracious bacteria, then they would have faced a pressing shortage of hydrogen. This would have given the methanogens a good reason to enter into an intimate partnership with a hydrogen-producing bacterium, such as *Rhodobacter*. So far, so good. But what forced the prototype eukaryote up into the oxygenated surface waters before it lost its genes for oxygen respiration? Again, it may have been the sulphate-reducing bacteria. This time, the competition could have been for nutrients like nitrates, phosphates, and some metals, which are more plentiful in the sunlit surface waters. If the prototype eukaryote were no longer tied to its waterhole, then it would benefit from moving up in the world. If so, competition may have pressed the first eukaryotic cells up into the oxygenated surface waters long before they lost their genes for oxygen respiration, where they would have found good use for them. What an ironic turn of events! It seems the majestic rise of the eukaryotes was contingent on unequal competition between incompatible tribes of bacteria, the glories of nature upon the flight of the weak. The Bible was right: the meek really did inherit the Earth.

Is this truly what happened? It's too early to say for sure. I'm reminded of that amiably cynical Italian turn of phrase, which translates roughly as 'It may not be true, but it is well contrived'. In my view, the hydrogen hypothesis is a radical hypothesis, which makes better use of the known evidence than any other theory; and it has about the right combination of probability and improbability to explain the fact that the eukaryotes arose only once.

Beyond that there is another consideration, which makes me believe the hydrogen hypothesis, or something like it, is basically correct—and this relates to a more profound advantage provided by mitochondria. It explains why *all* known eukaryotes either have, or once had (then lost) mitochondria. As we noted earlier, the eukaryotic lifestyle is energetically profligate. Changing shape and engulfing food is highly energetic. The only eukaryotes that can do it without mitochondria are parasites that live in the lap of luxury, and they barely need to do anything *but* change their shape. In the next few chapters, we'll see that virtually every aspect of the eukaryotic lifestyle—changing shape with a dynamic cytoskeleton, becoming large, building a nucleus, hoarding reams of DNA, sex, multicellularity—all these depend on the existence of mito-chondria, and so can't, or are at least highly unlikely to, happen in bacteria.

The reason relates to the precise mechanism of energy production across a membrane. Energy is generated in essentially the same way in both bacteria and mitochondria, but the mitochondria are internalized within cells, whereas bacteria use their cell membrane. Such internalization not only explains the success of the eukaryotes, but it even throws light on the origin of life itself. In Part 2, we'll consider how the mechanism of energy-generation in bacteria and mitochondria shows how life might have originated on earth, and why it gave the eukaryotes, and only the eukaryotes, the opportunity to inherit the world.

PART 2

The Vital Force
Proton Power and the Origin of Life

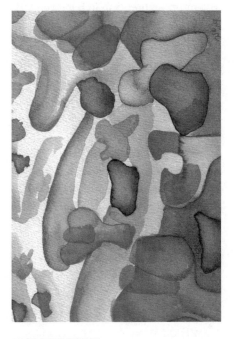

The way in which mitochondria generate energy is one of the most bizarre mechanisms in biology. Its discovery has been compared with those of Darwin and Einstein. Mitochondria pump protons across a membrane to generate an electric charge with the power, over a few nanometres, of a bolt of lightning. This proton power is harnessed by the elementary particles of life— mushroom-shaped proteins in the membranes—to generate energy in the form of ATP. This radical mechanism is as fundamental to life as DNA itself, and gives an insight into the origin of life on Earth.

The elementary particles of life—energy-generating proteins in the mitochondrial membranes

Energy and life go hand in hand. If you stop breathing, you will not be able to generate the energy you need for staying alive and you'll be dead in a few minutes. Keep breathing. Now the oxygen in your breath is being transported to virtually every one of the 15 trillion cells in your body, where it is used to burn glucose in cellular respiration. You are a fantastically energetic machine. Gram per gram, even when sitting comfortably, you are converting 10 000 times more energy than *the sun* every second.

This sounds improbable, to put it mildly, so let's consider the numbers. The sun's luminosity is about 4×10^{26} watts and its total mass is 2×10^{30} kg. Over its projected lifetime, about 10 billion years, each gram of solar material will produce about 60 million kilojoules of energy. The generation of this energy is not explosive, however, but slow and steady, providing a uniform and long-lived rate of energy production. At any one moment, only a small proportion of the sun's vast mass is involved in nuclear fusions, and these reactions take place only in the dense core. This is why the sun can burn for so long. If you divide the luminosity of the sun by its mass, each gram of solar mass yields about 0.0002 milliwatts of energy, which is 0.0000002 joules of energy per gram per second (0.2 µJ/g/sec). Now let's assume that you weigh 70 kg, and if you are anything like me you will eat about 12 600 kilojoules (about 3000 calories) per day. Assuming barely 30 per cent efficiency, converting this amount of energy (into heat or work or fat deposits) averages 2 millijoules per gram per second (2 mJ/g/sec) or about 2 milliwatts per gram—a factor of 10 000 greater than the sun. Some energetic bacteria, such as *Azotobacter*, generate as much as 10 joules per gram per second, out-performing the sun by a factor of 50 million.

At the microscopic level of cells, all life is animated, even the apparently sessile plants, fungi and bacteria. Cells whirr along, machine-like in the way that they channel energy into particular tasks, whether these are locomotion, replication, constructing cellular materials, or pumping molecules in and out of the cell. Like machines, cells are full of moving parts, and to move they need energy. Any form of life that can't generate its own energy is hard to distinguish from inanimate matter, at least in philosophical terms. Viruses only 'look' alive because they are organized in a way that suggests the hand of a designer, but they occupy a shadowy landscape between the living and the nonliving. They have all the information they need to replicate themselves, but must remain inert until they infect a cell, as they can only replicate themselves using the energy and cellular machinery of the infected cell. This means that viruses could not have been the first living things on Earth, nor could they have delivered life from outer space to our planet: they depend utterly on other living organisms and cannot exist without them. Their simplicity is not primitive, but a refined, pared-down complexity.

Despite its obvious importance to life, biological energy receives far less attention than it deserves. According to molecular biologists, life is all about information. Information is encoded in the genes, which spell out the instructions for building proteins, cells, and bodies. The double helix of DNA, the stuff of genes, is an icon of our information age, and the discoverers of its structure, Watson and Crick, are household names. The reasons for this status are a mixture of the personal, the practical, and the symbolic. Crick and Watson were brilliant and flamboyant, and unveiled the structure of DNA with the aplomb of conjurors. Watson's famous book narrating the discovery, *The Double Helix*, defined a generation and changed the way that science is perceived by the general public; and he has been an outspoken and passionate advocate of genetic research ever since. In practical terms, sequencing the codes of genes enables us to compare ourselves with other organisms and to peer into our own past, as well as the story of life. The human genome project is set to reveal untold secrets of the human condition, and gene therapy holds a candle of hope for people with crippling genetic diseases. But most of all, the gene is a potent symbol. We may argue over nature versus nurture, and rebel against the power of the genes; we may worry about genetically modified crops and the evils of cloning or designer babies; but whatever the rights and wrongs, we worry because we know deep down, viscerally, that genes are important.

Perhaps because molecular biology is so central to modern biology we pay lip service to the energy of life in the same way that we acknowledge the industrial revolution as a necessary precursor of the modern information age. Electrical power is so obviously essential for a computer to function that the point is almost too banal to be worth making. Computers are important because of their data-processing capacity, not because they are electronic. We may only appreciate the importance of a power supply when the batteries run out, and there's no plug to be seen. In the same way, energy is important to supply the needs of cells, but is plainly secondary to the information systems that control it and draw on it. Life without energy is dead, but energy without information to control it might seem as destructive as a volcano, an earthquake, or an explosion. Or is it? The flood of life-giving rays from the sun suggests an uncontrolled flow of energy is not inevitably destructive.

In contrast to our worries over genetics, I wonder how many people exercise themselves over the sinister implications of bioenergetics. Its terminology is what the Soviets used to call obscurantist, as full of mysterious symbols as a wizard's robes. Even willing students of biochemistry are wary of terms like 'chemiosmotics' and 'proton-motive force'. Although the implications of these ideas may turn out to be as important as those of genetics, they are little known. The hero of bioenergetics, Peter Mitchell, who won the Nobel Prize for chemistry in 1978, is hardly a household name, even though he ought to be as well known as Watson and Crick. Unlike Watson and Crick, Mitchell was an eccentric and reclusive genius, who set up his own laboratory in an old country house in Cornwall, which he had renovated himself, following his own designs. At one time, his research was funded in part by the proceeds from a herd of dairy cows, and he even won a prize for the quality of his cream. His writings did not compete with Watson's *Double Helix*—besides the usual run of dry academic papers (even more obscure than usual in Mitchell's own case), he expounded his theories in two 'little grey books', published privately and circulated among a few interested professionals. His ideas can't be encapsulated in a visually arresting emblem like the double helix, redolent of the standing of science in society. Yet Mitchell was largely responsible for articulating and proving one of the very greatest insights in biology, a genuine and bizarre revolution that overturned long-cherished ideas. As the eminent molecular biologist Leslie Orgel put it: 'Not since Darwin has biology come up with an idea as counterintuitive as those of, say, Einstein, Heisenberg or Schrödinger. . . his contemporaries might well have asked "Are you serious, Dr Mitchell?"'

Part 2 of this book is broadly about Mitchell's discovery of the way that life generates its energy, and the implications of his ideas for the origin of life. In later chapters, these ideas will enable us to see what the mitochondria did for us: why they are essential for the evolution of all higher forms of life. We'll see that the precise mechanism of energy generation is vital: it constrains the opportunities open to life, and it does so very differently in bacteria and eukaryotic cells. We'll see that the precise mechanism of energy generation precluded bacteria from ever evolving beyond bacteria—from ever becoming complex multicellular organisms—while at the same time it gave the eukaryotes unlimited possibilities to grow in size and sophistication, propelling them up a ramp of ascending complexity to the marvels that we see all around us today. But this same mechanism of energy generation constrained the eukaryotes, too, albeit in utterly different ways. We'll see that sex, and even the origin of two sexes, is explained by the constraints of this same form of energy generation. And beyond that we'll see that our terminal decline into old age and death also stems from the small print of the contract that we signed with our mitochondria two billion years ago.

To understand all this, we first need to grasp the importance of Mitchell's insights into the energy of life. His ideas are simple enough in outline, but to feel their full force we'll need to look a little deeper into their details. To do this, we'll take a historical perspective, and as we go along we can savour the dilemmas, and the great minds that wrestled with them in the golden age of biochemistry, littered with Nobel Prizes. We'll follow the shining path of discovery, which showed how cells generate so much energy that they put the sun in the shade.

4

The Meaning of Respiration

Metaphysicians and poets used to write earnestly about the flame of life. The sixteenth century alchemist Paracelsus even explicitly declared: 'Man dies like a fire when deprived of air.' While metaphors are supposed to illuminate truths, I suspect that the metaphysicians would have been contemptuous of Lavoisier, the 'father of modern chemistry', who argued that the flame of life was not merely a metaphor, but exactly analogous to a real flame. Combustion and respiration are one and the same process, Lavoisier said, in the kind of literal scientific spoiler that poets have protested about ever since. In a paper addressed to the French Royal Academy in 1790, Lavoisier wrote:

Respiration is a slow combustion of carbon and hydrogen, similar in every way to that which takes place in a lamp or lighted candle and, in that respect, breathing animals are active combustible bodies that are burning and wasting away. . . it is the very substance of the animal, the blood, which transports the fuel. If the animal did not habitually replace, through nourishing themselves, what they lose through respiration, the lamp would very soon run out of oil and the animal would perish, just as the lamp goes out when it lacks fuel.

Both carbon and hydrogen are extracted from the organic fuels present in food, such as glucose, so Lavoisier was correct in saying that the respiratory fuels are replenished by food. Sadly, he never got much further. Lavoisier lost his head to the guillotine in the French Revolution four years later. In his book, *Crucibles*, Bernard Jaffe assigns the 'judgement of posterity' to this deed: 'Until it is realized that the gravest crime of the French Revolution was not the execution of the King, but of Lavoisier, there is no right measure of values; for Lavoisier was one of the three or four greatest men France has produced.' A century after the Revolution, in the 1890s, a public statue of Lavoisier was unveiled. It later transpired that the sculptor had used the face, not of Lavoisier, but of Condorcet, the Secretary of the Academy during Lavoisier's last years. The French pragmatically decided that 'all men in wigs look alike anyway', and the statue remained until it was melted down during the Second World War.

Though Lavoisier revolutionized our understanding of the chemistry of respiration, even he didn't know where it took place—he believed it must happen in the blood as it passed through the lungs. In fact, the site of respiration remained controversial through much of the nineteenth century, and it was not until 1870 that the German physiologist Eduard Pflüger finally persuaded biologists that respiration takes place within the individual cells of the body, and is a general property of all living cells. Even then, nobody knew exactly whereabouts in the cell respiration took place; it was commonly ascribed to the nucleus. In 1912, B. F. Kingsbury argued that respiration actually took place in the mitochondria, but this was not generally accepted until 1949, when Eugene Kennedy and Albert Lehninger first demonstrated that the respiratory enzymes are located in the mitochondria.

The combustion of glucose in respiration is an electrochemical reaction—an *oxidation* to be precise. By today's definition, a substance is *oxidized* if it loses electrons. Oxygen (O_2) is a strong oxidizing agent because it has a strong chemical 'hunger' for electrons, and tends to extract them from substances such as glucose or iron. Conversely, a substance is *reduced* if it gains electrons. Because oxygen gains the electrons extracted from glucose or iron, it is said to be reduced to water (H_2O). Notice that in forming water each atom of the oxygen molecule also picks up two protons (H^+) to balance the charges. Overall, then, the oxidation of glucose equates to the transfer of two electrons and two protons—which together make up two whole hydrogen atoms—from glucose to oxygen.

Oxidation and reduction reactions are always coupled, because electrons are not stable in isolation—they must be extracted from another compound. Any reaction that transfers electrons from one molecule to another is called a *redox reaction*, because one partner is oxidized and the other is simultaneously reduced. Essentially all the energy-generating reactions of life are redox reactions. Oxygen isn't always necessary. Many chemical reactions are redox reactions, as electrons are transferred, but they don't all involve oxygen. Even the flow of electricity in a battery can be regarded as a redox reaction, because electrons flow from a source (which becomes progressively oxidized) to an acceptor (which becomes reduced).

Lavoisier was *chemically* correct, then, when he said that respiration was a combustion, or oxidation, reaction. However, he erred not just about the site of respiration, but also about its function: he believed that respiration was needed to generate heat, which he thought of as an indestructible fluid. But clearly we don't function like a candle. When we burn fuel, we don't simply radiate the energy as heat, we use it to run, to think, to build muscles, to cook a meal, to make love, or for that matter, candles. All these tasks can be defined as 'work', in the sense that they require an input of energy to take place—they don't occur spontaneously. An understanding of respiration that reflected all this awaited a better appreciation of the nature of energy itself, which only came with the science of thermodynamics in the mid nineteenth century. The most revealing discovery, by British scientists James Prescott Joule and William Thompson (Lord Kelvin), in 1843, was that heat and mechanical work are interchangeable—the principle of the steam engine. This led to a more general realization, later referred to as the first law of thermodynamics, that energy can be converted from one form into another, but never created nor destroyed. In 1847, the German physician and physicist Hermann von Helmholtz applied these ideas to biology, when he showed that the energy released from food molecules in respiration was used partly to generate the force in the muscles. This appliance of thermodynamics to muscle contraction was a remarkably mechanical insight in an age still besieged by 'vitalism'—the belief that life was animated by special forces, or spirits, which could not be reproduced by mere chemistry.

The new understanding of energy eventually fostered an appreciation that the bonds of molecules contain an implicit 'potential' energy that can be released when they react. Some of this energy can be captured, or *conserved* in a different form, by living things, and then channelled into work, such as the contraction of muscles. For this reason, we can't talk about 'energy generation' in living things, although it is such a convenient phrase that I have occasionally transgressed. When I say energy generation I mean the conversion of potential energy, implicit in the bonds of fuels like glucose, into the biological energy 'currencies' that organisms use to power the various forms of work; in other words, I mean the generation of more working currency. And it is to these energy currencies that our story now turns.

Colours in the cell

By the end of the nineteenth century, scientists knew that respiration took place in cells, and was the source of energy for every aspect of life. But how it actually worked—how the energy released by the oxidation of glucose was coupled to the energetic demands of life—was anybody's guess.

Clearly glucose does not ignite spontaneously in the presence of oxygen. Chemists say that oxygen is thermodynamically reactive but *kinetically* stable: it doesn't react quickly. This is because oxygen must be 'activated' before it is able to react. Such activation requires either an input of energy (like a match), or a catalyst, which is to say a substance that lowers the activation energy needed for the reaction to take place. For scientists of the Victorian era, it seemed likely that any catalyst involved in respiration would contain iron, for iron has a high affinity for oxygen—as in the formation of rust—but can also bind to oxygen reversibly. One compound that was known to contain iron, and to bind to oxygen reversibly, was haemoglobin, the pigment that imparts the colour to red blood cells; and it was the colour of blood that gave the first clue to how respiration actually works in living cells.

Pigments such as haemoglobin are coloured because they absorb light of particular colours (bands of light, as in a rainbow) and reflect back light of other colours. The pattern of light absorbed by a compound is known as its absorption spectrum. When binding oxygen, haemoglobin absorbs light in the blue-green and yellow parts of the spectrum, but reflects back red light, and this is the reason why we perceive arterial blood as a vivid red colour. The absorption spectrum changes when oxygen dissociates from haemoglobin in venous blood. Deoxyhaemoglobin absorbs light across the green part of the spectrum, and reflects back red and blue light. This gives venous blood its purple colour.

Given that respiration takes place inside cells, researchers started looking for similar pigments in animal tissues rather than in the blood. The first success came from a practicing Irish physician named Charles MacMunn, who worked in a small laboratory in the hay loft over his stables, carrying out research in his spare time. He used to keep watch for patients coming up the path through a small hole in the wall, and would ring through to his housekeeper if he didn't wish to be disturbed. In 1884, MacMunn found a pigment inside tissues, whose absorption spectrum varied in a similar manner to haemoglobin. He claimed that this pigment must be the sought-after 'respiratory pigment', but unfortunately McMunn could not explain its complex absorption spectrum, or even show that the spectrum was attributable to it at all. His findings were quietly forgotten until David Keilin, a Polish biologist at Cambridge, rediscovered the pigment in 1925. By all accounts Keilin was a brilliant researcher, an inspiring lecturer, and a kindly man, and he made a point of deferring priority to MacMunn. In fact, though, Keilin went well beyond MacMunn's observations, showing that the spectrum was not attributable to one pigment, but to three. This enabled him to explain the complex absorption spectrum that had stumped MacMunn. Keilin named the pigments *cytochromes* (for cellular pigments) and labelled them a, b, and c, according to the position of the bands on their absorption spectra. These labels are still in use today.

Curiously, however, none of Keilin's cytochromes reacted directly with oxygen. Clearly something was missing. This missing link was elucidated by the German chemist Otto Warburg (in Berlin) who received the Nobel Prize for his work in 1931. I say elucidated, because Warburg's observations were indirect, and quite ingenious. They had to be, for the respiratory pigments, unlike haemoglobin, are present in vanishingly tiny amounts within cells, and could not be isolated and studied directly using the rough and ready techniques of the time. Instead, Warburg drew on a quirky chemical property—the binding of carbon monoxide to iron compounds in the dark, and its dissociation from them when illuminated—to work out the absorption spectrum of what he called the 'respiratory ferment'.[1] The spectrum turned out to be that of a haemin compound, similar to haemoglobin and chlorophyll (the green pigment of plants that absorbs sunlight in photosynthesis).

Interestingly, the respiratory ferment absorbed light strongly in the blue part of the spectrum, reflecting back green, yellow, and red light. This imparted a brownish shade, not red, like haemoglobin, nor green like chlorophyll. However, Warburg found that simple chemical changes could turn the ferment red or green, with spectra closely resembling those of haemoglobin or chlorophyll. This raised a suspicion, expressed in his Nobel lecture, that the 'blood pigment and the leaf pigments have both arisen from the ferment. . . for evidently, the ferment existed earlier than haemoglobin and chlorophyll.' His words imply that respiration had evolved before photosynthesis, a visionary conclusion, as we shall see.

The respiratory chain

Despite these great strides forward, Warburg still could not grasp how respiration actually took place. At the time of his Nobel Prize, he seemed inclined to believe that respiration was a one-step process (releasing all the energy bound up in glucose at once) and was unable to explain how David Keilin's cytochromes fitted into the overall picture. Keilin, in the meantime, was developing the idea of a *respiratory chain*. He imagined that hydrogen atoms, or at least their constituent components, protons and electrons, were stripped from glucose, and passed down a chain of cytochromes, from one to the next, like firemen passing buckets hand to hand, until they were finally reacted with oxygen to form water. What advantage might such a series of small steps offer? Anyone who has seen pictures from the 1930s of the disastrous end of the Hindenburg, the largest zeppelin ever built, will appreciate the great amount of energy released by the reaction of hydrogen with oxygen. By breaking up this reaction into a number of intermediate steps, a small and manageable amount of energy could be released at each step, said Keilin. This energy could be used later on (in a manner then still unknown) for work such as muscle contraction.

---

[1] Respiration can be stopped by exposing cells to carbon monoxide in the dark, and started again by illuminating the cells, which causes carbon monoxide to dissociate. Warburg reasoned that the speed of respiration would depend on the speed at which carbon monoxide (CO) dissociated after illumination. If he shone light at a wavelength readily absorbed by the ferment, CO would dissociate quickly, and he would measure a quick rate of respiration. On the other hand, if the ferment did not absorb light at a particular wavelength, CO would not dissociate and respiration would remain blocked. By illuminating the ferment with 31 different wavelengths of light (generated by flames and vapour lamps), and measuring the rate of respiration in each case, Warburg pieced together the absorption spectrum of the ferment.

Keilin and Warburg maintained a lively correspondence throughout the 1920s and 1930s, disagreeing on many particulars. Ironically, Keilin's concept of a respiratory chain was given more credence by Warburg himself, who discovered additional non-protein components of the chain during the 1930s, which are now referred to as coenzymes. For his new discoveries Warburg was offered a second Nobel Prize in 1944, but, being Jewish, he was refused permission to receive it by Hitler (who nonetheless allowed himself to be swayed by Warburg's international prestige, and did not have him imprisoned, or worse). Sadly, Keilin's own profound insights into the structure and function of the respiratory chain were never honoured with a Nobel Prize, surely an oversight on the part of the Nobel committee.

The overall picture then emerging was this. Glucose is broken down into smaller fragments, which are fed into an asset-stripping merry-go-round of linked reactions, known as the Krebs Cycle.[2] These reactions strip out the carbon and oxygen atoms, and discharge them as carbon dioxide waste. The hydrogen atoms bind to Warburg's coenzymes and enter the respiratory chain. There, the hydrogen atoms are split into their constituent electrons and protons, the further passage of which differs. We'll look into what happens to the protons later on; for now, we'll concentrate on the electrons. These are passed down the full length of the chain by the string of electron carriers. Each of the carriers is successively reduced (gaining electrons) and then oxidized (losing electrons) by the next link in the chain. This means that the respiratory chain forms a succession of linked redox reactions, and so behaves like a tiny electrical wire. Electrons are transferred down the wire from carrier to carrier at a rate of about 1 electron every 5 to 20 milliseconds. Each of the redox reactions is exergonic—in other words, each releases energy that can be used for work. In the final step, the electrons pass from cytochrome c to oxygen, where they are reunited with protons to form water. This last reaction takes place in Warburg's respiratory ferment, which had been re-named *cytochrome oxidase* by Keilin, because it uses oxygen to oxidize cytochrome c. Keilin's term is still in use today.

Today we know that the respiratory chain is organized into four gigantic molecular complexes, embedded in the inner membrane of the mitochondria (Figure 5). Each complex is millions of times the size of a carbon atom, but even so they are barely visible down the electron microscope. The individual complexes are composed of numerous proteins, coenzymes, and cytochromes,

---

[2] Sir Hans Krebs received the Nobel Prize in 1953 for elucidating the cycle, although many others contributed to a detailed understanding. Krebs' seminal paper on the cycle in 1937 was rejected by *Nature*, a personal set-back that has since encouraged generations of disappointed biochemists. In addition to its central role in respiration the Krebs Cycle is also the cell's starting point for making amino acids, fats, haems, and other important molecules. I regret this is not the place to discuss it.

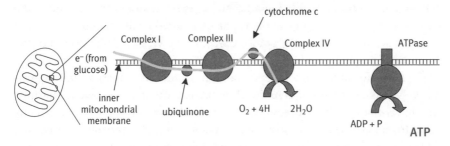

5 Simplified representation of the respiratory chain, showing complexes I, III, and IV, and the ATPase. Complex II is not shown here, as electrons (e⁻) enter the chain at either complex I or complex II, and are passed on from either of these complexes to complex III by the carrier ubiquinone (also known as Coenzyme Q, sold in supermarkets as a health food supplement, though with questionable efficacy). The passage of electrons down the chain is illustrated by the curvy line. Cytochrome c carries electrons from complex III to complex IV (cytochrome oxidase) where they react with protons and oxygen to form water. Notice that all of the complexes are embedded separately in the membrane. Whereas ubiquinone and cytochrome c shuttle electrons between the complexes, the nature of the intermediate that connected electron flow down the respiratory chain with ATP synthesis in the ATPase was a mystery that confounded the field for an entire generation.

including those discovered by Keilin and Warburg. Curiously, mitochondrial genes encode some of the proteins, while nuclear genes encode others, so the complexes are an amalgam encoded by two separate genomes. There are tens of thousands of complete respiratory chains embedded in the inner membrane of a single mitochondrion. It seems that these chains are physically separated from each other, and indeed even the complexes within individual chains seem to be physically independent.

ATP: the universal energy currency

Although Keilin's early concept of the respiratory chain was correct in its essentials, perhaps the most important question remained unanswered—how was energy conserved, rather than being dissipated on the spot? Energy is released by the passage of electrons down the respiratory chains to oxygen, but is consumed elsewhere in the cell, usually outside the mitochondria, at a later time. There had to be some kind of intermediate, presumably a molecule of some sort, that could conserve the energy released in respiration, then transfer it to other compartments of the cell, and couple it to some kind of work. Whatever this intermediate was, it had to be sufficiently adaptable to be used for the various different types of work carried out by the cell, and it had to be stable enough to remain intact until needed (as even moving around the short distances of the cell takes some time). In other words, it had to be the molecular equivalent of a common currency, or coins that could be exchanged for services rendered. The respiratory chain is then the Mint, where the new currency is produced. So what could this currency be?

The first glimmer of an answer came from studies of fermentation. The age-old importance of fermentation in wine-making and brewing belied how little was known about the process. The beginnings of a chemical understanding came from Lavoisier, once again, who measured the weight of all the products, and declared fermentation to be no more than a chemical splitting of sugar to give alcohol and carbon dioxide. He was quite right, of course, but in a sense he missed the point, for Lavoisier thought of fermentation as just a chemical process with no inherent function. For Lavoisier, yeast was merely sediment that happened to catalyse the chemical breakdown of sugar.

By the nineteenth century, the students of fermentation split into two camps—those who thought that fermentation was a living process with a function (mostly the vitalists, who believed in a special vital force, irreducible to 'mere' chemistry), and those who considered fermentation to be purely a chemical process (mostly the chemists themselves). The century-long feud appeared to be settled by Louis Pasteur, a vitalist, who demonstrated that yeast was composed of living cells, and that fermentation was carried out by these cells in the absence of oxygen. Indeed Pasteur famously described fermentation as 'life without oxygen'. As a vitalist, Pasteur was convinced that fermentation must have a purpose, which is to say, a function that was beneficial in some way for yeast, but even he admitted to being 'completely in the dark' about what this purpose might have been.

Only two years after Pasteur's death in 1895, the belief that living yeast was necessary for fermentation was overturned by Eduard Buchner, who received the Nobel Prize for his work in 1907. Buchner used German brewers' yeast rather than Pasteur's French vintners' yeast. Clearly the German yeast was more robust, for unlike Pasteur, Buchner did succeed in grinding it up with sand in a mortar to form a paste, then squeezing juice from the paste using a hydraulic press. If sugar was added to this 'pressed yeast juice', and the mixture incubated, fermentation began within a few minutes. The mixture evanesced alcohol and carbon dioxide in the same proportions as live yeast, albeit in lesser volume. Buchner proposed that fermentation was carried out by biological catalysts that he named *enzymes* (from the Greek *en zyme*, meaning *in yeast*). He concluded that living cells are chemical factories, in which enzymes manufacture the various products. For the first time, Buchner had shown that these chemical factories could be reconstituted even after the demise of the cells themselves, so long as the conditions were suitable. This discovery heralded the end of vitalism, and a new sense that all living processes could ultimately be explained by similarly reductionist principles—the dominant theme of twentieth century biomolecular sciences. But Buchner's legacy also reduced living cells to little more than a bag full of enzymes, which to this day dulls our perception of the importance of membranes in biology, as we shall see.

Using Buchner's yeast juices, Sir Arthur Harden in England and Hans von Euler in Germany (and others) gradually pieced together the succession of steps in fermentation during the first decades of the twentieth century. They unravelled about a dozen steps in all, each catalysed by its own enzyme. The steps are linked together like a factory production line, in which the product of one reaction is the starting point for the next. For their work, Harden and von Euler shared the Nobel Prize in 1929. But the biggest surprise came in 1924, when yet another Nobel laureate, Otto Meyerhof, showed that almost exactly the same process takes place in muscle cells. In muscles, admittedly, the final product was lactic acid, which produces cramps rather than the enjoyable inebriation of alcohol, but Meyerhof showed that almost all the twelve factory production-line steps are the same. This was a striking demonstration of the fundamental unity of life, implying that even simple yeasts are related by descent to human beings, as postulated by Darwin.

By the end of the 1920s it was becoming clear that cells use fermentation to generate energy. Fermentation acts as a backup power supply (indeed the only power supply in some cells), which is usually switched on when the main energy generator, oxygen respiration, fails. Thus, fermentation and respiration came to be seen as parallel processes, both of which served to provide energy for cells, one in the absence of oxygen, the other in its presence. But the larger question remained: how are the individual steps coupled to the conservation of energy for use in other parts of the cell, at other times? Did fermentation, like respiration, generate some kind of energy currency?

The answer came in 1929 with the discovery of ATP by Karl Lohman in Heidelberg. Lohman showed that fermentation is linked to the synthesis of ATP (adenosine triphosphate), which can be stored in the cell for use over a period of hours. ATP is composed of adenosine bound to three phosphate groups linked end to end in a chain, a somewhat precarious arrangement. Splitting off the terminal phosphate group from ATP releases a large amount of energy that can be used to power work—indeed, is *required* to power much biological work. In the 1930s the Russian biochemist Vladimir Engelhardt showed that ATP is necessary for muscle contraction—muscles tense in a state of rigor, as in rigor mortis, when deprived of ATP. Muscle fibres split ATP to liberate the energy that they need to contract and relax again, leaving adenosine diphosphate (ADP) and phosphate (P):

$$ATP \rightarrow ADP + P + energy.$$

Because the cell has a limited supply of ATP, fresh supplies must be regenerated continually from ADP and phosphate—and to do so, of course, requires an input of energy, as can be seen if the equation above is followed backwards. This is the function of fermentation: to provide the energy that is needed to regenerate ATP. Fermenting one molecule of glucose regenerates two molecules of ATP.

Engelhardt immediately squared up to the next question. ATP is *needed* for muscle contraction, but is only produced by fermentation when oxygen levels are low. If the muscles are to contract in the *presence* of oxygen, then presumably some other process must generate the ATP needed; this, said Engelhardt, must be the function of oxygen respiration. In other words, oxygen respiration also serves to generate ATP. Engelhardt set about trying to prove his assertion. The difficulty faced by researchers at the time was technical: muscles are not easy to grind up to use for studies of respiration—they are damaged and leak. Engelhardt resorted to an unusual experimental model, which could be manipulated rather more easily—the red blood cells of birds. Using them, he showed that respiration really does generate ATP, and in far greater quantities than fermentation. Soon afterwards, the Spaniard Severo Ochoa showed that as many as 38 molecules of ATP could be generated by respiration from a single molecule of glucose, a finding for which he too received the Nobel Prize, in 1959. This means that oxygen respiration can produce 19 times more ATP per molecule of glucose than does fermentation. The total production is astonishing. In an average person, ATP is produced at a rate of 9×10^{20} molecules per second, which equates to a turnover rate (the rate at which it is produced and consumed) of about *65 kg* every day.

Few people accepted the universal significance of ATP at first, but work by Fritz Lipmann and Herman Kalckar in Copenhagen in the 1930s confirmed it, and by 1941 (now in the US), they proclaimed ATP to be 'the universal energy currency' of life. In the 1940s this must have been an audacious claim, the kind that can easily backfire and cost the advocates their careers. Yet astonishingly, given the flamboyance and variety of life, it is basically true. ATP has been found in every type of cell ever studied, whether plant, animal, fungal, or bacterial. In the 1940s, ATP was known to be the product of both fermentation and respiration, and by the 1950s, photosynthesis was added to the list—it, too, generates ATP, in this case by trapping the energy of sunlight. So the three great energy highways of life, respiration, fermentation, and photosynthesis, all generate ATP, another profound example of the fundamental unity of life.

The elusive squiggle

ATP is often said to have a 'high-energy' bond, which is denoted with a 'squiggle' (~) rather than a simple hyphen. When broken, this bond is supposed to release a large amount of energy that can be used to power various forms of work about the cell. Unfortunately, this easy representation is not actually true, for there is nothing particularly unusual about the chemical bonds in ATP. What is unusual is the *equilibrium* between ATP and ADP. There is far, *far* more ATP in the cell, relative to ADP, than there would be if the reaction on page 79 were left to find its natural equilibrium. If ATP and ADP were mixed together in a test tube and left for a few days, then virtually the entire mixture would break down into ADP and phosphate. What we see in the cell is the absolute reverse: the ADP and phosphate is converted almost totally into ATP. This is a little like pumping water uphill—it costs a lot of energy to pump the water up, but once you have a reservoir on top of a hill, there is a lot of potential energy that can be tapped into later when the water is allowed to rush down again. Some hydro-electric schemes work this way. Water is pumped up to a high reservoir at night when demand is low. It is then released when there is a surge in demand. In England, apparently, there is a massive surge in demand for electricity after popular soap operas, when millions of people go to the kitchen at the same time, and put the kettle on for a nice cup of tea. This surge in demand is met by opening the floodgates of Welsh mountain reservoirs, which are refilled at night after the demand has settled, ready for the next mass teatime.

In the cell, ADP is continually pumped 'uphill' to generate a reservoir of potential energy in the form of ATP. This reservoir of ATP awaits the opening of the floodgates, whereupon it is used to power various tasks about the cell, just as the flow of water back downhill is used to power electrical devices. Of course, a lot of energy is needed to produce such a high concentration of ATP, just as a lot of energy is needed to pump water uphill. Providing this energy is the function of respiration and fermentation. The energy released from these processes is used to generate very high cellular levels of ATP in the cell, against the normal chemical equilibrium.

These ideas help us to understand how ATP is used to power work in the cell, but they don't explain how the ATP is actually formed. The answer seemed to lie in the studies of fermentation by Efraim Racker in the 1940s. Racker was one of the giants of bioenergetics. A Pole by birth, raised in Vienna, he fled the Nazis to Britain at the end of the 1930s, like many of his contemporaries. After internment on the Isle of Man at the outbreak of the war, he moved to the US, and there settled in New York for some years. Deciphering the mechanism of ATP synthesis in fermentation was the first of his many important contributions over fifty years. Racker discovered that, in fermentation, the energy released by breaking down sugars into smaller fragments is used to attach phosphate groups onto the fragments, against a chemical equilibrium. In other words, fermentation generates high-energy phosphate intermediates, and these in turn transfer their phosphates to form ATP. The overall change is energetically favourable, just as water flowing downhill can be used to turn a waterwheel—the flow of water is *coupled* to the turning of the waterwheel. The formation of ATP likewise takes place via coupled chemical reactions, so that the energy released by fermentation drives a coupled energy-consuming reaction, the formation of ATP. Presumably, thought Racker, and the entire field, a similar model of chemical coupling would also explain how ATP is formed in respiration. Quite the contrary! Rather than offering an insight, it started a wild-goose chase that was to run on for decades. On the other hand, the eventual resolution gave more powerful insights into the nature of life and complexity than anything else in molecular biology, bar the structure of the DNA double helix itself.

The problem hinged on the identity of the high-energy intermediates. In respiration, ATP is produced by a giant enzyme complex called the ATPase (or ATP synthase), which was also discovered by Racker and his colleagues in New York. As many as 30 *thousand* ATPase complexes stud the inner mitochondrial membrane, and can be made out faintly down the electron microscope, sprouting like mushrooms from the membrane (Figure 6). When first visualized in 1964, Racker described them as the 'fundamental particles of biology', an epithet that seems even more apposite today, as we shall see. The ATPase complexes share the inner mitochondrial membrane with the complexes of the respiratory chains, but they are not physically connected to them: they are embedded separately in the membrane. Herein lies the root of the problem. How do these separate complexes communicate with each other across the physical gap? More specifically, how do the respiratory chains transfer the energy released by the flow of electrons to the ATPase, to generate ATP?

In respiration, the *only* known reactions were the redox reactions taking place when electrons were transported down the respiratory chain. The complexes were known to be oxidized and reduced in turn, but that was it: they did not seem to interact with any other molecules. All the reactions were physically separated from the ATPase. Presumably, thought researchers, there must be a high-energy intermediate, as in fermentation, which was formed using the energy released by respiration. This intermediate would then move physically across to the ATPase. After all, chemistry requires contact; action at a distance to a chemist is voodoo. The proposed high-energy intermediate would need to contain a bond equivalent to the sugar-phosphate formed in fermentation, which, when broken, would pass on the energy needed for the high-energy bond of ATP. The ATPase presumably catalysed this reaction.

As so often happens in science, on the threshold of a revolution, the broad outlines seemed to be understood in full. All that remained to be done was to fill in a few details, such as the identity of the high-energy intermediate, which came to be known simply as the squiggle, at least in polite company. Certainly

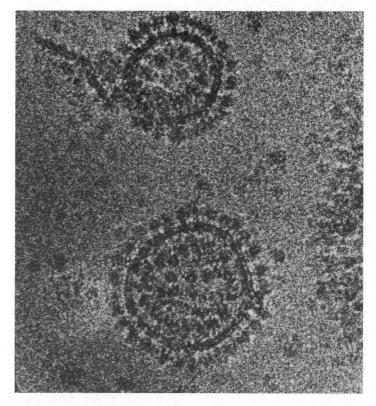

6 The 'elementary particles of life' as christened by Efraim Racker. The ATPase proteins sprout like mushrooms on stalks from the membrane vesicles.

the intermediate was elusive—an entire generation of the finest minds and cleverest experimentalists spent two decades searching for it; they proposed and rejected at least twenty candidates, but even so, finding it only seemed to be a matter of time. Its existence was prescribed by the chemical nature of the cell, little more than a bag of enzymes, as the disciples of Eduard Buchner knew only too well. Enzymes did chemistry, and chemistry was all about the bonds between atoms in molecules.

But one detail was troubling and nagged at the chemistry of respiration: the number of ATP molecules produced varied. Somewhere between 28 and 38 molecules of ATP are formed from a single glucose molecule. The actual number varies over time, and although it can be as high as 38, it is typically at the lower end of this range. But the important point is the lack of consistency. Because ATP is formed from the passage of electrons down the respiratory chain, the passage of one pair of electrons down the chain generates between 2 and 3 ATPs: not a round number. Chemistry, of course, is all about round numbers, as anyone knows who has ever struggled to balance chemical equations. It's not possible to have half a molecule react with two thirds of another molecule. So how could the production of ATP require a variable and non-integer number of electrons?

Another detail also nagged. Respiration *requires* a membrane, and can't take place at all without it. The membrane is more than just a bag to contain the respiratory complexes. If the membrane is disrupted, respiration is said to become uncoupled, like a bicycle that loses its chain: however furiously we pedal the wheels will not turn. When respiration is uncoupled, the oxidation of glucose via the respiratory chain proceeds apace, but no ATP is formed. In other words, the input is uncoupled from the output and the energy released is dissipated as heat. This curious phenomenon is not simply a matter of mechanical damage to the membrane: it can also be induced by a number of apparently unrelated chemicals, known as *uncouplers*, which do not mechanically disrupt the membrane. All these chemicals (including, interestingly, aspirin and, indeed, ecstasy) uncouple the oxidation of glucose from the production of ATP in a similar fashion, but did not seem to share any kind of chemical common denominator. Uncoupling could not be explained in conventional terms.

By the early 1960s, the field had begun to sink into a slough of despond. As Racker put it (in words reminiscent of Richard Feynman's celebrated dictum on quantum mechanics): 'Anyone who is not thoroughly confused just doesn't understand the problem.' Respiration generated energy in the form of ATP, but in a manner that did not defer to the basic rules of chemistry, indeed that seemed to flout them. What was going on? Even though these strange findings were crying out for a radical rethink, nobody was prepared for the shocking answer supplied by Peter Mitchell in 1961.

5

Proton Power

Peter Mitchell was an outsider to the field of bioenergetics. He had studied bio-chemistry at Cambridge during the war, and began his PhD there in 1943, as he had been injured in a sporting accident before the war and was not enlisted for service. Mitchell was a flamboyant character in the war years, well known about town for his artistic and creative flair, and impish sense of humour. He was an accomplished musician, and liked to wear his hair long in the style of the young Beethoven. Mitchell, too, later fell deaf. His mien was embellished by private means, and he was one of the select few who could afford to drive a Rolls Royce in the drab post-war years; his uncle Godfrey Mitchell owned the Wimpey construction empire. Mitchell's shares in the company later helped to keep his private research laboratory, the Glynn Institute, afloat. Despite being recog-nized as one of the brightest young scientists, it took him seven years to finish his PhD, in part because his research was diverted towards wartime goals (the production of antibiotics), and in part because he was asked to resubmit his thesis; one of his examiners had complained that 'the discussion seemed silly, not a presentation'. David Keilin, who knew Mitchell better, remarked: 'The trouble is that Peter is too original for his examiners.'

Mitchell's work concerned bacteria, and especially the problem of how bacteria import and export particular molecules in and out of the cell, very often against a concentration gradient. In more general terms, Mitchell was interested in *vectorial* metabolism, which is to say, reactions that have a direc-tion in space as well as time. The key to bacterial transport systems, for Mitchell, lay in the outer membrane of the bacterial cell. This was plainly not just an inert physical barrier, as all living cells require a continuous and selective exchange of materials across this barrier. At the least, food must be taken up and waste products removed. The membrane acts as a semi-permeable barrier, restricting the passage of molecules and controlling their concentration inside the cell. Mitchell was fascinated by the molecular mech-anics of active transport across membranes. He appreciated that many mem-brane proteins are as specific for the molecules that they transport as enzymes are for their raw materials. Also like enzymes, active transport grinds to a halt as the gradient opposing it gathers strength. The force acting to dispel the gradient strengthens, just as it gets harder to blow air into a balloon as it fills up.

Mitchell developed many of his ideas while at Cambridge in the 1940s and early 1950s, and later in Edinburgh in the late 1950s. He saw active transport as an aspect of physiology, concerning the operation of living bacteria. At that time, there was little intercourse between physiologists and biochemists. Clearly, though, active transport across a membrane requires an input of energy, and this in turn led Mitchell to ponder on bioenergetics, an aspect of biochemistry. He soon realized that if a membrane pump establishes a concentration gradient, then the gradient itself could in principle act as a driving force. Cells could harness such a force in the same way that the escape of air from a balloon can propel it across the room, or the escape of steam can propel a piston in an engine.

These considerations were enough for Mitchell to put forward a radical new hypothesis in *Nature* in 1961, while still at Edinburgh. He proposed that respiration in cells worked by *chemiosmotic coupling*, by which term he meant a chemical reaction that could drive an osmotic gradient, or vice versa. *Osmosis* is a familiar term from schooldays, even if we can't quite remember what it means. It usually means the flow of water across a membrane from a less concentrated to a more concentrated solution, but Mitchell, characteristically, didn't mean it in that sense at all. By 'chemiosmosis' we might imagine that he was referring to a flow of chemicals other than water across a membrane, but this was not what he meant either. He actually used the word 'osmotic' in the original Greek sense, meaning 'push'. Chemiosmosis, for Mitchell, was the pushing of molecules across a membrane *against* a concentration gradient—it is therefore, in a sense, the exact opposite of osmosis, which follows the concentration gradient. The purpose of the respiratory chain, said Mitchell, is no more nor less than to push protons over the membrane, creating a reservoir of protons on the other side. The membrane is little more than a dam. The pent-up force of protons, trapped behind this dam, can be released a little at a time to drive the formation of ATP.

It works like this. Recall from the previous chapter that the complexes of the respiratory chain are plugged through the membrane. The hydrogen atoms that enter the respiratory chain are split into protons and electrons. The electrons pass down the chain like the current in a wire, via a succession of redox reactions (Figure 7). The energy released, said Mitchell, doesn't form a high-energy chemical intermediate at all; the squiggle is elusive because it does not exist. Rather, the energy released by electron flow is used to pump protons across the membrane. Three of the four respiratory complexes use the energy released by electron flow to thrust protons across the membrane. The membrane is otherwise impermeable to protons, so any backflow is restricted, and

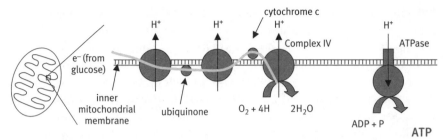

7 Simplified representation of the respiratory chain, as in Figure 5, but now showing the nature of the intermediate—the proton. Electrons (e^-) pass down the chain from complex I to complex IV, and the energy released at each step is coupled to the expulsion of a proton across the membrane. This leads to a difference in proton concentration across the membrane, which can be measured as a difference in acidity (pH, or acidity, is defined as the proton concentration) and as an electrical potential difference, as protons carry a single positive charge. The reservoir of protons acts as a reservoir of potential energy, just as a reservoir of water on top of a hill acts as a reservoir of potential energy that can be used to generate hydroelectric power. Likewise, the flow of protons down the concentration gradient can be used to power mechanical tasks, in this case ATP synthesis. The flow of protons through the ATPase is called the 'proton-motive force' and it turns the tiny molecular motor of the ATPase, to generate ATP from ADP and phosphate.

a reservoir develops. Protons carry a positive charge, which means that the proton gradient has both an electrical and a concentration component. The electrical component generates a potential difference across the membrane, while the concentration component generates a difference in pH, or acidity (acidity is defined as proton concentration) with the outside more acid than the inside. The combination of pH and potential difference across the membrane constitutes what Mitchell called the 'proton-motive force.' It is this force that drives the synthesis of ATP. Because ATP is synthesized by the ATPase, Mitchell predicted that the ATPase would need to be powered by the proton-motive force—a current of protons flowing down the proton gradient from the pent-up reservoir; what Mitchell liked to call proton electricity, or proticity.

Mitchell's ideas were ignored, regarded with hostility, dismissed as mildly insane, or claimed as derivative. Racker later wrote: 'Given the general attitude of the establishment, these formulations sounded like the pronouncements of a court jester or of a prophet of doom.' The theory was couched in the strange, almost mystical, terminology of electrochemists, and made use of concepts that were unfamiliar to most enzymologists at the time. Only Racker and Bill Slater in Amsterdam (another protégé of Keilin) took it seriously at first, albeit with open-minded scepticism; and Slater soon lost his patience.

Mitchell, who was by turns brilliant, argumentative, irascible, and grandiloquent, exacerbated the situation. He could infuriate his opponents. In an argument with Mitchell, Slater was seen hopping around on one foot in rage, illuminating the expression 'hopping mad.' These arguments took their toll on Mitchell too, who was forced to resign from Edinburgh University with stomach ulcers. In a two-year interim out of science, he restored the decaying eighteenth-century manor, Glynn House, near Bodmin in Cornwall, as a family home and private research institute. He returned to the research front in 1965, having girded up his loins for the battles to come. And come they did. The raging disputes of the next two decades were dubbed the 'ox phos' wars (from oxidative phosphorylation, the mechanism of ATP production in respiration).

The explanatory power of proton power

Mitchell's hypothesis neatly solved the nagging difficulties that dogged the older theories. It explained why a membrane was necessary, and why it had to be intact—a leaky membrane would allow a drizzle of protons back through, and so dissipate the proton-motive force as heat. A porous dam is no use to anyone.

It also explained how the mysterious uncoupling agents worked. Recall that 'uncoupling' refers to the loss of correspondence between glucose oxidation and ATP production, like a bike that loses its chain—the energy put into peddling is no longer connected to a useful function. Uncoupling agents all disconnect the energy input from the output, but otherwise seemed to have little else in common. Mitchell showed that they did have something in common—they are weak acids, which dissolve in the lipids of the membrane. Because weak acids can either bind or release protons, according to the acidity of their surroundings, they can shuttle protons across the membrane. In alkaline or weakly acidic conditions they lose a proton and gain a negative charge. Drawn by the electric charge they cross to the positive, acidic side of the membrane. Then, being a weak acid in strongly acidic conditions, they pick up a proton again. This neutralizes the electrical charge, so they become subject to the concentration gradient again. The weak acid traverses the membrane to the less acidic side, whereupon it loses the proton and once more becomes subject to the electrical tug. This kind of cycling can only happen if the uncoupling agent dissolves in the membrane regardless of whether it has bound a proton or not; and it was this subtle requirement that confounded earlier attempts at an explanation. (Some weak acids are soluble in lipids, but only when they have bound to a proton or vice versa; when they have released their proton they are no longer soluble in lipids, and so can't cross back over the membrane; they therefore can't uncouple respiration.)

Even more fundamentally, the chemiosmotic hypothesis explained the voodoo 'action at a distance' that seemed to beg a high-energy intermediate, the elusive squiggle. Protons pumped across a membrane in one spot generate a force that acts equally anywhere on the surface of the membrane, just as the pressure of water behind a dam depends on the overall volume of water, not on the location of the pump. So protons are pumped over the membrane in one place, but can return through an ATPase anywhere else in the membrane with a force that depends on the overall proton pressure. In other words, there was no *chemical* intermediate, but the proton-motive force itself acted as an intermediate—the energy released by respiration was conserved as the reservoir of protons. This also explained how a non-integer number of electrons could generate ATP—although a fixed number of protons are pumped across the membrane for every electron transported, some of the protons leak back through the dam, whereas others are tapped off for other purposes: they are not used to power the ATPase (we'll return to this in the next section).

Perhaps most importantly, the chemiosmotic theory made a number of explicit predictions, which could be tested. Over the following decade, Mitchell, working in the refurbished Glynn House with his life-long research colleague Jennifer Moyle, and others, proved that mitochondria do indeed generate a pH gradient as well as an electrical charge (of about 150 millivolts) across the inner membrane. This voltage might not sound like a lot (it's only about a tenth of that available from a torch battery) but we need to think of it in molecular terms. The membrane is barely 5 nm (10^{-9} m) thick, so the voltage experienced from one side to the other is in the order of 30 million volts per metre—a similar voltage to a bolt of lightning, and a thousand times the capacity of normal household wiring. Mitchell and Moyle went on to show that a sudden rise in oxygen levels elicited a transient rise in the number of protons pumped over the membrane; they showed that respiratory 'uncouplers' really did work by shuttling protons back across the membrane; and they showed that the proton-motive force did indeed power the ATPase. They also demonstrated that proton pumping is coupled to the passage of electrons down the respiratory chain, and slows or even stops if any raw materials (hydrogen atoms, oxygen, ADP, or phosphate) run short.

By then, Mitchell and Moyle were not the only experimentalists working on chemiosmotics. Racker himself helped to convince the field by showing that if the respiratory complexes were isolated and then added to artificial lipid vesicles, they could still produce a proton gradient. But perhaps the one experiment that did more than any other to convince researchers, or botanists at least, of the veracity of the theory was carried out by André Jagendorf and Ernest Uribe at Cornell University in 1966. Jagendorf's initial reaction to the chemiosmotic hypothesis had been hostile. He wrote: 'I had heard Peter

Mitchell talk about chemiosmosis at a bioenergetics meeting in Sweden. His words went into one of my ears and out the other, leaving me feeling annoyed they had allowed such a ridiculous and incomprehensible speaker in.' But his own experiments convinced him otherwise.

Working with chloroplast membranes, Jagendorf and Uribe suspended the membranes in acid, at pH 4, and gave the acid time to equilibrate across the membrane. Then they injected an alkali, at pH 8, into the preparation, creating a pH difference of 4 units across the membrane. They found that large amounts of ATP were created by this process, without the need for light or any other energy source: ATP synthesis was powered by the proton difference alone. Notice that I'm talking about *photosynthetic* membranes here. A striking feature of Mitchell's theory is that it reconciles quite distinct modes of energy production that seem to be unrelated, like photosynthesis and respiration—both produce ATP via a proton-motive force across a membrane.

By the mid 1970s, most of the field had come round to Mitchell's point of view—Mitchell even maintained a chart showing the dates when his rivals 'converted', to their fury—even though many molecular details still needed to be worked out, and remained controversial. Mitchell was sole recipient of the Nobel Prize for chemistry in 1978, another source of acrimony, although I believe his conceptual leap justified it. He had been through a personally traumatic decade, fighting poor health as well as a hostile bioenergetic establishment, but lived to see the conversion of his fiercest critics. In thanking them for their intellectual generosity in his Nobel lecture, Mitchell quoted the great physicist Max Planck—'a new scientific idea does not triumph by convincing its opponents, but rather because its opponents eventually die.' To have falsified this pessimistic dictum, said Mitchell, was a 'singularly happy achievement.'

Since 1978, researchers have whittled away at the detailed mechanisms of electron transport, proton pumping, and ATP formation. The crowning glory was John Walker's determination of the structure of the ATPase in atomic detail, for which he shared the Nobel Prize for chemistry in 1997 with Paul Boyer, who had suggested the basic mechanism many years earlier. (This was broadly similar in principle, but differed in detail, from the mechanism favoured by Mitchell.) The ATPase is a marvellous example of nature's nano-technology: it works as a rotary motor, and as such is the smallest known machine, constructed from tiny moving protein parts. It has two main components, a drive shaft, which is plugged straight through the membrane from one side to the other, and a rotating head, which is attached to the drive shaft, resembling a mushroom head when seen down the electron microscope. The pressure of the proton reservoir on the outside of the membrane forces protons through the drive shaft to rotate the head; for each three protons that pass through the drive shaft, the head cranks around by 120°, so three cranks complete a turn. There are three binding sites on the head, and these are where the ATP is assembled. Each time the head rotates, the tensions exerted force chemical bonds to form or break. The first site binds ADP; the next crank of the head attaches the phosphate onto the ADP to form ATP; and the third releases the ATP. In humans, a complete turn of the head requires 9 protons and releases 3 molecules of ATP. Just to complicate matters, in other species, the ATPase often requires different numbers of protons to rotate the head.

The ATPase is freely reversible. Under some circumstances it can go into reverse, whereupon it splits ATP, and uses the energy released to pump protons *up* the drive shaft, back across the membrane against the pressure of the reservoir. In fact the very name ATPase (rather than ATP synthase) signifies this action, which was discovered first. This bizarre trait hides a deep secret of life, and we'll return to it in a moment.

The deeper meaning of respiration

In a broad sense, respiration generates energy using proton pumps. The energy released by redox reactions is used to pump protons across a membrane. The proton difference across the membrane corresponds to an electric charge of about 150 mV. This is the proton-motive force, which drives the ATPase motor to generate ATP, the universal energy currency of life.

Something very similar happens in photosynthesis. In this case, the sun's energy is used to pump protons across the chloroplast membrane in an analogous fashion to respiration. Bacteria, too, function in the same way as mitochondria, by generating a proton-motive force across their outer cell membrane. For anyone who is not a microbiologist, there is no field of biology more confusing than the astonishing versatility with which bacteria generate energy. They seem to be able to glean energy from virtually anything, from methane, to sulphur, to concrete. This extraordinary diversity is related at a deeper level. In each case, the principle is exactly the same: the electrons pass down a redox chain to a terminal electron acceptor (which may be CO_2, NO_3^-, NO_2^-, NO, SO_4^{2-}, SO_3^-, O_2, Fe^{2+}, and others). In each case the energy derived from the redox reactions is used to pump protons across a membrane.

Such a deep unity is noteworthy not just for its universality, but perhaps even more because it is such a peculiar and roundabout way of generating energy. As Leslie Orgel put it, 'Few would have laid money on cells generating energy with proton pumps.' Yet proton pumping is the secret of photosynthesis, and all forms of respiration. In all of them, the energy released by redox reactions is used to pump protons across a membrane, to generate a proton-motive force. It seems that pumping protons across a membrane is as much a signature of life on earth as DNA. It is fundamental.

In fact the proton-motive force has a much broader significance than just generating ATP, as Mitchell realized. It acts as a kind of force field, enveloping bacteria with an impalpable source of power. Proton power is involved in several fundamental aspects of life, most notably the active transport of molecules in and out of the cell across the external membrane. Bacteria have dozens of membrane transporters, many of which use the proton-motive force to pump nutrients into the cell, or waste products out. Instead of using ATP to power active transport, bacteria use protons: they hive off a little energy from the proton gradient to power active transport. For example, the sugar lactose is transported into the cell against a concentration gradient by coupling its transport to the proton gradient: the membrane pump binds one lactose molecule and one proton, so the energetic cost of importing lactose is met by the dissipation of the proton gradient, not by ATP. Similarly, to maintain low sodium levels inside the cell, the removal of one sodium ion is paid for by the import of one proton, again dissipating the proton gradient without consuming ATP.

Sometimes the proton gradient is dissipated for its own sake, to produce heat. In these circumstances, respiration is said to be uncoupled, for electron flow and proton pumping continue as normal, but without ATP production. Instead, the protons pass back through pores in the membrane, thereby dissipating the energy bound up in the proton gradient as heat. This can be useful in itself, as a means of producing heat, as we shall see in Part 4, but it also helps to maintain electron flow during times of low demand, when 'stagnant' electrons are prone to escape from the respiratory chain to react with oxygen, producing destructive oxygen free radicals. Think of this like a hydroelectric dam on a river. At times of low demand there is a risk of flooding, which can be lowered by having an over-flow channel. Similarly, in the respiratory chain, a through-flow of electrons can be maintained by uncoupling electron flow from ATP synthesis. Instead of flowing through the main hydroelectric dam gates (the ATPase), some protons are diverted through the overflow channels (the membrane pores). This through-flow helps to prevent any problems that may arise from having an overflowing reservoir of electrons, ready to form free radicals; and there are important health consequences, as we shall see in later chapters.

Besides active transport, the proton force can be put to other forms of work. For example, bacterial locomotion also depends on the proton-motive force as shown by the American microbiologist Franklin Harold and his colleagues in the 1970s. Many bacteria move around by rotating a rigid corkscrew-like flagella attached to the cell surface. They can achieve speeds of up to several hundred cell-lengths per second by this process. The protein that rotates the flagellum is a tiny rotary motor, not dissimilar to the ATPase itself, which is powered by the proton current through a drive shaft.

In short, bacteria are basically proton-powered. Even though ATP is said to be the universal energy currency, it isn't used for all aspects of the cell. Both bacterial homeostasis (the active transport of molecules in and out of the cell) and locomotion (flagellar propulsion) depend on proton power rather than ATP. Taken together, these vital uses of the proton gradient explain why the respiratory chain pumps more protons than are required for ATP synthesis alone, and why it is hard to specify the number of ATP molecules that are formed from the passage of one electron—the proton gradient is fundamental to many aspects of life besides ATP formation, all of which tap off a little.

The importance of the proton gradient also explains the odd propensity of the ATPase to go into reverse, pumping protons at the cost of burning up ATP. On the face of it, such a reversal of the ATPase might seem to be a liability, because it swiftly drains the cell of its ATP reserves. This only begins to make sense when we appreciate that the proton gradient is more important than ATP. Bacteria need a fully charged proton-motive force to survive, just as much as a galactic cruiser in Star Wars needs its protective force field fully operational before attacking the Empire's star fleet. The proton-motive force is usually charged up by respiration. However, if respiration fails, then bacteria generate ATP by fermentation. Now everything goes into reverse. The ATPase immediately breaks down the freshly made ATP and uses the energy released to pump protons across the membrane, maintaining the charge—which amounts to an emergency repair of the force field. All other ATP-dependent tasks, even those as essential as DNA replication and reproduction, must wait. In these circumstances, it might be said that the main purpose of fermentation is to maintain the proton-motive force. It is more important for a cell to maintain its proton charge than it is to have an ATP pool available for other critical tasks such as reproduction.

To me, all this hints at the deep antiquity of proton pumping. It is the first and foremost need of the bacterial cell, its life-support machine. It is a deeply unifying mechanism, common to all three domains of life, and central to all forms of respiration, to photosynthesis, and to other aspects of bacterial life, including homeostasis and locomotion. It is in short a fundamental property of life. And in line with this idea, there are good reasons to think that the origin of life itself was tied to the natural energy of proton gradients.

6

The Origin of Life

How life began on Earth is one of the most exhilarating fields of science today—a wild west of ideas, theories, speculations, and even data. It is too large a subject to embark on in detail here, so I will limit myself to a few observations on the importance of chemiosmotics. But for perspective let me paint a quick picture of the problem.

The evolution of life depends in very large measure on the power of natural selection—and this in turn depends on the inheritance of characteristics that can be subjected to natural selection. Today we inherit genes made of DNA; but DNA is a complicated molecule and can't have just 'popped' into existence. Moreover, DNA is chemically inert, as we noted in the Introduction. Recall that DNA does little more than code for proteins, and even this is achieved by way of a more active intermediary, RNA, which in various forms physically translates the DNA code into the sequence of amino acids in a protein. In general, proteins are the active ingredients that make life possible—they alone have the versatility of structure and function needed to fulfil the multifarious requirements of even the simplest forms of life. Individual proteins are honed to the requirements of their particular tasks by natural selection. First among these tasks, proteins are needed to replicate DNA and to form RNA from the DNA template, for without heredity natural selection is not possible; and for all their glories proteins are not repetitive enough in structure to form a good heritable code. The origin of the genetic code is therefore a chicken and egg problem. Proteins need DNA to evolve, but DNA needs proteins to evolve. How did it all get started?

The answer agreed by most of the field today is that the intermediary, RNA, used to be central. RNA is simpler than DNA, and can even be put together in a test tube by chemists, so we can bring ourselves to believe that it may once have formed spontaneously on the early Earth or in space. Plenty of organic molecules, including some of the building blocks of RNA, have been found on comets. RNA can replicate itself in a similar manner to DNA, and so forms a replicating unit that natural selection can act upon. It can also code for proteins directly, as indeed it does today, and so provides a link between template and function. Unlike DNA, RNA is not chemically inert—it folds into complex shapes and is able to catalyse some chemical reactions in the same way as enzymes (RNA catalysts are called ribozymes). Thus, researchers into the origin of life point to a primordial 'RNA world', in which natural selection acts upon independently self-replicating RNA molecules, which slowly accrue complexity, until being displaced by the more robust and efficient combination of DNA and proteins. If this whistlestop tour whets your appetite for more, I can recommend *Life Evolving* by Christian de Duve as a good place to start.

Elegant as it is, there are two serious problems with the 'RNA world'. First, ribozymes are not very versatile catalysts, and even allowing for the most rudimentary catalytic efficiency, there is a big question mark over whether they could have brought a complex world into existence. To me they are rather less suitable than minerals as the original catalysts. Metals and minerals are found at the heart of many enzymes today, including iron, sulphur, manganese, copper, magnesium, and zinc. In all these cases, the enzyme reaction is catalysed by the mineral (technically, the prosthetic group), not the protein, which improves the efficiency rather than the nature of the reaction.

Second, more importantly, there is an accounting problem with energy and thermodynamics. The replication of RNA is work, and therefore requires an input of energy. The requirement for energy is constant, because RNA is not very stable, and is easily broken down. Where did this energy come from? There were plenty of sources of energy on the early earth, mooted by astrobiologists—the impact of meteorites, electrical storms, the intense heat of volcanic eruptions, or underwater hydrothermal vents, to name but a few. But how these diverse forms of energy were converted into something that life could use is rarely described—none of them is used directly, even today. Probably the most sensible suggestion, which has been in and out of favour over several decades, is the fermentation of a 'primordial soup', cooked up by a combination of all the various forms of energy.

The idea of a primordial soup gained experimental support in the 1950s, when Stanley Miller and Harold Urey passed electric sparks, to simulate lightning, through a mixture of gases believed to represent the earth's early atmosphere—hydrogen, methane, and ammonia. They succeeded in producing a rich mixture of organic molecules, including some precursors of life, such as amino acids. Their ideas fell out of favour because there is no evidence that the earth's atmosphere ever contained these gases in sufficient quantities; and organic molecules are far harder to form in the more oxidizing atmosphere now thought to have existed. But the existence of plentiful organic material on comets has brought us round full circle. Many astrobiologists, keen to link life with space, argue that the primordial soup could have been cooked up in outer space. The earth then received generous helpings in the huge asteroid bombardment that pockmarked the moon and earth for half a billion years from 4.5

to 4 billion years ago. If the soup really did exist, then perhaps life could have started out by fermenting a soup after all.

But there are several problems with fermentation as the original source of energy. First, as we have seen, fermentation stands apart from both respiration and photosynthesis, in that it does not pump protons across a membrane. This leads to a discontinuity and a problem with time. If all the fermentable organic compounds came from outer space, then the nutrient supply should have begun to run out after the great asteroid bombardment drew to a close 4 billion years ago. Life would dribble away to extinction unless it could invent photosynthesis, or some other way of producing organic molecules from the elements, before the fermentable substrates ran out. And this is where we run into a problem with time. Traces of fossil evidence suggest that life on Earth began at least 3.85 billion years ago, and that photosynthesis evolved some time between 3.5 billion and 2.7 billion years ago (although this evidence has been questioned lately). Given the discontinuity between fermentation and photosynthesis—not a single intermediary step brings us any loser to the evolution of photosynthesis—then the gap of at least several hundred million years, and perhaps a billion years, looks very awkward. With no other source of energy, could the organic molecules delivered by asteroids really nourish life for that long? It doesn't sound very likely to me, especially given the tendency of ultraviolet radiation to break down complex organic molecules in the days before the ozone layer.

Second, the perception of fermentation as simple and primitive is wrong. It reflects our prejudice that microbes are biochemically simple, which is untrue, and dates back to the ideas of Louis Pasteur, who described fermentation as 'life without oxygen', implying simplicity. But Pasteur, as we have seen, admitted to being 'completely in the dark' about the function of fermentation, so he could hardly conclude that it was simple. Fermentation requires at least a dozen enzymes, and, as the first and so only means of providing energy, can be seen as irreducibly complex. I use this term deliberately, for it has been presented by some biochemists to argue that the evolution of life required the guiding hand of a Creator—that life is only possible following 'intelligent design'. I disagree with this position, as any evolutionary biologist would, but the objection nonetheless must be tackled, and can present problems in some cases. In the case of fermentation, it is genuinely hard to see how all these interlinked enzymes could have evolved as a functional unit in an RNA world that was not supplied by any other form of energy. But notice that I am specifying, 'a world not supplied by any other form of energy'. What we need is a means of generating energy that is 'reducibly complex'. So the problem we must wrestle with is not how fermentation could have evolved without any other source of energy, but where the energy necessary for its evolution came from. If photosynthesis evolved later, and fermentation is too complex to evolve without an energy supply, we are still left with respiration as a further possibility. Could respiration have evolved on the early earth? The usual objection is that there was very little oxygen available on the primordial earth (see my book, *Oxygen: The Molecule that Made the World* for a discussion of this), but this is not an obstacle. Other forms of respiration use sulphate or nitrate, or even iron, instead of oxygen—and all of them pump protons across a membrane. They are therefore far closer in their basic mechanism to photosynthesis and hint at possible intermediary steps. Notice that this places the evolution of respiration before photosynthesis, as Otto Warburg suggested in 1931. So we are faced with the question: is respiration, too, 'irreducibly complex'? I shall argue that it is not—on the contrary, it is almost an inevitable outcome of the conditions on the primordial earth—but before we think about this we need to consider a final, fatal, objection to fermentation as primitive.

This third objection relates to the properties of LUCA, the Last Universal Common Ancestor of all known life on Earth. Some very interesting data suggest that classical fermentation did not exist in LUCA; and if fermentation did not exist in LUCA, then presumably it did not exist in the earlier forms of life, dating back to the very origin of life, either. These data come from Bill Martin, whom we met in Part 1. There we considered the three domains of life—the archaea, the bacteria, and the eukaryotes. We saw that the eukaryotes were almost certainly formed by the union of an archaeon and a bacterium. If so, then the eukaryotes must have evolved relatively recently, and LUCA must have been the last ancestor common to the bacteria and the archaea. Martin employs this logic in considering the origin of fermentation. Up to a point, we can assume that any basic properties shared by bacteria and archaea, such as the universal genetic code, were inherited from this common ancestor, whereas any major differences presumably evolved later. For example, photosynthesis (to generate oxygen) is found only in the cyanobacteria, the green algae and the plants. Both the plants and algae are impostors—they rely on their chloroplasts for photosynthesis, and these are derived from cyanobacteria. Thus we can say that photosynthesis evolved in the cyanobacteria. Crucially, it is not found in any archaea at all, or in any other group of bacteria besides the cyanobacteria, so we can infer that photosynthesis evolved in the cyanobacteria alone, and that this happened *after* the split between the bacteria and archaea.

Returning to fermentation, let's apply the same argument. If fermentation were the first means of generating energy, then we should find a similar pathway in both the archaea and bacteria, just as we find the universal genetic code in both—both inherited it from their common ancestor. Conversely, if fermentation only evolved later, like photosynthesis, then we would not expect to find fermentation in both the archaea and the bacteria, but only in some groups. So what do we find? The answer is interesting, for both the archaea and bacteria *do* ferment, but do so by using different enzymes to catalyse the steps. Several of them are completely unrelated. Presumably, if the archaea and bacteria do not share the same enzymes for fermentation, then the classical fermentation pathway must have evolved later on, independently, in the two domains. This means that LUCA could not ferment, at least as we know it today. And if she could not ferment, then she must have got her energy from somewhere else. We are forced to draw the same conclusion for a third time—fermentation was not the primordial source of energy on Earth. Life must have started another way, and the idea of a primordial soup is wrong, or at best irrelevant.

The first cell

If proton pumping across a membrane is fundamental to life, as I have argued, then on the same basis it should be present in both bacteria and archaea. It is. Both have respiratory chains with similar components. Both use the respiratory chain to pump protons across a membrane, generating a proton-motive force. Both share an ATPase that is basically similar in its structure and function.

Although respiration is far more complex than fermentation today, when pared down to its essentials it is actually far simpler: respiration requires electron transport (basically just a redox reaction), a membrane, a proton pump, and an ATPase, whereas fermentation requires at least a dozen enzymes working in sequence. The main problem with respiration evolving early in the history of life is the need for a membrane, as Mitchell himself appreciated (he discussed it in a presentation in Moscow in 1956). Modern cell membranes are complex, and it's hard to imagine them evolving in an RNA world. Of course, simpler alternatives exist. The problem with them is that they are largely impermeable to anything. An impermeable membrane obstructs exchange with the outside world, which in turn prevents metabolism, and so life itself. Given that a membrane does seem to have figured in LUCA's respiration, can we infer from the modern archaea and bacteria what kind of membrane it might have been?

The answer betrays an extraordinary evolutionary divide with the most profound implications, which were elucidated by Bill Martin and Mike Russell (from the University of Glasgow) at the Royal Society of London in 2002. The membranes of bacteria and archaea are both composed of lipids, but beyond that they have very little in common. The membrane lipids of bacteria are made up of hydrophobic (oily) fatty acids bound to a hydrophilic (water-loving) head, by way of a chemical bond known as an *ester* bond. In contrast, the membrane lipids of archaea are built up from branching 5-carbon units called isoprenes, joined together as a polymer. The isoprene units form numerous cross-links, giving the archaeal membrane a rigidity not found in bacteria. Furthermore, the isoprene chains are bound to the hydrophilic head by a different type of chemical bond, known as an *ether* bond. The hydrophilic heads of both bacteria and archaea are made of glycerol phosphate—but each uses a different mirror-image form. These are no more interchangeable than the left and right hands of a pair of gloves. Lest such differences may seem modest, bear in mind that all the components of the lipid membranes are made by the cell using specific enzymes via complex biochemical pathways. As the components are different, the enzymes needed to make them are also different, and so too are the genes that code for the enzymes.

The differences in both the construction and the final composition of bacterial and archaeal membranes are so fundamental that Martin and Russell came to the conclusion that LUCA (the last universal common ancestor) could not have had a lipid membrane. Her descendents must surely have evolved lipid membranes independently later on. But if she practiced chemiosmotics, as we have seen she almost certainly did, then LUCA must equally surely have had some sort of membrane across which she could pump protons. If this membrane was not made of lipids, what else could it be made of? Martin and Russell gave a radical answer: LUCA they say, might have had an inorganic membrane, a thin, bubbly layer of iron-sulphur minerals, enclosing a microscopic cell, filled with organic molecules.

According to Martin and Russell, iron-sulphur minerals catalysed the first organic reactions, to produce sugars, amino acids and nucleotides, and eventually perhaps the 'RNA world' discussed earlier, in which natural selection could take over. Their Royal Society paper gives a detailed insight into the kind of reactions that might have taken place; we will confine ourselves to the energetic implications, profound enough in themselves.

Full metal jacket

The idea that iron-sulphur minerals, such as iron pyrites (fool's gold), may have played a role in the origin of life dates back to the discovery of 'black smokers', three kilometres under the ocean, in the late 1970s. The black smokers are hydrothermal vents—large, tottering black towers, superheated at the high pressures of the sea floor, billowing 'black smoke' into the surrounding ocean. The 'smoke' is composed of volcanic gases and minerals, including iron and hydrogen sulphide, which precipitate out in the surrounding waters as iron-sulphur minerals. The greatest surprise was that the smokers are full of life, despite the high temperatures and pressures, and the absolute darkness. An entire ecological community thrives there, gaining its energy directly from the hydro-thermal vents, apparently independent of the sun.[1]

Iron-sulphur minerals have an ability to catalyse organic reactions—as indeed they still do today in the prosthetic groups of many enzymes, such as iron-sulphur proteins. The possibility that iron-sulphur minerals might have been the midwife of life itself, catalysing the reduction of carbon dioxide to form a plethora of organic molecules in the hellish black smokers was developed by the German chemist and patent attorney Gunter Wächtershäuser, in a brilliant series of papers in the late 1980s and 1990s. One researcher exclaimed that read-ing them felt like a stumbling across a scientific paper that had fallen through a time warp from the end of the twenty-first century.

Wächtershäuser conceived of these first organic reactions taking place on the surface of iron-sulphur minerals. His ideas seemed to be supported by genetic trees, which suggested that the hyperthermophiles (microbes that thrive at high pressure and searing temperatures) are among the most ancient groups in both bacteria and archaea. However, this genetic evidence has been ques-tioned recently (see de Duve, for example) and Wächtershäuser's postulated reactions have been criticized on thermodynamic grounds. Perhaps most importantly, the black smoker story also suffers from a dilution problem. Once the precursors have reacted on the two-dimensional surface of a crystal, they dissociate and are free to diffuse away into the widest reaches of ocean. There is nothing to contain them unless they remain attached; and it is hard to envisage the fluid cycles of biochemistry evolving in a fixed position on the surface of a mineral.

Mike Russell put forward an alternative set of ideas in the late 1980s, and has been refining them ever since, most recently in collaboration with Bill Martin. Russell is less interested in the large, menacing black smokers than in more modest volcanic seepage sites. One such site is the 350 million-year-old iron pyrites deposits at Tynagh in Ireland. The minerals form huge numbers of tubular structures, about the size of pen lids, as well as bubbly deposits, which Russell postulates may have been similar to the hatchery of life itself. Such bubbles were probably formed, he says, by the mixing of two chemically different fluids: hot, reduced, alkaline waters that seeped up from deep in

---

[1] No microbes are truly independent of the sun's energy. All life on earth, even the microbes in the deep hot biosphere, gains its energy from redox reactions. These are only made possible because the oceans and air are out of chemical equilibrium with the earth itself—an imbal-ance that depends on the oxidizing power of the sun. The microbes of the deep hot biosphere make use of redox reactions that would not be possible were it not for the relative oxidation of the oceans, ultimately attributable to the sun. One reason that their metabolism and turnover is so slow—a single cell may take a million years to reproduce—is that they are dependent on the desperately slow trickle-down of oxidized minerals from further up.

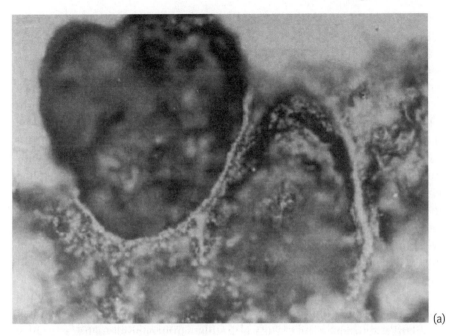

(a)

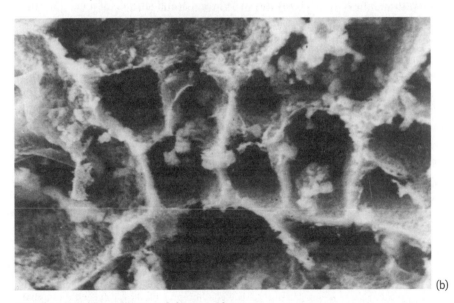

(b)

8 Primordial cells with iron-sulphur membranes.
(a) Electron micrograph of a thin section of iron-sulphur mineral (pyrites) from Tynagh, Ireland, 360 million years old. (b) Electron micrograph of structures formed in a laboratory by injecting sodium sulphide (NaS) solution, representing hydrothermal fluid, into iron chloride ($FeCl_2$) solution, representing the iron-rich early oceans.

the crust, and the more oxidized and acidic ocean waters above, containing carbon dioxide and iron salts. Iron-sulphur minerals such as mackinawite (FeS) would have precipitated into microscopic bubbly membranes at the mixing zone.

This is not just speculation. Russell and his long-term collaborator, Alan Hall, have simulated it in the laboratory. By injecting sodium sulphide solution (representing the hydrothermal fluid oozing up from the bowels of the earth) into iron chloride solution (representing the early oceans) Russell and Hall produced a host of tiny, microscopic bubbles, bounded by iron-sulphide membranes (Figure 8). These bubbles have two remarkable traits, which make me believe that Russell and Hall are thinking along the right lines. First, the cells are naturally chemiosmotic, with the outside more acidic than the inside. This situation is similar to the Jagendorf-Uribe experiment, in which a pH difference across the membranes was enough to generate ATP. Because Russell's cells come with a natural pH gradient, all that the cells need to do to generate ATP is to plug an ATPase through the membrane, surely orders of magnitude easier to evolve than an entire functional fermentation pathway! If the first step on the way to the origin of life required little more than an ATPase, then Racker's prescient description of the ATPase as the 'elementary particle of life' may have had more substance than even he could have known.

Second, the iron-sulphur crystals in the bubbly membranes conduct electrons (as indeed do the iron-sulphur proteins that still exist in mitochondrial membranes today). The reduced fluids that well up from the mantle are rich in electrons, whereas the relatively oxidized oceans are poor in electrons, setting up a potential difference across the membrane of several hundred millivolts—quite similar to the voltage across bacterial membranes today. This voltage stimulates the flow of electrons from one compartment to the other through the membrane. What's more, the flow of negatively charged electrons draws positively charged protons from inside, giving rise to a rudimentary proton pumping mechanism.

Not only do the iron-sulphur cells provide a continuous supply of energy, but they also act as miniature electrochemical reactors, catalysing fundamental biochemical reactions, and concentrating the reaction products. The basic building blocks of life, including RNA, ADP, simple amino acids, small peptides, and so on, could all have been formed by virtue of the catalytic properties of the iron-sulphur minerals, and perhaps sedimental clays, in the reactions described by Gunter Wächtershäuser, but with two great advantages—they are concentrated by the membranes (preventing them from diffusing away into the oceans), and they are powered by a natural source of energy, the proton gradient.

Life itself

Does all this sound improbable? In the previous chapter, I suggested that the origin of life was not as improbable as the evolution of the eukaryotes. Think about what is happening here. Such conditions could not have been rare on the early earth. Volcanic activity has been estimated to be fifteenfold greater than today. The crust was thinner, the oceans shallower, and the tectonic plates were only just forming. Volcanic seepage sites must have existed across much of the surface of the earth, to say nothing of more violent volcanic activity. The formation of many millions of tiny cells, bounded by iron-sulphur membranes, requires no more than a difference in redox state and acidity between the oceans and the volcanic fluids emanating from deep in the crust—a difference that certainly existed.

The early earth, as envisaged by Russell, is a giant electrochemical cell, which depends on the power of the sun to oxidize the oceans. UV rays split water and oxidize iron. Hydrogen, released from water, is so light that it is not retained by gravity, and evaporates off into space. The ocean becomes gradually oxidized, relative to the more reduced conditions in the mantle. According to the basic rules of chemistry, the mixing zone inevitably forms natural cells, replete with their own chemiosmotic and redox gradients. Mixing would have been assisted by the high tidal range, drawn the tug of the newly formed moon, which was closer to the earth then than today. We can be almost certain that such cells would really have formed, perhaps on a massive scale. And of course we can see their remains in the geology of places like Tynagh. There is a long way to go from here to make even a bacterium, but these conditions are a good first step.

Not only would the requisite conditions have been probable, but they would have been stable and continuous. They depend only on the power of the sun, without requiring problematic inventions like photosynthesis or fermentation. The sun is only needed to oxidize the oceans, as we know it must. Of all the forms of energy mooted by astrobiologists—meteorite impacts, volcanic heat, lightning—the power of the sun has often been curiously overlooked by scientists, if not by prehistoric mythologies. As the distinguished microbiologist Franklin Harold put it in his classic text, *The Vital Force* (the title of which I honour in the title of this Part): 'One cannot help but suspect that the great stream of energy that passes across the earth plays a larger role in biology than our current philosophy knows: that perhaps the flood of power not only permitted life to evolve, but called it into being.'

For hundreds of millions of years, the sun provided the constant source of energy needed to pay the debt to the second law of thermodynamics. It created chemical disequilibria, and promoted the formation of naturally chemiosmotic cells. The primordial conditions are still faithfully replicated in the fundamental properties of all cells today. Both organic and inorganic cells are bounded by a membrane, which physically contains the cell's organic constituents, preventing them from diffusing away into the oceans. In both organic and inorganic cells, biochemical reactions are catalysed by minerals (today embedded as the prosthetic groups of enzymes). In both cases, the membrane is the barrier as well as carrier of energy. In both cases, energy is captured by a chemiosmotic gradient, with a positive charge and acidic conditions on the outside, and relatively negative, alkaline conditions on the inside. In both cases, redox reactions, electron transport and proton pumping regenerate the gradient. When the bacteria and archaea finally emerged from their nursery, to venture into the open oceans, they took with them an unmistakable seal of their origin. They parade it still today.

But this imprint, echoing the origin of life itself, was also life's primary limitation. Why, we may ask, did bacteria never evolve beyond bacteria? Why did four billion years of bacterial evolution never succeed in producing a truly multicellular, intelligent bacterium? More specifically, why did the evolution of the eukaryotes require a union between an archaeon and a bacterium, rather than just the gradual accrual of complexity by a favoured line of bacteria or archaea? In Part 3, we'll see that the answer to this long-standing riddle, and an explanation for the marvellous flowering of the eukaryotic line into plants and animals, lies in the fundamental nature of energy production by chemiosmotics across a bounding membrane.

PART 3

Insider Deal
The Foundations of Complexity

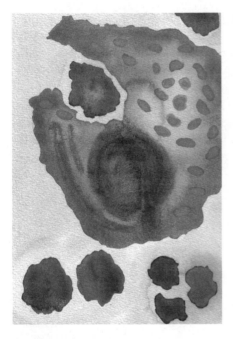

Bacteria ruled supreme on Earth for two billion years. They evolved almost unlimited biochemical versatility but never discovered the secrets of greater size or morphological complexity. Life on other planets may get stuck in the same rut. On Earth, large size and complexity only became possible once energy generation had been internalized in mitochondria. But why did bacteria never internalize their own energy generation? The answer lies in the tenacious survival of mitochondrial DNA, a two-billion-year-old paradox.

A large cell with things inside—in eukaryotes, energy generation is internalized in mitochondria

Here is a list of words to make an evolutionary biologist spill their beer: purpose, teleology, ramp of ascending complexity, non-Darwinian. All these terms are associated with a religious view of evolution—the sense that life was 'programmed' to evolve, to become more complex, to give rise to humanity on a smooth curve from the lowest animals to the angels, each approaching closer to God—the great 'chain of being'. Such a view is popular not just with religious theorists, but nowadays with astrobiologists too. The idea that the laws of physics virtually summon life forth in the universe that we see around us is a comforting one, and evokes the idea that even human sentience may be an inevitable outcome of the workings of physics. I disagreed in Part 1, and we will consider the theme further in Part 3 by looking at the origin of biological complexity.

In Part 1, we observed that all complex multicellular organisms on earth are composed of eukaryotic cells; in contrast, bacteria have remained resolutely bacterial for the best part of four billion years. There is a chasm between bacterial and eukaryotic cells, and life elsewhere in the universe might well get stuck in the bacterial rut. We have seen that the eukaryotic cell was first formed in an unusual union between a bacterium and an archaeon. The question we'll look into now is the 'seeding' of complexity in eukaryotes: what exactly is it about the eukaryotic cell that seems to encourage the evolution of complexity? However misleading the impression may be, surveying the grand canvas of evolution *after* the appearance of the eukaryotic cell *does* engender a sense of purpose. The idea of a great chain of being, striving to approach closer to God, is not accidental, even if it is wrong. In Part 3, we'll see that the seeds of complexity were sown by mitochondria, for once mitochondria existed, life was almost bound to become more complex. The drive towards greater complexity came from within, not from on high.

In his celebrated book *Chance and Necessity*, the committed atheist and Nobel Prize-winning molecular biologist Jacques Monod tackled the theme of purpose. Plainly, he said, it is pointless to discuss the heart without mentioning that it is a pump, whose function is to pump blood around the body. But that is to ascribe purpose. Worse, if we were to say that the heart evolved *to* pump blood, we would be committing the ultimate sin of teleology—the assignment of a forward-looking purpose, a predetermined end-point to an evolutionary trajectory. But the heart could hardly have evolved 'for' anything else; if it didn't evolve *to* pump blood, then it is truly a miracle that it happened to become so fine a pump. Monod's point was that biology is full of purpose and apparent trajectories, and it is perverse to pretend they don't exist; rather, we must explain them. The question we must answer is this: how does the operation of blind chance, a random mechanism without foresight, bring about the exquisitely refined and purposeful biological machines that we see all around us?

Darwin's answer, of course, was natural selection. Blind chance serves only to generate random variation within a population. Selection is not blind, or at least not random: it selects for the overall fitness of an organism in its particular environment—the survival of the fittest. The survivors pass on their successful genetic constitution to their offspring. Thus any changes that improve the function of the heart at pumping blood will be passed on, while any that undermine it will be eliminated by selection. In each generation (in the wild) only a few per cent might survive to reproduce, and they will tend to be the luckiest or best adapted. Over many generations luck no doubt balances out, so natural selection tends to select the best adapted of the best adapted, inevitably refining function until other selective pressures balance out the tendency to change. Natural selection therefore works as a ratchet, which turns the operation of random variation into a trajectory. In retrospect this may well *look* like a ramp of ascending complexity.

Ultimately, biological fitness is written in the sequence of the genes, for they alone are passed on to the next generation (well, almost alone: mitochondria are, too). Over evolutionary time, alterations in the genetic sequence, subjected to round after round of natural selection, build tiny refinement upon tiny refinement, until finally erecting the dizzying cathedral of biological complexity. Although Darwin knew nothing of genes, the genetic code at once suggests a mechanism for producing random variation in a population: mutations in the sequence of 'letters' in DNA can change the sequence of amino acids in proteins, which might have a positive, or a negative, or a neutral, effect on their function. Copying errors alone generate such variation. Each generation produces perhaps several hundred small changes in the DNA sequence (out of several billion letters), which may or may not affect fitness. Such small changes undoubtedly occur, and generate some of the raw material for the slow evolutionary change anticipated by Darwin. The gradual divergence in the sequence of genes of different species, over hundreds of millions of years, shows this process in action.

But small mutations are not the only way to bring about change in the genome (the complete library of genes in one organism), and the more we learn about genomics (the study of genomes), the less important small mutations seem to be. At the least, greater complexity demands more genes—the small bacterial genome could hardly code for a whole human being, still less the myriad genetic differences between individuals. Surveying species, there is a general correlation between the degree of complexity and the number of genes, if not the total DNA content. So where do all these extra genes come from? The answer is duplications of existing genes, or whole genomes, or from the union of two or more different genomes, or from the spread of repetitive DNA sequences—apparently 'selfish' replicators, which copy themselves throughout the genome, but may later be co-opted to serve some useful function (useful, that is, to the organism as a whole).

None of these processes is strictly Darwinian, in the sense of gradual, small refinements to an existing genome. Rather, they are large-scale, dramatic changes in the total DNA content—giant leaps across genetic space, transforming existing gene sequences at a single stroke—even if they generate the raw material for new genes, rather than the new genes themselves. Excepting these leaps across genetic space, the process is otherwise Darwinian. Changes to the genome are brought about in an essentially random manner, and then subjected to rounds of natural selection. Small changes hone the sequence of new genes to new tasks. So long as the big jumps in DNA content do not generate an unworkable monster, they can be tolerated. If there is no benefit in having twice as much DNA, then we can be sure that natural selection will jettison it again—but if complex organisms need a lot of genes, then the elimination of superfluous DNA surely puts a ceiling on the maximum possible complexity, for it eliminates the raw material needed to form new genes.

This brings us back to the ramp of complexity. We have seen that there is a big discontinuity between bacteria and eukaryotes. It is remarkable that bacteria are *still* bacteria: while enormously varied and sophisticated in biochemical terms, they have resolutely failed to generate real *morphological* complexity in four billion years of evolution. In their size, shape, and appearance, they can hardly be said to have evolved in any direction at all. In contrast, in half the time open to bacteria, the eukaryotes unquestionably ascended a ramp of complexity—they developed elaborate internal membrane systems, specialized organelles, complex cell cycles (rather than simple cell division), sex, huge genomes, phagocytosis, predatory behaviour, multicellularity, differentiation, large size, and finally spectacular feats of mechanical engineering: flight, sight, hearing, echolocation, brains, sentience. Insofar as this progression happened over time, it can reasonably be plotted out as a ramp of ascending complexity. So we are faced with bacteria, which have nearly unlimited biochemical diversity but no drive towards complexity, and eukaryotes, which have little biochemical diversity, but a marvellous flowering in the realm of bodily design.

When confronted with the divide between bacteria and eukaryotes, the Darwinian might reply: 'Ah, but the bacteria *did* generate complexity—they gave rise to the more complex eukaryotes, which in turn gave rise to many organisms of inordinately greater complexity.' This is true, but only in a sense, and here is the rub. The mitochondria, I shall argue, could *only* be derived by endosymbiosis—a union of two genomes in the same cell, or a giant leap across genetic space—and without mitochondria, the complex eukaryotic cell simply could *not* evolve. This viewpoint stems from the idea that the eukaryotic cell itself was forged in the merger that gave rise to mitochondria, and that the possession of mitochondria is, or was in the past, a *sine qua non* of the eukaryotic condition. This picture differs from the mainstream view of the eukaryotic cell, so let's remind ourselves quickly why it matters.

In Part 1, we examined the origin of the eukaryotic cell, as surmised by Tom Cavalier-Smith, which best represents the mainstream view. To recapitulate, a prokaryotic cell (without a nucleus) lost its cell wall, perhaps through the action of an antibiotic produced by other bacteria, but survived the loss, as it already had an internal protein skeleton (cytoskeleton). The loss of the cell wall had profound consequences for the cell in terms of its lifestyle and manner of reproduction. It developed a nucleus and a complicated life cycle. Using its cytoskeleton to move around and change shape like an amoeba, it developed a new, predatory lifestyle, engulfing large particles of food such as whole bacteria by phagocytosis. In short, the first eukaryotic cell evolved its nucleus and its eukaryotic lifestyle by standard Darwinian evolution. At a relatively late stage, one such eukaryotic cell happened to engulf a purple bacterium, perhaps a parasite like *Rickettsia*. The internalized bacteria survived and eventually transmuted, by standard Darwinian evolution, into mitochondria.

Notice two things about this line of reasoning: first, it exhibits what we might call a Darwinian bias, in that it limits the importance attributed to the union of two dissimilar genomes, a basically non-Darwinian mode of evolution; and second, it limits the importance of mitochondria in this process. Mitochondria are incorporated into a fully functional eukaryotic cell, and are readily lost again in many primitive lines such as *Giardia*. Mitochondria, in this view, are an efficient means of generating energy, but no more nor less than that. The new cell simply had a Porsche engine fitted, in place of its old-fashioned milk-cart motor. I think this view gives little real insight into why all complex cells possess mitochondria, or conversely, why mitochondria are *needed* for the evolution of complexity.

Now consider the hydrogen hypothesis of Bill Martin and Miklos Müller, which we also discussed in Part 1. According to this radical hypothesis, a mutual chemical dependency between two very different prokaryotic cells led to a close relationship between the two. Eventually one cell physically engulfed the other, combining two genomes within a single cell: a giant leap across genetic space to create a 'hopeful monster'. This genetic leap, in turn, set up a series of

Darwinian selection pressures on the new entity, leading to a transfer of genes from the guest to the host. The critical point of the hydrogen hypothesis is that there *never was* a primitive eukaryote, one that supposedly possessed a nucleus and had a predatory lifestyle, but did not have any mitochondria. Rather, the first eukaryote was born of the union between two prokaryotes, a fundamentally non-Darwinian process—there was no halfway house.

Just look at Figure 9, a tree of life drawn in 1905 by the Russian biologist Konstantine Merezhkovskii, to see what an uncomfortable reversal of the standard branching tree of life this creates. There has been plenty of controversy over trees of life in the past, notably from Stephen Jay Gould, who claimed that the Cambrian explosion inverted the usual tree. The Cambrian explosion refers to the great, and geologically sudden, proliferation of life around 560 million years ago. Later on, most of the major branches were ruthlessly pollarded, as whole phyla fell extinct. Daniel Dennett, in *Darwin's Dangerous Idea*, lambasts Gould's apparently radical evolutionary trees for being the same as any other evolutionary tree, except with distorted axes—a low-lying scrub bush, throwing up a few scraggly shoots, rather than a lofty tree of life. But there is no danger of this in Merezhkovskii's case. His evolutionary tree is a genuinely upside down variety. Here, the branches fuse, rather than bifurcate, to generate a new domain of life.

I'm not trying to cry revolution. There is nothing exceptional about these arguments, and symbiosis is part of the standard evolutionary canon, even if it is played down as a mechanism of generating novelty. For example, the late, great John Maynard Smith and Eörs Szathmáry, in their stimulating book *The Origins of Life*, argue that biological symbiosis is analogous to a motorbike, which is a symbiosis between a bicycle and the internal combustion engine. Even if we view this symbiosis as an advance, they say, with rather crusty humour, someone still had to invent the bicycle and the internal combustion engine first. Likewise in life, natural selection must invent the parts first, and symbiosis just makes creative use of the available parts. Thus symbiosis is best explained in Darwinian terms.

All this is true, but it obscures the fact that some of the most profound evolutionary novelties are made possible *only* by symbiosis. Presumably, if we follow Maynard Smith and Szathmáry, if a bicycle and an internal combustion engine can evolve independently by natural selection, then so too, in principle, could the motorcycle. No doubt it's faster to evolve a motorcycle by shuffling existing components, but there is no fundamental reason why it should not have evolved anyway, given enough time, in the absence of symbiosis. In the case of the eukaryotic cell, I disagree. Left to themselves, I will argue, bacteria could not evolve into eukaryotes by natural selection alone: symbiosis was *needed* to bridge the gulf between bacteria and eukaryotes, and in particular a mitochon-

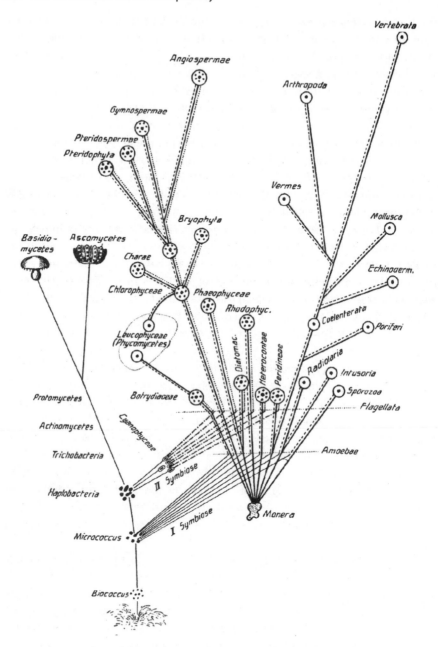

9 Merezhkovskii's inverted tree of life, showing fusion of branches. The standard 'Darwinian' tree of life is strictly bifurcating: branches branch but do not fuse. The origin of the eukaryotic cell was endosymbiotic. On the tree of life, this is represented as bifurcation backwards: branches fuse together, inverting part of the tree of life.

drial merger was *necessary* to sow the seeds of complexity. Without mito-chondria, complex life is simply not possible, and without symbiosis, mito-chondria are not possible—without the mitochondrial merger we would be left with bacteria and nothing but. Regardless of whether we consider symbiosis Darwinian or not, an understanding of why symbiotic mitochondria are neces-sary is paramount to an understanding of our own past, and our place in the universe.[1]

In Part 3, we will see why there is such a yawning chasm between the prokary-otes and the eukaryotes, and why this deep divide can only be bridged by symbiosis—it is next to impossible, given the mechanism of chemiosmotic ener-gy production (discussed in Part 2) for eukaryotes to evolve by natural selection from prokaryotes. This is why bacteria are still bacteria, and why it is unlikely that life as we know it, based on cells, carbon chemistry, and chemiosmosis, will progress beyond the bacterial level of complexity anywhere else in the universe. In Part 3, we'll see why mitochondria seeded complexity in the eukaryotes, placing them at the beginning of the ramp of ascending complexity; and in Part 4, we'll see why mitochondria impelled the eukaryotes onwards up the ramp.

---

[1] In his excellent book *Mendel's Demon*, Mark Ridley muses about the need for a merger in the evolution of the eukaryotic cell: was it a fluke, along with the retention of a contingent of mitochondrial genes? Could the eukaryotes have evolved without such a merger? Ridley argues that both the merger and the retention of genes were probably flukes. I disagree, but for an alternative view I can strongly recommend his book.

7

Why Bacteria are Simple

The great French molecular biologist François Jacob once remarked that the dream of every cell is to become two cells. In our own bodies, this dream is held very carefully in check; otherwise the result would be cancer. But Jacob was trained as a microbiologist, and for bacteria, one cell becoming two is more than a dream. Bacteria replicate at colossal speed. When well fed, *E. coli* bacteria divide once every 20 minutes, or 72 times a day. A single *E. coli* bacterium weighs about a trillionth of a gram (10^{-12} g). Seventy-two cell divisions in a day corresponds to an amplification of 2^{72} ($= 10^{72 \times \log 2} = 10^{21.6}$), which is an increase in weight from 10^{-12} grams to 4000 metric tons. In two days, the mass of exponentially doubling *E. coli* would be 2664 times larger than the mass of the Earth (which weighs 5.977×10^{21} metric tons)!

Luckily this does not happen, and the reason is that bacteria are normally half starved. They swiftly consume all available food, whereupon their growth is limited once again by the lack of nutrients. Most bacteria spend most of their lives in stasis, waiting for a meal. Nonetheless, the speed at which bacteria do mobilize themselves to replicate upon feeding illustrates the overwhelming strength of the selection pressures at work. Amazingly, *E. coli* cells divide in two faster than they can actually replicate their own DNA, which takes about 40 minutes (or twice the time required for cell division). They can do this because they begin a new round of DNA replication well before the previous round is finished. During rapid cell division, several copies of the full bacterial genome are produced simultaneously in any one cell.

Bacteria are gripped by the naked tyranny of natural selection. Speed is paramount, and herein lies the secret to why bacteria are still bacteria. Imagine a population of bacteria whose growth is restricted by nutrient availability. Feed it. The bacterial cells proliferate. The cells that replicate fastest swiftly dominate the population, whilst those that replicate more slowly are displaced. When the nutrient supply runs out, we are left with a new population held in limbo, at least until the next meal comes along. As long as the fastest replicators are robust, and so able to survive in the wild, the new population mix inevitably comprises mostly the fastest replicators. This is as plain as the fact that the Chinese will come to dominate the world's population unless their stringent birth-control laws succeed in limiting families to one or two children.

Because cell division is faster than DNA replication, the speed at which bacteria can possibly divide is limited by the speed at which they can replicate their DNA. Even though bacteria can speed up their DNA replication by making more than one copy per cell division, there is a limit to the number of copies that can be made at once. In principle, the speed of DNA replication depends on the size of the genome, and the resources available for copying it. A suitable energy reserve in the form of ATP is necessary, if not sufficient, for replication. Cells that are not energetically efficient, or starved of resources, make less ATP and so tend to be slower to copy their genome. In other words, to thrive, bacteria must replicate their genome faster than the competition, and to do so requires either a smaller genome or more effective energy production. If two bacterial cells generate ATP at the same speed, then the cell with the smallest genome will tend to replicate fastest, and will eventually come to dominate the population.

A bacterial cell can tolerate a larger genome if the extra genes enable it to produce ATP more effectively than its competitors during times of lean resources. In a fascinating study, Konstantinos Konstantinidis and James Tiedje, at Michigan State University, examined all 115 fully sequenced bacterial genomes. They found that the bacteria with the largest genomes (about 9 or 10 million letters, encoding 9000 genes) dominate in environments where resources are scarce but diverse, and where there is little penalty for slow growth, in particular the soil. Many soil bacteria only manage to produce about three generations in a year, so there is less selection for speed than for any replication whatsoever. Under these conditions, an ability to take advantage of scant resources is important—and this in turn requires more genes to code for the extra metabolic flexibility needed. So versatility pays dividends if it offers a clear advantage in terms of reproductive speed. It is no accident that bacteria such as *Streptomyces avermitilis*, which are ubiquitous in the soil, are metabolically versatile with big genomes to match.

Thus, in bacteria, larger genomes can be tolerated when growth is slow, and versatility is at a premium. Even so, there is still selection for small size in relation to other versatile bacteria, and this apparently sets a bacterial genome 'ceiling' of about 10 million DNA letters. These are the largest genomes among bacteria, and most have far fewer genes. In general, it is probably fair to say that bacterial genomes are small in size because larger genomes take more time and energy to replicate, and so are selected against. Yet even the most versatile bacteria have small genomes in comparison with the eukaryotic cells living in the same environment. How the eukaryotes were released from a selection pressure that stifles even the most versatile bacteria is the subject of this chapter.

Gene loss as an evolutionary trajectory

To maintain a small genome, bacteria could either remain passively unchanging, always with the same hand of genes, like a gambler with cold feet, or they could be more dynamic, constantly losing and winning genes—playing their hand and taking another. Perhaps surprisingly, at least for anyone who likes to think of evolution as a steady progression towards greater sophistication (and so more genes), bacteria are quick to gamble with their genes. They lose as often as not: gene loss is common in bacteria.

One of the most extreme examples of gene loss is *Rickettsia prowazekii*, the cause of typhus, a terrible epidemic that preys on overcrowded populations in filthy conditions, rife with rats and lice. Over history, epidemic typhus has wiped out whole armies, including Napoleon's armies in Russia, the vestiges of which escaped from Russia in 1812 ridden with typhus, along with many refugees from Poland and Lithuania. *Rickettsia prowazekii* is named after two pioneering investigators in the early years of the twentieth century, the American Howard Ricketts and the Czech Stanislaus von Prowazek. Along with the French Tunisian Charles Nicolle, Ricketts and Prowazek discovered that the disease is transmitted through the faeces of the human body louse. Sadly, by the time a vaccine had finally been developed in 1930, almost all these early pioneers had died of typhus, including both Ricketts and Prowazek. The sole survivor, Nicolle, received the Nobel Prize for his dedicated work in 1928. Nicolle's discoveries were put to use in the First and Second World Wars, when hygienic measures, such as shaving, washing, and burning clothes, helped limit the spread of the disease.

Rickettsia is a tiny bacterium, almost as small as a virus, which lives as a parasite inside other cells. It is so well adapted to this lifestyle that it can no longer survive outside its host cells. Its genome was first sequenced by Siv Andersson and her colleagues at the University of Uppsala in Sweden, and was published in *Nature* in 1998 to great clamour. The genome of *Rickettsia* has been streamlined by its intracellular lifestyle in a similar manner to our own mitochondria— and its remaining genes also share many sequence similarities, prompting Andersson and colleagues to declare *Rickettsia* the closest living relative to mitochondria, though as we saw in Part 1, others disagree.

Here it is *Rickettsia's* propensity to lose genes that concerns us. Over evolutionary time *Rickettsia* has lost most of its genes, and now has a mere 834 protein-coding genes left. While this is an order of magnitude more than the mitochondria of most species, *Rickettsia* has barely a quarter of the number of genes of its closest relatives in the wild. It was able to lose most of its genes in this way simply because they were not needed: life inside other cells, if you can survive there at all, is a spoon-fed existence. The parasites live in the kitchen of a lavish chef, and need make little for themselves. Ironically, instead of becoming fatter, they lose weight: they throw away superfluous genes.

Let's pause here for a moment, and think about the pressure to lose genes. Genetic damage is random, and might happen to any gene at any time; but *gene loss* is not random. Any cell or organism that loses an essential gene (or has it damaged such that its function is lost) will perish: it can no longer survive in the wild, and so will be eliminated by natural selection. On the other hand, if a gene is not essential, then its loss or damage, by definition, cannot be catastrophic. In our own case, our primate ancestors lost the gene for making vitamin C a few million years ago, but did not perish because their diet was rich in fruit, which provided them with plenty of vitamin C. They survived and prospered. We know because most of the gene is still there in our 'junk' DNA, as eloquent as the wreck of a ship with a hole beneath the waterline. The remaining sequence corresponds closely to the functional gene in other species.

At a biochemical level *Rickettsia* is an extreme equivalent of our primate ancestors. It doesn't need the genes for making many essential cellular chemicals (such as amino acids and nucleotides) from scratch any more than we need a gene for making vitamin C: it can simply import them all from its host cell. If the genes for making such chemicals in *Rickettsia* happen to be damaged, so what?—they can be lost with impunity. Unusually among bacteria, nearly a quarter of the total genome of *Rickettsia* is composed of 'junk' DNA. This 'junk' is the recognizable relic of recently sunken genes. These shipwrecked genes lie broken, their memory not yet obliterated: their hulks are still rotting in the genome. Such junk DNA will almost certainly be lost altogether in time, as it slows the replication of *Rickettsia*. Mutations that delete unnecessary DNA will be selected for when they happen, as they speed up replication. So damage is a first step, followed by the complete loss of genes. *Rickettsia* has already lost four-fifths of its genome in this way and the process is continuing today. As Siv Andersson put it: 'genome sequences are only snapshots in evolutionary time and space.' Here the snapshot is a moment in the evolutionary degeneration of a parasite that is losing its unnecessary genes.

Balancing gene loss and gain in bacteria

Most bacteria, of course, are not intracellular parasites, but live in the outside world. They need many more genes than *Rickettsia*. Nonetheless, they face a similar pressure to lose superfluous genes—they just can't afford to lose as many. The tendency of free-living bacteria to lose genes can be measured in the laboratory. In 1998, the Hungarian researchers Tibor Vellai, Krisztina Takács, and Gábor Vida, then all at the Eötvös Loránd University in Budapest, reported some simple (conceptually if not technically) but revealing experiments. They engineered three bacterial gene 'rings', or plasmids (the genetic 'loose change' we met in Chapter 1). Each plasmid contained a gene conferring resistance to an antibiotic, and the only important difference between them was their size— each plasmid contained different amounts of non-coding DNA. The plasmids were then added to cultures of *E. coli* bacteria, which were allowed to grow. The bacteria take up the plasmid—they are *transfected*—and can call upon the gene if necessary.

In the first set of experiments, the Hungarian investigators grew the three transfected cultures in the presence of the antibiotic. Any bacterium that lost its plasmid would thereby lose its resistance to the antibiotic, and so be killed. Under this selection pressure, the colonies with the largest plasmids grew the slowest, because they had to spend more time and effort copying their DNA. After only 12 hours in culture the cells with the smallest plasmids had outgrown their lumbering cousins tenfold. In the second set of experiments, bacteria were cultured without antibiotic. Now all three cultures grew at similar speeds, regardless of the size of the plasmid. How so? When the cultures were double-checked for the presence of the plasmids, it turned out that the superfluous plasmids were being lost. All three cultures of bacteria were able to grow at a similar rate because they jettisoned the genes for antibiotic resistance, which are not essential when bacteria are cultured in the absence of the antibiotic. The bacteria simply threw away the unnecessary genes in their rush to replicate faster—a case of 'use it or lose it!'

These studies show that bacteria can lose superfluous genes in a matter of hours or days. Such fast gene loss means that bacterial species tend to retain the smallest number of genes compatible with viability at any one moment. Natural selection is like an ostrich with its head buried in the sand—it doesn't matter how stupid an action might be in the long term, so long as it provides some momentary respite. In this case, if the gene for antibiotic resistance is not necessary, it is lost from most cells in a population even if it may turn out to be needed again at some point in the future. Just as bacteria lose the genes for antibiotic resistance, they also lose other genes that are not essential at a particular moment. Such genes are more easily lost from a portable chromosome such as a plasmid, but bacteria can also lose genes that are part of their main chromosome, albeit more slowly. Any gene that is not used regularly will tend to be lost as a result of random mutations and selection for faster replication. The efficiency of these mechanisms acting on the main bacterial chromosome is illustrated by their low amount of 'junk' DNA, as well as the low number of genes in relation to eukaryotes. Bacteria are small and streamlined because they bin any excess baggage at the first opportunity.

Throwing away genes is not as foolhardy as it may sound, however, for bacteria can also pick up the same genes again, as well as others. The existence of lateral gene transfer—the uptake of DNA from the surroundings (from dead cells) or other bacteria, by a form of copulation known as bacterial conjugation—shows that bacteria can and do accumulate new genes. Active gain of genes compensates for gene loss. In a fluctuating environment, it is unlikely that *all* redundant genes will be lost from *all* bacteria in a population before the conditions change again (for example, with changing seasons), as gene loss is a random process. At least a fraction of the bacterial population is likely to have retained the redundant genes in full working order, and when the conditions change again they can pass them around the population by lateral gene transfer. Such open-handedness with genes explains how antibiotic resistance can spread so swiftly through an entire population of bacteria.

Although the importance of lateral gene transfer in bacteria has been recognized since the 1970s, we have only recently begun to appreciate the degree to which it can confound evolutionary trees. In some bacterial species, more than 90 per cent of observed variation in a population comes from lateral gene transfer, rather than the conventional selection of cells in clones or colonies. The transfer of genes between different species, genera, and even domains means that bacteria do not pass on a consistent core of genes by vertical inheritance, as we do to our children. This makes it embarrassingly difficult to define the term 'species' in bacteria. In plants and animals, a species is defined as a population of individuals that can interbreed to produce fertile offspring. This definition does not apply to bacteria, which divide asexually to form clones of supposedly identical cells. In theory, the clones drift apart over time as a result of mutations, leading to genetic and morphological differences sufficient to call 'speciation'. But lateral gene transfer often confounds this outcome. Genes can be switched so quickly and so comprehensively that the cacophony obliterates all traces of ancestry—no gene is passed on to daughter cells for more than a few generations before being replaced by an equivalent gene from another cell with a different ancestry. The current champion is *Neisseria gonorrhoeae*: this recombines genes so quickly that it is impossible to detect any clonal groups at all: even the gene for ribosomal RNA, often claimed to represent the true phylogeny (lineage) of bacteria, is swapped so often that it gives no indication of ancestry.

Over time, such gene transfers make a big difference. Just to give a single example, gene transfer has produced two different strains of the bacterial 'species' *E. coli* that differ more radically in their gene content (a third of their genome, or nearly 2000 different genes) than all the mammals put together, perhaps even all the vertebrates! The importance of vertical inheritance, descent with modification, in which the genes are *only* passed on to the daughter cells during cell division, is often ambivalent among bacteria. Imagine trying to work out our own provenance by examining the heirlooms passed down in the family, only to discover that our ancestors were compulsive klepto-maniacs, forever pilfering each other's family silver. As the branching 'tree of life' is based strictly on vertical inheritance—the erroneous assumption that the heirlooms only pass from parents to children—its veracity is open to question. Among the bacteria at least a network may be a better analogy. As one despairing expert put it, reflecting on the troubles of constructing a tree of life, 'only God can make a tree'.

So why are bacteria so open-handed with their genes? It might sound like altruistic behaviour, sharing genetic resources for the good of the population as a whole, but it is not; it is still a form of selfishness, what Maynard Smith described as an 'evolutionarily stable strategy'. Compare lateral transfer with conventional 'vertical' inheritance. In the latter case, if an antibiotic threatens a population of bacteria, and only a few cells have retained the genes needed to save their lives, then the rest of the unprotected population will die, and only the offspring of the tattered survivors can thrive to replenish the population. If conditions then change again, favouring a different gene, this surviving popu-lation too may be decimated. In swiftly changing conditions, only the cells that retain an enormous repertoire of genes will survive most exigencies, and they will be so large and unwieldy that they can be out-competed by bacteria able to replicate faster in the interim. Such streamlined bacteria, of course, may be threatened by any exigencies at all—but not if they are able to pick up genes from the environment; then they can combine speedy replication with the genetic resilience to cope with almost anything thrown at them. Bacteria that lose and gain genes in this way will thrive in place of either lumbering genetic giants, or bacteria that refuse to pick up any new genes at all. Presumably, the most effective way of picking up new genes is by conjugation, rather than from the dead bacteria whose genes may be damaged, so ultimately an apparently altruistic, though individually selfish, sharing of genes is favoured. Overall, then, we see the dynamic balance of two different trends in bacteria—the tendency to gene loss, which reduces the bacterial genome to the smallest possible size in the prevailing conditions; and the accumulation of new genes by means of lateral gene transfer, according to need.

I have cited examples of gene loss in bacteria like *Rickettsia*, and in the lab, but beyond the sparseness of their genome (the small number of genes and the lack of junk DNA), it is difficult to prove that gene loss is important in bacteria in 'the wild'. But the importance of lateral gene transfer among bacteria also testifies to the strength and pervasiveness of the selection pressure for bacteria to lose any superfluous genes—otherwise they would not be under such an obligation to pick them up again. Despite taking up new genes, bacteria don't expand their genomes, so presumably they must lose genes at the same rate. And they lose genes at this rate because the competition between cells within a species (and between cells in different species) must continually reduce the genome to the smallest size possible in the prevailing conditions.

The upper limit of any known bacterial genome is about 9 or 10 million letters, encoding some 9000 genes. Presumably, any bacteria that acquire more genes than this tend to lose them again, as the time needed to copy the extra genes slows down replication without providing any countering benefits. This is a stark contrast between bacteria and eukaryotes. The more we learn about bacteria, the harder it becomes to make valid generalizations about them. In recent years, we have discovered bacteria with straight chromosomes, with nuclei, cytoskeletons, and internal membranes, all traits once considered to be unique prerogatives of the eukaryotes. One of the few definitive differences that hasn't evaporated on closer inspection is gene number. Why is it that there are no bacteria with more than 10 million DNA letters, when, as we noted in Chapter 1, the single-celled eukaryote *Amoeba dubia* has managed to accumulate 670 *billion* letters—67 000 times more letters than the largest bacteria, and for that matter 200 times more than humans? How did the eukaryotes manage to evade the reproductive constraints imposed on bacteria? The answer that I think gets to the heart of the matter was put forward by Tibor Vellai and Gábor Vida in 1999, and is disarmingly simple. Bacteria are limited in their physical size, genome content, and complexity, they say, because they are forced to respire across their *external* cell membrane. Let's see why that matters.

The stumbling block of geometry

Recall from Part 2 how respiration works. Redox reactions generate a proton gradient across a membrane, which is then used to power the synthesis of ATP. An intact membrane is necessary for energy generation. Eukaryotic cells use the inner mitochondrial membrane to generate ATP, while bacteria, which do not have organelles, must use their external cell membrane.

The limitation for bacteria is geometric. For simplicity, imagine a bacterium shaped like a cube, then double its dimensions. A cube has six sides, so if our cubic bacterium had dimensions of one thousandth of a millimetre each way (1 μm), doubling its size would quadruple the surface area, from 6 μm^2 (1 × 1 × 6) to 24μm^2 (2 × 2 × 6) μm^2. The volume of the cube, however, depends on its length multiplied by its breadth by its depth, and this rises eightfold, from 1 μm^3 (1 × 1 × 1) to 8 μm^3 (2 × 2 × 2). When the cube has dimensions of 1 μm each way, the surface area to volume ratio is 6/1 = 6; with dimensions of 2 μm each way, the surface area to volume ratio is 24/8 = 3. The cubic bacterium now has half as much surface area in relation to its volume. The same thing happens if we double the dimensions of the cube again. The surface area to volume ratio now falls to $96/64 = 1.5$. Because the respiratory efficiency of bacteria depends on the ratio of surface area (the external membrane used for generating energy) to volume (the mass of the cell using up the available energy) this means that as bacteria become larger their respiratory efficiency declines hyperbolically (or more technically, with mass to the power of $2/3$, as we'll see in the next Part).

This decline in respiratory efficiency is coupled to a related problem in absorbing nutrients: the falling surface area to volume ratio restricts the rate at which food can be absorbed relative to the requirement. These problems can be mitigated to some extent by altering the shape of the cell (for example, a rod has a larger surface-area-to-volume ratio than a sphere) or by folding the membrane into sheets or villi (as in our own intestinal wall, which is subject to the same need to maximize absorption). Presumably, however, there comes a point when complex shapes are selected against, simply because they are too fragile, or too difficult to replicate with any accuracy. As any spatially challenged plasticine modeller knows, an imperfect sphere is much the most robust and replicable shape. We aren't alone: most bacteria are spherical (cocci) or rod-like (bacilli) in shape.

In terms of energy, a bacterial cell with double the 'normal' dimensions will produce half as much ATP per unit volume, while being obliged to divert more energy towards replicating the cellular constituents, such as proteins, lipids, and carbohydrates, that make up the extra cell volume. Smaller variants, with smaller genomes, will almost invariably be favoured by selection. It is therefore hardly surprising that only a handful of bacteria have achieved a size comparable with eukaryotes, and these exceptions merely prove the rule. For example, the giant sulfur bacterium *Thiomargarita namibiensis* (the 'sulfur pearl of Namibia'), discovered in the late 1990s, is eukaryotic in size: 100 to 300 microns in diameter (0.1 to 0.3 mm). Although this caused some excitement, it is actually composed almost entirely of a large vacuole. This vacuole accumulates raw materials for respiration, which are continually washed up and swept away by the upwelling currents off the Namibian coast. Their giant size is a sham—they amount to no more than a thin layer covering the surface of a spherical vacuole, like the rubber skin of a water-filled balloon.

Geometry is not the only stumbling block for bacteria. Think again about proton pumping. To generate energy, bacteria need to pump protons across their external cell membrane, into the space outside the cell. This space is known as the periplasm, because it is itself bounded by the cell wall.[1] The cell

---

[1] Technically the periplasm refers to the space between the inner and outer cell membranes of Gram-negative bacteria. These are named after the way in which they are coloured by a particular stain known as the Gram stain. Bacteria that are coloured by this stain are called Gram-positive; bacteria that are not stained are called Gram-negative. This odd behaviour

wall presumably helps to keep protons from dissipating altogether. Peter Mitchell himself observed that bacteria acidify their medium during active respiration, and presumably more protons are free to disperse if the cell wall is lost. Such considerations may help to explain why bacteria that lose their cell wall become fragile: they not only lose their structural support but also lose the outer boundary to their periplasmic space (of course they retain the inner boundary, the cell membrane itself). Without this outer boundary, the proton gradient is more likely to dissipate, at least to some extent—some protons appear to be 'tethered' to the membrane by electrostatic forces. Any dispersal of proton gradient is likely to disrupt chemiosmotic energy production: energy is not produced efficiently. As energy production runs down, all other aspects of a cell's housekeeping are forced to run down too. Fragility is the least of what we would expect; it's more surprising that the denuded cells can survive at all.

How to lose the cell wall without dying

While many types of bacteria do lose their cell wall during parts of their life cycle only two groups of prokaryotes have succeeded in losing their cell walls permanently, yet lived to tell the tale. It's interesting to consider the extenuating circumstances that permitted them to do so.

One group, the *Mycoplasma*, comprises mostly parasites, many of which live inside other cells. *Mycoplasma* cells are tiny, with very small genomes. *M. genitalium*, discovered in 1981, has the smallest known genome of any bacterial cell, encoding fewer than 500 genes. Despite its simplicity, it ranks among the most common of sexually transmitted diseases, producing symptoms similar to *Chlamydia* infection. It is so small (less than a third of a micron in diameter, or an order of magnitude smaller than most bacteria) that it must normally be viewed under the electron microscope; and difficulties culturing it meant its significance was not appreciated until the important advances in gene sequencing in the early 1990s. Like *Rickettsia*, *Mycoplasma* have lost virtually all the genes required for making nucleotides, amino acids, and so forth. Unlike

---

actually reflects differences in the cell wall and the cell membrane. Gram-negative bacteria have two outer cell membranes and a thin cell wall, which is contiguous with the outer cell membrane. In contrast, Gram-positive bacteria have a thicker cell wall, but only one cell membrane. Technically, then, only Gram-negative bacteria have a periplasm, because only Gram-negative bacteria have a space between their two cell membranes. However, both types of bacteria have a cell wall, which encloses a space that lies outside the cell but inside the wall. For simplicity I shall refer to this space as the periplasm, because it fulfils many of the same purposes in all bacteria, despite their differing structures.

Rickettsia, however, *Mycoplasma* have also lost all the genes for oxygen respiration, or indeed any other form of membrane respiration: they have no cytochromes, and so must rely on fermentation for energy. As we saw in the previous chapter, fermentation does not involve pumping protons across a membrane, and this might explain how *Mycoplasma* can survive without a cell wall. But fermentation produces up to 19 times fewer ATPs from a molecule of glucose than does oxygen respiration, and this in turn helps to account for the regressive character of *Mycoplasma*—their tiny size and genome content. They live like hermits, with little to call their own.

The second group of prokaryotes that thrive without a cell wall is the *Thermoplasma*, which are extremophile archaea that live in hot springs at 60°C and an optimal acidity of pH 2. They would probably fare well in a British fish and chip shop, as their preferred living conditions are equivalent to hot vinegar. Lynn Margulis once argued that *Thermoplasma* may be the archaeal ancestors of the eukaryotic cell, on the grounds that they can survive 'in the wild' without a cell wall; but, as we saw in Part 1, stronger evidence supports the methanogens as the putative original host. When the complete genome sequence of *Thermoplasma acidophilum* was reported in *Nature* in 2000, it provided no evidence of a close link to the eukaryotes.

How do *Thermoplasma* survive without a cell wall? Simple: their acidic surroundings fulfil the role of the periplasm, so they have no need of a periplasm of their own. Normally, bacteria pump protons across the external cell membrane into the periplasm outside the cell, which is bounded by the cell wall. This small periplasmic space is therefore acidic, and its acidity is essential for chemiosmosis. In other words, bacteria normally carry around with them a portable acid bath. In contrast, *Thermoplasma* already live in an acid bath, which is effectively a giant communal periplasm, so they can relinquish their own portable acid bath. As long as they can maintain neutral conditions inside the cell, they can take advantage of the natural chemiosmotic gradient across the cell membrane. So how do they stay neutral inside? Again, the answer is simple: they actively pump protons out of the cell in the same way as any other bacteria, by cell respiration. In other words, as in most prokaryotes, the energy released from food is used to pump protons out of the cell against a concentration gradient; and the backflow of protons into the cell is used to power the ATPase, driving ATP synthesis.

In principle, the absence of a cell wall should not undermine the energetic efficiency or genome size of *Thermoplasma* but in practice the cells are somewhat regressive. Although they can measure up to 5 microns in diameter, their genome, of 1 to 2 million letters, encodes only 1500 genes, and is among the smallest of bacterial genomes; indeed, it is the smallest non-parasitic genome known. Perhaps the extra effort needed to keep out a high concentration of protons saps the energy that *Thermoplasma* can afford to divert to replicating its genome.[2]

Let's round this up. The exceptions of *Mycoplasma* and *Thermoplasma* only go to prove the rule: the complexity of both bacteria and archaea is curtailed by their need to generate energy across the outer cell membrane. In general, bacteria can't grow larger because their energetic efficiency falls off quickly as their cell volume increases. If they lose their cell wall, the outer boundary to the periplasm, the proton gradient is more likely to dissipate away, sapping the energy supply and rendering the bacteria fragile. The only prokaryotes that have survived without the cell wall are tiny regressive hermits, such as *Mycoplasma*, which live by parasitism and fermentation, or specialists like *Thermoplasma*, which can only survive in acid. Despite losing their cell walls, and so in principle being able to consume particles, neither group shows any tendency towards the predatory eukaryotic habit of engulfing food by phagocytosis. Neither do they show any tendency to develop a nucleus, or for that matter any other eukaryotic traits. These traits, I shall argue, depend on the possession of mitochondria.

Why insider dealing pays

The advantage of mitochondria is that they reside physically inside their host cell. Recall that mitochondria are bounded by two membranes, an outer and an inner membrane, which enclose two distinct spaces, the inner matrix and the inter-membrane space. The respiratory chains and the ATPase complexes are all embedded in the inner mitochondrial membrane, and pump protons from the inner matrix to the inter-membrane space (see Figure 1, page 12). The acid environment needed for chemiosmosis is therefore contained within the mitochondria and does not affect other aspects of cellular function. (Technically it is not actually acidic, as the protons are buffered, but this doesn't alter the thrust of the argument.)

Internalization of energy generation within the cell means that an external cell wall is no longer needed, and so can be lost without inducing fragility. Loss of the cell wall frees up the external cell membrane to specialize in other tasks, such as signalling, movement, and phagocytosis. Most importantly of all, internalization releases the eukaryotic cell from the geometric constraints that

[2] *Thermoplasma* are variable in size, but usually quite large, spherical cells with a small genome. If they live in strong acid, and need to restrict the entry of protons into the cell, they could do this by lowering their surface-area to volume ratio—i.e., by being large and spherical in shape. Large size, of course, undermines the efficiency of respiration, which might explain the small genome size. It would be interesting to know whether cell volume in *Thermoplasma* correlates with the acidity of their surroundings.

oppress bacteria. Eukaryotes are on average 10 000 to 100 000 times the volume of bacteria, but as they become larger, their respiratory efficiency doesn't slope off in the same way. To increase energetic efficiency, all that eukaryotic cells need to do is to increase the surface area of mitochondrial membranes within the cell; and this can be done simply by having a few more mitochondria. Internalization of energy production therefore enables both the loss of the cell wall and a much greater cell volume. In the fossil record, the sheer size of eukaryotic cells often helps to distinguish them from bacteria—and this greater size appeared quite suddenly, in geological terms, with the internalization of energy generation in the cell. Suddenly, some 2 billion years ago, large eukaryotic cells appear in the fossil record; presumably this must date with some accuracy the origin of the mitochondria, although they themselves can't be made out in the fossils.

So bacteria are under a strong selection pressure for small size whereas eukaryotes are not. As eukaryotic cells grow larger, they can maintain their energy balance simply by keeping more mitochondria inside—herding more pigs, as it were. So long as they can find enough food to oxidize—enough to feed the pigs—they are not constrained by geometry. Whereas large size is penalized in bacteria, it actually pays dividends in eukaryotes. For example, large size enables a change in behaviour or lifestyle. A large energetic cell does not have to spend all its time replicating its DNA, but can instead spend time and energy developing an arsenal of protein weapons. It can behave like a fungal cell, and squirt lethal enzymes onto neighbouring cells to digest them before absorbing their juices. Or it can turn predator and live by engulfing smaller cells whole, digesting them inside itself. Either way, it doesn't need to replicate quickly to stay ahead of the competition—it can simply eat the competition. Predation, the archetypal eukaryotic lifestyle, is born of large size, and it depends on overcoming the energetic barriers to being larger. A parallel with human society is the larger communities made possible by farming: with more manpower, it was possible to satisfy food production and still have enough people left over to form an army, or invent lethal new weapons. The hunter-gathers couldn't sustain such a high population and were bound to lose out to the numerous and specialized competition.

Among cells, it is interesting that predation and parasitism tend to pull in opposite directions. As a rule of thumb, parasites are regressive in character, and in this regard the eukaryotic parasites are no exception. The very word 'parasite' conveys something contemptible. Conversely the term 'predator' can send shivers up and down the spine. Predation tends to drive evolutionary arms races, in which the predator and prey compete to grow ever larger: the *red queen* effect, whereby both sides must run to stay in the same place, relative to each other. I know of no bacterial cells that are predatory in the eukaryotic fashion of physically engulfing their prey. Perhaps this should not be surprising. A predatory lifestyle requires a very substantial energetic investment before anything is caught and eaten. At the cellular level, engulfing food by phagocytosis, in particular, demands a dynamic cytoskeleton and an ability to change shape vigorously, both of which consume copious quantities of ATP. So phagocytosis is made possible by three factors: the ability to change shape (which requires losing the cell wall, then developing a far more dynamic cytoskeleton); sufficiently large size to physically engulf prey; and a plentiful supply of energy.

Bacteria can lose their cell wall but have never developed phagocytosis. Vellai and Vida, whom we met earlier, argue that the additional requirements of phagocytosis for large size *and* plentiful ATP may have prevented bacteria from ever becoming effective predators in the eukaryotic style. Respiring over the outer membrane means that bacteria are obliged to generate less energy, relative to their size, as they become bigger. When they become large enough to physically engulf other bacteria they are less likely to have the energy needed to do so. Worse, if the cell membrane is specialized for energy generation, then phagocytosis would also be detrimental, for it would disrupt the proton gradient. It is possible that bacteria could circumvent such problems by relying on fermentation, rather than respiration, as this does not require a membrane. But fermentation also generates substantially less energy than respiration, and this may limit the ability of cells to survive by phagocytosis. Vellai and Vida note that all the eukaryotic cells that live by the combination of fermentation and phagocytosis are parasites, and so might be able to make energy savings in other areas (for example, not synthesizing their own nucleotides and amino acids, the building blocks of DNA and proteins).[3] By sacrificing their energetic expenses in some areas they might be able to justify the energetic costs of phagocytosis. But I'm not aware of any research that looks into this hypothesis systematically, and unfortunately Tibor Vellai has moved on from this field of research.

---

[3] Fermentation presents some interesting dilemmas, for although it is far less efficient than respiration, in terms of the quantity of ATP generated from one molecule of glucose, it is also faster—it produces more ATP in a short space of time. This means that cells growing by fermentation can out-compete those growing by respiration for the same resources. Exactly how this works out in reality, however, is less certain, as fermentation can't complete the oxidation of molecules like glucose, but rather releases waste products like alcohol into the surroundings, to our own benefit. Of course, this is also to the benefit of any cells that can burn alcohol, which is to say, are capable of respiration. So, like the hare and the tortoise, it may be that respiration, though slower, pays dividends in the end. In terms of phagocytosis, running out of energy in 'mid-bite' may be more detrimental than biting slowly. A second interesting possibility is that respiration actually *encouraged* the evolution of multicellular organisms, as they were large enough to hoard any raw materials, to prevent the fermenting cells from frittering them away first.

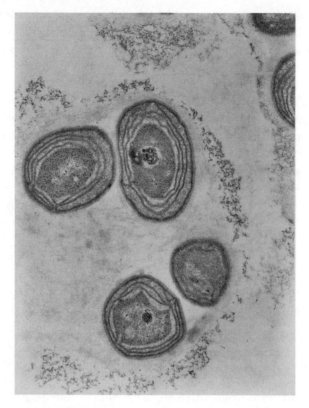

10 Internal bioenergetic membranes of the bacterium
Nitrosomonas, giving it a 'eukaryotic' look.

These ideas are interesting and may go some way towards explaining the
dichotomy between bacteria and eukaryotes, but they leave a suspicion at the
back of my mind. Why are bacteria *invariably* penalized if they get bigger?
Bacteria are so inventive that it is remarkable that none of them have ever
solved the challenge of simultaneously increasing their size *and* their energy
status. It doesn't sound so difficult: all they needed to do was grow some internal membranes for generating energy. If internalization of energy production
inside the cell enabled eukaryotes to make their quantum leap in size and
behaviour, what was to stop bacteria from having internal membranes themselves? Some bacteria, such as *Nitrosomonas* and *Nitrosococcus* do in fact have
quite complex internal membrane systems, devoted to generating energy
(Figure 10). They have a eukaryotic 'look' about them. The cell membranes are
extensively infolded, creating a large periplasmic compartment. It seems to be
a small step from here to a fully compartmentalized eukaryotic cell; so why did
it never happen?

In the next chapter, we'll take up the story of the first chimeric eukaryote that we abandoned without a nucleus at the end of Part 1, and look into what may have become of it next. Guided by the principles of energy generation, which we explored in Part 2, we'll see why a symbiosis between two cells was successful, and why, by the same token, it was not possible for bacteria to compartmentalize themselves in the same way as eukaryotes, by natural selection alone. We'll see why only eukaryotes could become giant predators in a bacterial world—indeed, why they overturned the bacterial world forever.

8

Why Mitochondria Make Complexity Possible

In the last chapter, we considered why bacteria have remained small and unsophisticated, at least in terms of their morphology. The reasons relate mostly to the selection pressures that face bacteria. These are different from eukaryotic cells because bacteria, for the most part, do not eat each other. Their success in a population therefore depends largely on the speed of their replication. This in turn depends on two critical factors: first, copying the bacterial genome is the slowest step of replication, so the larger the genome, the slower is replication; and second, cell division costs energy, so the least energetically efficient bacteria replicate the slowest. Bacteria with large genomes will always tend to lose out in a race against those with smaller genomes, because bacteria swap genes, by way of lateral gene transfer, and so can keep loading up cassettes of useful genes, and throwing them away again as soon as they become burdensome. Bacteria are therefore faster and more competitive if genetically unburdened.

If two cells have the same number of genes, and have equally efficient energy-generating systems, then the cell that can replicate the fastest will be the smaller of the two. This is because bacteria depend on their outer cell membrane to generate energy, as well as absorbing food. As bacteria become larger in size, their surface area rises more slowly than their internal volume, so their energetic efficiency tails away. Larger bacteria are energetically less efficient, and always likely to lose out in competition with smaller bacteria. Such an energetic penalty against large size precludes phagocytosis, for physically engulfing prey demands both large size and plenty of energy to change shape. Eukaryotic-style predation—catching and physically eating prey—is therefore absent among bacteria. It seems that eukaryotes escape this problem because they generate their energy internally, which makes them relatively independent of their surface area, and enables them to become many thousands of times larger without losing energetic efficiency.

As a distinction between the bacteria and eukaryotes, this reason sounds flimsy. Some bacteria have quite complex internal membrane systems and could be released from the surface-area constraint, yet still don't approach eukaryotes in size and complexity. Why not? We'll look into a possible answer in this chapter, and it is this: mitochondria need genes to control respiration over a large area of internal membranes. All known mitochondria have retained a contingent of their own genes. The genes that mitochondria retain are specific, and the mitochondria were able to retain them because of the nature of their symbiotic relationship with their host cell. Bacteria do not have this advantage. Their tendency to throw away any superfluous genes has prevented them from ever harnessing the correct core contingent of genes to govern energy generation, and this has always prevented them from developing the size and complexity of the eukaryotes.

To understand the reasons why mitochondrial genes are important, and why bacteria can't acquire the correct set of genes for themselves, we'll need to penetrate further into the intimate relationship between the cells that took part in the original eukaryotic union, two billion years ago. We'll take up the story where we left off in Part 1. There, we parked the chimeric eukaryote as a cell that had mitochondria but had not yet developed a nucleus. Because a eukaryotic cell is, by definition, a cell that has a 'true' nucleus, we can't really refer to our chimera as a eukaryote. So let's think now about the selection pressures that turned our strange chimeric cell into a proper eukaryotic cell. These pressures hold the key not just to the origin of the eukaryotic cell, but also to the origin of real complexity, for they explain why bacteria have always remained bacteria: why they could never evolve into complex eukaryotes by natural selection alone, but required symbiosis.

Recall from Part 1 that the key to the hydrogen hypothesis is the transfer of genes from the symbiont to the host cell. No evolutionary novelties were called for, beyond those that already existed in the two collaborating cells entered in an intimate partnership. We know that genes were transferred from the mitochondria to the nucleus, because today mitochondria have few remaining genes, and there are many genes in the nucleus that undoubtedly have a mitochondrial origin, for they can be found in the mitochondria of other species that lost a different selection of genes. In all species, mitochondria lost the overwhelming majority of their genes—probably several thousand. Exactly how many of these genes made it to the nucleus, and how many were just lost, is a moot point among researchers, but it seems likely that many hundreds did make it to the nucleus.

For those not familiar with the 'stickiness' and resilience of DNA, it may seem akin to a conjuring trick for genes from the mitochondria to suddenly appear in the nucleus, like a rabbit produced from a top hat. How on earth did they do that? In fact such gene hopping is commonplace among bacteria. We have already noted that lateral gene transfer is widespread, and that bacteria routinely take up genes from their environment. Although we normally think of the 'environment' as outside the cell, acquiring spare genes from inside the cell is even easier.

Let's assume that the first mitochondria were able to divide within their host cell. Today, we have tens or hundreds of mitochondria in a single cell, and even after two billion years of adaptation to living within another cell they still divide more or less independently. At the beginning, then, it's not hard to picture the host cell as having two or more mitochondria. Now imagine that one dies, perhaps because it can't get access to enough food. As it dies, it releases its genes into the cytoplasm of the host cell. Some of these genes will be lost altogether, but a handful might be incorporated into the nucleus, by means of normal gene transfer. This process could, in principle, be repeated every time a mitochondrion dies, each time potentially transferring a few more genes to the host cell.

Such transfer of genes might sound a little tenuous or theoretical, but it is not. Just how rapid and continuous the process can be in evolutionary terms was demonstrated by Jeremy Timmis and his colleagues at the University of Adelaide in Australia, in a *Nature* paper of 2003. The researchers were interested in chloroplasts (the plant organelles responsible for photosynthesis), rather than mitochondria, but in many respects chloroplasts and mitochondria are similar: both are semi-autonomous energy-producing organelles, which were once free-living bacteria, and both have retained their own genome, albeit dwindling in size. Timmis and colleagues found that chloroplast genes are transferred to the nucleus at a rate of about 1 transfer in every 16 000 seeds in the tobacco plant *Nicotiana tabacum*. This may not sound impressive, but a single tobacco plant produces as many as a million seeds in a single year, which adds up to more than 60 seeds in which at least one chloroplast gene has been transferred to the nucleus—in every plant, in every generation.

Very similar transfers take place from the mitochondria to the nucleus. The reality of such gene transfers in nature is attested by the discovery of duplications of chloroplast and mitochondrial genes in the nuclear genomes of many species—in other words the same gene is found in both the mitochondria or chloroplast *and* in the nucleus. The human genome project has revealed that there have been at least 354 separate, independent transfers of mitochondrial DNA to the nucleus in humans. These DNA sequences are called *numts*, or nuclear-mitochondrial sequences. They represent the entire mitochondrial genome, in bits and pieces: some bits repeatedly, others not. In primates and other mammals, such numts have been transferred regularly over the last 58 million years, and presumably the process goes back further, as far as we care to look. Because DNA in mitochondria evolves faster than DNA in the nucleus, the sequence of letters in numts can act as a time capsule, giving an impression of what mitochondrial DNA might have looked like in the distant past. Such alien sequences can cause serious confusion, however, and were once mistaken for dinosaur DNA, leaving one team of researchers with red faces.

Gene transfer continues today, occasionally making itself noticed. For example, in 2003, Clesson Turner, then at the Walter Reed Army Medical Center in Washington, and collaborators, showed that a spontaneous transfer of mitochondrial DNA to the nucleus was responsible for causing the rare genetic disease Pallister-Hall syndrome in one unfortunate patient. How common such genetic transfers are in the pantheon of inherited disease is unknown.

Gene transfers occur predominantly in one direction. Think back again to the first chimeric eukaryote. If the host cell were to die, it would release its symbionts, the proto-mitochondria, back into the environment, where they may or may not perish—but regardless of their fate, the experiment in chimeric co-existence would certainly have perished. On the other hand, if a single mitochondrion were to die, but a second viable mitochondrion survived in the host cell, then the chimera as a whole would still be viable. To get back to square one, the surviving mitochondrion would just have to divide. Each time a mitochondrion died, the genes released into the host cell could potentially be integrated into its chromosome by normal genetic recombination. This means there is a gene ratchet, favouring the transfer of genes from the mitochondria to the host cell, but not the other way around.

The origin of the nucleus

What happens to the genes that are transferred? According to Bill Martin, whom we met in both Parts 1 and 2, such a process might account for the origin of the eukaryotic nucleus. To understand how, we need to recall two points that we have discussed in earlier chapters. First, recall that Martin's hydrogen hypothesis argues that the eukaryotic cell was first forged from the union of an archaeon and a bacterium. And second, recall from Chapter 6 (page 98) that archaea and bacteria have different types of lipid in their cell membranes. The details don't matter here, but consider the kind of membranes we would expect to find in that first, chimeric eukaryote. The host cell, being an archaeon, should have had archaeal membranes. The mitochondria, being bacterial, should have had bacterial membranes. So what do we actually see today? Eukaryotic membranes are uniformly bacterial in nature—both in their lipid structure and in many details of their embedded proteins (like the proteins that make up the respiratory chain, and similar proteins found in the nuclear membrane). The bacterial-style membranes of the eukaryotes include the cell membrane, the mitochondrial membranes, other internal membrane structures, and the double nuclear membrane. In fact there is no trace of the original archaeal membranes in the eukaryotes, despite the fact that other features make it virtually certain that the original host cell was indeed an archaeon.

Such basic consistency, when we would expect to find disparity, has led some researchers to question the hydrogen hypothesis, but Martin considers the apparent anomaly to be a strength. He suggests that the genes for making bacterial lipids were transferred to the host cell, along with many other genes. Presumably, if functional, the genes went ahead with their normal tasks, such as making lipids; there is no reason why they should not function normally as before. But there may have been one difference—the host cell may have lost the ability to target protein products to particular locations in the cell (protein targeting relies on an 'address' sequence that differs in different species). The host cell may therefore have been able to *make* bacterial products, such as lipids, but not known exactly what to do with them; in particular, where to send them. Lipids, of course, don't dissolve in water, and so if not targeted to an existing membrane would simply precipitate as lipid vesicles—spherical droplets enclosing a hollow watery space. Such droplets fuse as easily as soap bubbles, extending into vacuoles, tubes, or flattened vesicles. In the first eukaryote, these vesicles might simply have coalesced where they were formed, around the chromosome, to form loose, baggy membrane structures. Now this is exactly the structure of the nuclear membrane today—it is not a continuous double membrane structure like the mitochondria or chloroplasts, but is composed of a series of flattened vesicles, and these are continuous with the other membrane systems within the cell. What's more, when modern eukaryotic cells divide, they dissolve the nuclear membrane, to separate the chromosomes destined for each of the daughter cells; and a fresh nuclear membrane forms around the chromosomes in each of these daughter cells. It does so by coalescing in a manner reminiscent of Martin's proposal, and remains continuous with the other membrane systems of the cell. Thus, in Martin's scenario, gene transfer accounts for the formation of the nuclear membrane, as well as all the other membrane systems of eukaryotic cells. All that was needed was a degree of orientational confusion, a map-reading hiatus.

There is still one step to go: we need to put together a cell with bacterial-style membranes throughout, in other words we need to replace the archaeal lipids of the cell membrane with bacterial lipids. How did this happen? Presumably, if bacterial lipids offered any advantage, such as fluidity, or adaptability to different environments, then any cell that expressed only the bacterial lipids would be at an advantage. Natural selection would ensure that the archaeal lipids were replaced, if such an advantage existed: there was little call for evolutionary 'novelty'; it was merely a matter of playing with existing parts. It remains possible, however, that some eukaryotes did not go the whole hog. It would be interesting to know if there are still any primitive eukaryotic cells that retain vestiges of archaeal lipids in their membranes. In support of the possibility, virtually all eukaryotes, including fungi, plants, and animals like ourselves still possess all the genes for making the basic carbon building blocks of archaeal lipids, the *isoprenes* (see page 99). We don't use them for building membranes any more, however, but for making an army of *isoprenoids*, otherwise known as terpenoids or terpenes. These include any structure composed of linked isoprene units, and together make up the single largest family of natural products known, totalling more than 23 000 catalogued structures. These include steroids, vitamins, hormones, fragrances, pigments, and some polymers. Many isoprenoids have potent biological effects, and are being used in pharmaceutical development; the anticancer drug *Taxol*, for example, a plant metabolite, is an isoprenoid. So we haven't lost the machinery for making archaeal lipids at all; if anything, we have enriched it.

If his theory is correct, then Martin has derived an essentially complete eukaryotic cell via a simple succession of steps: it has a nucleus enveloped by a discontinuous double membrane; it has internal membrane structures; and it has organelles such as mitochondria. The cell is free to lose its cell wall (but not, of course, its external cell membrane), as it no longer needs a periplasm to generate energy. Being derived from a methanogen, it wraps its genes in histone proteins and has a basically eukaryotic system of transcribing its genes and building proteins (see Part 1). On the other hand, this hypothetical progenitor eukaryotic cell probably did not engulf its food whole by phagocytosis—despite having a cytoskeleton (inherited from the archaea or the bacteria), it has not yet derived the dynamic cytoskeleton characteristic of mobile protozoa like amoeba. Rather, the first eukaryotes may have resembled unicellular fungi, which secrete various digestive enzymes into their surroundings, to break down food externally. This conclusion is corroborated by some recent genetic studies, but we won't look into these here, for too many uncertainties remain.

Why did mitochondria retain any genes at all?

So the transfer of genes from the mitochondria to the host cell is capable of explaining the origin of the eukaryotic cell, without requiring any evolutionary innovations (new genes with different functions) whatsoever. Yet the sheer *ease* of gene transfer raises another suspicious question. Why are there any genes left in the mitochondria at all? Why were they not all transferred to the nucleus?

There are big disadvantages to retaining genes in the mitochondria. First, there are hundreds, even thousands of copies of the mitochondrial genome in each cell (usually 5 to 10 copies in every mitochondrion). This enormous copy-number is one of the reasons that mitochondrial DNA is so important in forensics, and in identifying ancient remains—from such an embarrassment of riches, it is usually possible to isolate at least a few mitochondrial genes. But by the same token, it also means that whenever the cell divides a vast number of ostensibly superfluous genes must be copied. Not only that, but every single mitochondrion is obliged to maintain its own genetic apparatus, enabling it to transcribe its genes and build its own proteins. By thrifty bacterial standards (which, as we have seen, eliminate any unnecessary DNA post haste) the existence of these supernumerary genetic outposts seems a costly extravagance. Second, as we shall see in Part 6, there are potentially destructive consequences of competition between different genomes within the same cell—natural selection can pit mitochondria against each other, or against the host cell, with no consideration of the long-term cost, merely the short-term gain for the individual genes. Third, storing genes, vulnerable informational systems, in the immediate vicinity of the mitochondrial respiratory chains, which leak destructive free radicals, is equivalent to storing a valuable library in the wooden shack of a registered pyromaniac. The vulnerability of mitochondrial genes to damage is reflected in their high evolution rate—in mammals, some twentyfold greater than the nuclear genes.

So there are serious costs to retaining mitochondrial genes. I repeat: if gene transfer is easy, why on earth are there any mitochondrial genes left? The first and most obvious reason is that the genes are not the problem: it is the *products* of the mitochondrial genes, the proteins, that need to function in the mitochondria. These are mostly involved in cellular respiration and so are vitally important to the life of the cell. If the genes are transported to the nucleus, then somehow their protein products need to be routed back to the mitochondria, and if they fail to get there the cell may well die. Even so, many proteins encoded in the nucleus *do* get back to the mitochondria: they are 'tagged' with a short chain of amino acids—an 'address' tag, pinpointing the final destination, as we discussed when considering lipids a few pages ago. The address tag is recognized by protein complexes in the mitochondrial membranes that act as customs posts, controlling import and export across the membranes. Many hundreds of proteins destined for the mitochondria are tagged in this way. But the simplicity of this system raises a question of its own—why can't *all* proteins that are destined for the mitochondria be tagged in this way?

The textbook answer is that they can—it just takes a long time to arrange, long even in terms of the vast stretches of evolutionary time. A number of chance events must be negotiated before a protein can successfully be targeted back to the mitochondria. First of all, the gene must be incorporated properly into the nucleus, which is to say that the entire gene (rather than a bit of it) must be transferred to the nucleus, and then integrated into the nuclear DNA. Once incorporated, it has to work: it must be switched on and transcribed to produce a protein. This may be difficult, as genes are inserted more or less randomly into the nuclear DNA, and can make a mess of the other genes already there, as well as regulatory sequences that govern genetic activity. Second, the protein must acquire the correct address tag, which again appears to be a chance event; otherwise it will not be targeted back to the mitochondria. Instead, it will be constructed in the cytoplasm and remain there, like a woebegone Trojan horse that failed to gain entrance to Troy. Acquiring the right address tag takes time, time that is measured in aeons. Thus, say theorists, the few remaining mitochondrial genes are just a shrinking residual. One day, perhaps a few hundred million years hence, no mitochondrial genes will be left at all. And the fact that different species have different numbers of genes that remain in their mitochondria lends support to the slow, random nature of this process.

The nucleus is not enough

But this answer is not quite convincing. *All* species have lost *almost* all their mitochondrial genomes but *not one species has lost them all.* None has more than a hundred genes left, having started out with probably several thousand some two billion years ago, so the process has run very nearly to completion in all species. This gene loss has occurred in parallel: different species have lost their mitochondrial genes independently. As a proportion of the genes lost, all species have now lost between 95 and 99.9 per cent of their mitochondrial genes. If chance alone were the dominating factor, we might expect that at least a few species would have gone the whole hog by now, and transferred all mitochondrial genes to the nucleus. Not one has done so. All known mitochondria have retained at least a few genes. What's more, mitochondria isolated from different species have invariably retained the same core of genes: they have independently lost the great majority of their genes but kept essentially the same handful, again implying that chance is not to blame. Interestingly, exactly the same applies to chloroplasts, which, as we have seen, are in a similar position: no chloroplast has lost all of its genes, and again, the same core of genes always figures among them. In contrast, other organelles related to mitochondria, such as hydrogenosomes and mitosomes, have almost invariably lost *all* their genes.

A number of reasons have been put forward to account for the fact that all known mitochondria have retained at least a few genes. Most are not terribly convincing. One idea once popular, for example, is that some proteins can't be targeted to the mitochondria because they are too large or too hydrophobic— but most of these proteins have in fact been successfully targeted to the mitochondria, either in one species or another, or by means of genetic engineering. Clearly the physical properties of proteins are not insurmountable obstacles to their parcelling and delivery to the mitochondria. Another idea is that the mitochondrial genetic systems harbour exceptions to the universal genetic code, and so mitochondrial genes are no longer strictly analogous to nuclear genes. If these genes were moved to the nucleus and read off according to the standard genetic code, the resulting protein would not be quite the same as that produced by the mitochondrial genetic system, and might not function correctly. But this can't be the full answer either, as in many species the mitochondrial genes *do* conform to the universal genetic code. There is no discrepancy in these cases, and therefore no reason why all the mitochondrial genes could not be transferred to the nucleus—and yet they remain stubbornly in the mitochondria. Likewise, there are no variations in the universal genetic code in chloroplast genes, and yet, like mitochondria, they always retain a core contingent of genes on site.

The answer that I believe to be correct is only now gaining credence among evolutionary biologists, despite being put forward by John Allen, then at the University of Lund in Sweden, as long ago as 1993. Allen argues that there are many good reasons why all the mitochondrial genes should have moved to the nucleus, and no clear 'technical' reasons why any should have stayed. Therefore, he says, there must be a very strong *positive* reason for their retention. They have not remained there by chance, but because natural selection has favoured their retention *despite* the manifold disadvantages. In the balance of pros and cons, the pros prevailed, at least in the case of the small number of genes that remain. But if the cons are so obvious and important, it is remarkable that we have overlooked the pros—they must be even weightier.

The reason, says Allen, is no less than the *raison d'être* of mitochondria: respiration. The speed of respiration is very sensitive to changing circumstances—whether we're awake or asleep, or doing aerobics, sitting around, writing books, or chasing a ball. These sudden shifts demand that mitochondria adapt their activity at a molecular level—a requirement that is too important, and abruptly swinging, to be controlled at a distance by the bureaucratic confederation of genes far away in the nucleus. Similar sudden shifts in requirements occur not just in animals but also in plants, fungi, and microbes, which are even more subject to the vicissitudes of the environment (such as changing oxygen levels, heat, or cold) at the molecular level. To respond effectively to these abrupt changes, Allen argues, mitochondria *need* to maintain a genetic outpost on site, as the redox reactions that take place in the mitochondrial membranes must be tightly regulated *by genes* on a local basis. Notice that I'm referring to the genes themselves here, and not to the proteins that they encode; we'll look into why the genes are important in a moment. But before we move on, let's note that the need for local genetic rapid-response units not only explains why mitochondria must retain a contingent of genes, but also, I believe, why the bacteria could not evolve into more complex eukaryotic cells by natural selection alone.

The problem of poise

Let's think back again to how respiration works. Electrons and protons are stripped from food, and react with oxygen to provide the energy that we need to live. The energy is released a bit at a time, by breaking the reaction into a series of small steps. These steps take place in the respiratory chain, down which electrons flow, as if down a tiny wire. At several points the energy released is used to pump protons across a membrane, trapping them on the other side, like water behind the dam of a reservoir. The flow of protons back from this reservoir, through special channels in the dam (the drive shafts of the ATPase motor) powers the formation of ATP, the energy 'currency' of the cell.

Let's consider briefly the *speed* of respiration. Everything is coupled like cogs, so the speed of one cog controls the speed of the rest. So what controls the overall speed of the cogs? The answer is *demand*, but let's think this through. If electrons flow quickly down the chain, then the protons are pumped quickly (for proton pumping depends on electron flow) and the proton reservoir 'fills up'. A full reservoir, in turn, provides a high pressure to form ATP quickly, as protons flow back through the dedicated drive shaft of the ATPase. Now think what happens if there is no demand for ATP. In Chapter 4, we saw that ATP is formed from ADP and phosphate, and when it is broken down again, to provide energy, it reverts to ADP and phosphate. When demand is low, ATP is not consumed by the cell. Respiration converts all the ADP and phosphate into ATP, and that's that: the raw materials are exhausted, and the ATPase must grind to a halt. If the ATPase motor is not turning, then protons can no longer pass through the drive shaft. The proton reservoir brims full. As a result, protons can no longer be pumped against the high pressure of the reservoir. And without proton pumping, electrons can't flow down the chain. In other words, if demand is low, everything backs up and the speed of respiration slows right down until fresh demand starts all the wheels turning again. So the speed of respiration ultimately depends on demand.

But this is what happens when everything is working well and the cogs are well greased. There are other reasons for respiration to slow down, and these are not related to demand but to supply. We have noted one instance: the supply of ADP and phosphate. Normally, the concentration of these raw materials reflects the consumption of ATP, but it is always possible that there is simply a shortage of ADP and phosphate. Then there is the supply of oxygen or glucose. If there is not enough oxygen around—if we are suffocating—electron flow down the chain must slow down because there is nothing to remove the electrons at the end. They are forced to back up in the chain, and everything else slows down just as if there were a shortage of ADP. What about glucose? Now the number of electrons and protons that enter the chain is restricted—as if we were starving—so the flow of electrons is forced to slow down, which is to say the volume of electrons flowing down the chain per second falls.

So, the overall speed of respiration should ideally reflect demand, which is to say *consumption* of ATP, but under difficult conditions, such as starvation or suffocation, or perhaps a metabolic shortage of raw materials, then the speed of respiration reflects the supply rather than the demand. In both cases, however, the overall speed of respiration is reflected in the speed that electrons flow down the respiratory chain. If electrons flow quickly, glucose and oxygen are consumed quickly, and by definition, respiration is fast. Now, after this little detour, we can return to the point. There is a third factor that causes respiration to slow down, and this relates neither to supply nor demand, but rather to the quality of the wiring: it relates to the components of the respiratory chain themselves.

The components of the electron-transport chains have a choice of two possible states: they can either be oxidized (they don't have an electron) or they can be reduced (they do have an electron). Obviously they can't be both at once— they either have an electron or they don't. If a carrier already has an electron, it can't receive another one until it has passed on the first to the next carrier in the chain. Respiration will be held up until it has passed on this electron. Conversely, if a carrier doesn't have an electron, it can't pass on anything to the next carrier until it has received an electron from an earlier carrier. Respiration will be held up until it receives one. The overall speed of respiration therefore depends on the dynamic equilibrium between oxidation and reduction. There are thousands of respiratory chains in a single mitochondrion. Respiration will proceed most rapidly when 50 per cent of the carriers within these chains are oxidized (ready to receive electrons from an earlier carrier), and 50 per cent are reduced (ready to pass on electrons to the next carrier). If the rate of respiration is plotted out mathematically, it fits the equation of a bell curve. Respiration is fast at the top of the bell curve and slows precipitously on either side, as the carriers become more oxidized or reduced. The point of optimal balance, the top of the bell curve at which respiration is fastest, is known as 'redox poise'. Straying from redox poise slows down energy production, and such inefficiency, as we have seen, is strongly selected against in bacteria.

But the penalty for straying from redox poise is worse than inefficiency: there is the devil to pay. All the carriers of the respiratory chain are potentially reactive—they 'want' to pass their electrons to a neighbour (they have a chemical propensity to do so). If respiration is progressing normally, each carrier is most likely to pass on its electrons to the next carrier in the chain, each one of which 'wants' the electron a bit more than did its predecessor; but if the next carrier is already full then the chain becomes blocked. There is now a greater risk that the reactive carriers will pass on their electrons to something else instead. The most likely candidate is oxygen itself, which easily forms toxic free radicals such as the superoxide radical. I discussed the damage caused by free radicals in *Oxygen*; here, the important point is that free radicals react indiscriminately to damage all kinds of biological molecules. Formation of free radicals by the respiratory chain has influenced life in profound and unexpected ways, including the evolution of warm-bloodedness, cell suicide, and ageing, as we'll see in later chapters. For now, though, let's just note that if the chain becomes blocked, it is more likely to leak free radicals, just as a blocked drainpipe is more likely to spring water from small cracks.

So there are two good reasons for sustaining poise: keeping respiration as fast as possible, and restricting the leak of reactive free radicals. But maintaining poise is not just a matter of keeping the correct balance of electrons entering the respiratory chains to those leaving at the other end: it also depends on the relative number of carriers within the chains, and this fluctuates because the carriers are continually replaced, like everything else in the body.

Let's think about this for a moment. What happens if there aren't enough carriers in the respiratory chains? A shortage of carriers means that the passage of electrons down the respiratory chain slows down, just as too few links in a human bucket chain means there is a slow supply of water getting to the fire. Such a slow transfer of water to the fire equates to a shortage of water: even if the reservoir is full, the house will burn down. Conversely, if there are too many carriers in the middle of the chain, these accumulate electrons faster than they can be passed on down the chain. In the bucket chain analogy, the buckets are being passed faster at the beginning of the chain than at the end—there is a build-up in the middle and everything goes haywire. In both cases, respiration slows down because there is an imbalance in the number of carriers in the respiratory chains, not in the levels of any raw materials. If the concentration of any of these carriers gets out of kilter with the requirements of respiration, respiration slows, and free radicals leak out to cause damage.

Why mitochondria need genes

Now we are in a position to see why the mitochondria (and chloroplasts too) must retain a contingent of genes of their own. Let's consider the last carrier of the respiratory chain, cytochrome oxidase, which we met in Chapter 4. Imagine that there are 100 mitochondria in a cell. One of these mitochondria does not have enough cytochrome oxidase. As a result, in this mitochondrion, respiration slows down, and electrons back-up in the chains, from where they can escape to form free radicals. The mitochondrion is inefficient and in danger of damaging itself. To rectify the situation, it needs to make more cytochrome oxidase, so it sends a message to the genes: *Make more cytochrome oxidase!* How would this message operate? The signal might well be the free radicals themselves: a sudden burst of free radicals can alter gene activity through the action of transcription factors that leap into action only when oxidized by free radicals (they are said to be 'redox-sensitive'). In other words, if there is not enough cytochrome oxidase, electrons back up in the chain and leak out as free radicals. The sudden appearance of free radicals is interpreted by the cell as a signal that there is not enough cytochrome oxidase. It responds accordingly by making some more.[1]

Let's imagine that the genes are in the nucleus. The message arrives, and the nucleus sends orders to make more copies of cytochrome oxidase. It directs the newly minted proteins to the mitochondria using the standard address tag— but this tag can't discriminate between different mitochondria. As far as the nucleus is concerned, 'mitochondria' is a concept and all the mitochondria in the cell share exactly the same address (and it's quite hard to see how this could be otherwise, given that the mitochondrial population is in a constant state of turnover). So the newly minted cytochrome oxidase is distributed to all 100 mitochondria. The mitochondrion that is short does not get enough. The rest receive too much and so immediately send a message back to the nucleus to say: *Switch off cytochrome oxidase production!* Clearly this situation is untenable. Mitochondria inevitably lose control over respiration, and over-produce free radicals. Cells that lose respiratory control would certainly be selected against. At the very least—and this is a critical point—an inability to control respiration ought to limit the number of mitochondria that a cell could profitably maintain.

Now think what happens the other way round. Imagine that the genes for cytochrome oxidase are retained in the mitochondria. When the signal *Make more cytochrome oxidase!* is sent, it only goes as far as the local gene contingent. These local genes produce more cytochrome oxidase, which is immediately incorporated into the respiratory chains, correcting the imbalance in electron

---

[1] One question is how the cell interprets a signal to 'know' that more cytochrome oxidase is needed. A free-radical signal is also produced if there is a low demand for ATP: electrons then back up in the respiratory chains, which leak radicals, but the situation is not improved by adding new complexes: there is still a low demand for ATP, and electron flow remains sluggish. But the cell can detect ATP levels, and so in principle could combine two signals: 'high ATP' with 'high free radicals'. An appropriate response would now be to dissipate the proton gradient, to maintain electron flow (see Part 2, page 92). There is evidence that this happens. In contrast, if there were not enough respiratory complexes, then ATP levels would decline and electrons again would back-up in the respiratory chains. Now the signal would combine 'low ATP' with 'high free radicals'. Such a system could in theory discriminate the need for more respiratory complexes from low demand.

flow and restoring redox poise. When the message is sent back: *Stop! Switch off cytochrome oxidase production!* it, too, goes only as far as the local gene contingent, and affects only that single mitochondrion. Such a rapid local response could take place in any of the cell's mitochondria and might in principle operate quite differently in different mitochondria in the same cell at the same time. The cell as a whole retains control over the speed of respiration, and so benefits despite the high costs of maintaining numerous genetic outposts. It would be worse to move the genes to the nucleus.

Professional biochemists or perceptive readers might object at this point. I mentioned in Part 2 that the respiratory complexes are constructed from a large number of subunits, as many as 45 separate proteins in complex I. Mitochondrial genes encode a handful of these subunits, but the great majority are encoded by nuclear genes. This means that the respiratory complexes are an amalgam encoded by two different genomes. How, then, could a few mitochondrial genes dominate? Surely any construction decisions would need to be shared with the nucleus? Not necessarily. It seems that the respiratory complexes assemble themselves around a few core subunits: once these core proteins have been implanted in the membrane they act as a beacon and a scaffold for the assembly of the rest of the subunits. So if the mitochondrial genes encoded these critical subunits, then they would control the number of new complexes being built. Effectively, the mitochondria make the construction decision, and plant a flag in the membrane; the nuclear components assemble themselves around the flag. Given that the nucleus serves hundreds of mitochondria at once, the overall number of flags in the cell as a whole, at any one time, might remain fairly constant. There would not need to be any change in the overall rate of nuclear transcription to compensate for fluctuations in individual mitochondria, but the effect would be to keep a tight grip on the rate of respiration in all the mitochondria in a cell at once.

If this is true, then Allen's theory makes some specific predictions about which genes should be retained in the mitochondria. They should encode mostly the core electron-transport proteins in the respiratory chains, such as cytochrome oxidase—those that implant in the membrane like a flag, as if to say *'Build here!'* This is indeed the case (see Figure 11). It is also the case for chloroplasts, which as we have seen are in a similar position. Of course, additional genes may also be retained (by chance or for other reasons) but both the mitochondrial and the chloroplast genes of *all* species *always* encode the critical electron-transport proteins, along with the necessary machinery to physically produce the proteins within the mitochondria (such as transfer RNA molecules). When gene loss has progressed to an extreme, it is only—and invariably—this core of respiratory genes that remains. For example, the mitochondria of *Plasmodium*, the cause of malaria, have retained just three protein-

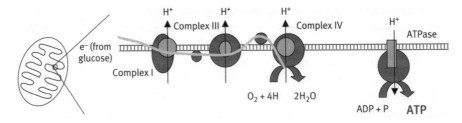

11 Very simplified representation of the respiratory chain, showing the coding of the subunits. Each complex is composed on numerous subunits, about 46 in the case of complex I. Some of these are encoded by mitochondrial genes and some by nuclear genes. John Allen's hypothesis argues that mitochondrial genes are necessary to control the rate of respiration on a local basis, and for this to work, the subunits encoded by mitochondrial genes should be the core subunits inserted in the membrane. The figure shows that this is broadly true: the subunits encoded by mitochondrial genes (shown in grey) are embedded in the core of the membrane, whereas the subunits encoded by nuclear genes (shown in black) assemble around them. Complex II is not shown here. It does not pump any protons, and does not have any subunits encoded by mitochondrial genes.

coding genes, and as a result they have had to keep all the complex machinery needed to make these proteins in each and every mitochondrion. All three of these genes encode cytochromes—the core electron-transporting proteins of the respiratory chain—exactly as predicted.

The theory makes another prediction, which also seems to be broadly true. This is that any organelles that do not need to conduct electrons will lose their genome. A good example of this is the hydrogenosome of some anaerobic eukaryotes (see Part 1, page 52). Hydrogenosomes are known to be related to mitochondria, and undoubtedly descend from bacteria. Their function is to carry out fermentations to generate hydrogen gas. They do not conduct electrons, and have no need to maintain redox poise. According to Allen's theory, they should have no need of a genome—and in virtually all cases they have indeed lost it.

Barriers to complexity in bacteria

If mitochondria *need* a core of genes to control the speed of respiration, might this explain why bacteria can't evolve into eukaryotes by natural selection alone? I believe so, although I should emphasize that this is my own speculation (which I have expanded on elsewhere: see Further Reading). Bacteria are about the same size as mitochondria, so clearly a single set of genes can control respiration over a certain area of energetic membranes. Presumably the same is true of bacteria that evolved extensive internal membrane systems, such as

Nitrosomonas and *Nitrosococcus*. They get by with a single gene set, so presumably that must be enough too. But let's expand our bacterium; let's double the area of internal membranes. Now, perhaps, we're beginning to lose control over some parts of the membrane. If you don't think so, double the area again. And again. We could double the internal membrane area of *Nitrosomonas* six or seven times before we're level with the eukaryotes. I doubt we could maintain control over the speed of respiration now. How might we regain control?

One way would be to copy a subset of genes and delegate it to regulate the extra membranes—but how could we choose the right genes? There is no way I can think of that does not involve some kind of foresight (an awareness of which genes to choose), and evolution has none. The only way such delegation might work would be to replicate the entire genome, and then whittle away at one of the two genomes until all the superfluous genes were gone (as actually happened in the mitochondria). But how would we know which genome should lose its genes? Both must be active for genetic control to work. In the meantime, however, we have a bacterium with two active genomes, each under a heavy selective pressure to throw away any excess genes. Either of the two genomes might be expected to lose some genes—but then the two dissimilar genomes would compete with each other, potentially leading to the destruction of the cell (more on this in Part 6), and certainly not stabilizing it in the selective battle against other cells.

Such competition between genomes might be stopped if it was possible to demark the sphere of influence of each genome. The eukaryotes solved the sphere-of-influence problem by sealing off the mitochondrial genomes within a double membrane. This is not possible in bacteria, however. If the spare set of genes were sealed off, there would be no way of getting food supplies in and ATP out. In particular, ATP exporters do not exist among bacteria—exporting energy in the form of ATP to their competitors in the outside world would be a suicidal behavioural trait for bacteria. The ATP exporters, along with the family of 150 mitochondrial transport proteins to which they belong, are a eukaryotic invention. We know this because the gene sequences of the ATP exporters are clearly related in plants, animals, and fungi, but there are no similar bacterial genes. This implies that the ATP exporters evolved in the last common ancestor of all the eukaryotes, before the divergence of the major groups, but after the formation of the chimeric ancestral eukaryotic cell.

The eukaryotes had time to evolve such niceties because the relationship between the two partners of the chimera was stable over evolutionary time. The two partners lived in harmony together, and didn't need anything else—there was ample time and stability for evolutionary change to take place. This stability was only possible because there were other advantages to the association between the collaborating partners. If the hydrogen hypothesis is correct, the

initial advantage was the mutual chemical dependency of two radically different cells, which lasted for long enough for the ATP exporters to evolve. In the
case of bacteria evolving simply by natural selection, however, there is no
corresponding stability. Simply duplicating a gene set and sealing it off within a
membrane could in itself provide no advantages in the interim. Far from it:
maintaining extra genes and membranes without any payback is energy sapping, and would no doubt be swiftly dumped by natural selection. Whichever
way we look at it, selection pressure is always likely to jettison the burdensome
additional genes needed for respiratory control over a wide area of membranes
in bacteria. The most stable state is always a small cell that respires across the
outer cell membrane. Such a cell will almost invariably be favoured by selection
in place of any larger, inefficient, free-radical-generating competitors.

So we can now finally appreciate the full set of barriers to large size and complexity in bacteria. Bacteria replicate as quickly as they can, and are limited at
least partly by the speed at which they can generate ATP. They generate ATP by
pumping protons across their external membrane. They can't grow larger
because their energetic efficiency tails away as their size increases. This fact in
itself makes the predatory eukaryotic lifestyle unlikely, because phagocytosis
requires a combination of large size with abundant energy that is precluded by
respiration across the outer membrane. Some bacteria developed complex
internal membrane systems. However, the area of these is several orders
of magnitude less than that of the mitochondrial membranes in a single
eukaryotic cell, because without gene outposts bacteria can't control the speed
of respiration over a wider area. Given the strong selection pressures for fast
reproduction and efficient energy generation, any of the possible transition
states *en route* to establishing such genetic outposts would likely have been
selected against whenever they arose. Only endosymbiosis was stable enough
to provide the long-term conditions necessary for respiratory control on a
wider scale.

Would things have happened differently somewhere else in an infinite
universe? Anything is possible, but it seems to me unlikely. Natural selection is
probabilistic: similar selection pressures are most likely to generate similar out-
comes anywhere in the universe. This explains why natural selection so often
converges on similar solutions, such as eyes and wings. Despite 4000 million
years of evolution, we know of no single example of bacteria that succeeded in
becoming eukaryotes by natural selection alone, or for that matter, of any mitochondria that lost all of their genes and still functioned as mitochondria. I
doubt whether such events would happen any more often anywhere else either.

What of a eukaryotic-style chimera? We saw in Part 1 that the eukaryotic cell
evolved here on earth just once, by way of what seems to have been a deeply
improbable chain of circumstances. Perhaps a similar concatenation would be

repeated elsewhere, but I see nothing in the laws of physics to suggest that the rise of complexity was inevitable. Physics is stymied by history. At best, the evolution of multicellular complexity seems to have been improbable; and without a kernel of complexity, intelligence is unthinkable. Yet once the loop that had kept bacteria simple was broken, the birth of that first large, complex cell, the first eukaryote, marked the beginning of a road that led, almost inexorably, to the spectacular feats of bioengineering that we see all around us today, including ourselves. This path was just as dependent on mitochondria as the origin of the eukaryotic cell itself, for the existence of mitochondria made the evolution of large size and greater complexity not just possible, but probable.

PART 4

Power Laws
Size and the Ramp of Ascending Complexity

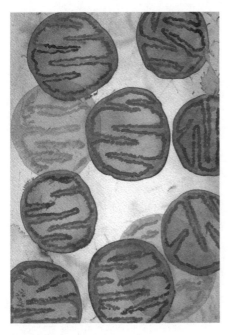

The more the merrier—mitochondrial numbers dictate the evolution of size and complexity

Does life inherently become more complex? There may be nothing in the genes to push life up a ramp of ascending complexity, but one force lies outside the genes. Size and complexity are usually linked, for larger size requires greater genetic and anatomical complexity. But there is an immediate advantage to being bigger: more mitochondria means more power and greater metabolic efficiency. It seems that two revolutions were powered by mitochondria—the accumulation of DNA and genes in eukaryotic cells, giving an impetus to complexity, and the evolution of warm-blooded animals, which inherited the earth.

 Size is a dominating bias in biology. By and large, we are mostly interested in the largest life-forms—the plants, animals, and fungi that we can actually see. Our interest in bacteria or viruses tends to be anthropocentric, a morbid curiosity, probing into the horrors of the diseases that they cause, and the more gruesome the better. Necrotizing bacteria that chew up whole limbs in a matter of days can hardly but attract more attention than the myriad microscopic plankton that exert such a profound influence on our planet's climate and atmosphere. Textbooks on microbiology tend to focus disproportionately on pathogens, despite the fact that only a tiny proportion of microbes actually cause disease. When we search for signs of life in space, we are really seeking extraterrestrial intelligence: we want proper aliens with twisting tentacles, not microscopic bacteria.

In the last few chapters, we have considered the origins of biological complexity: why it was that bacteria gave rise to our own remotest ancestors, the first eukaryotes—morphologically complex cells with nuclei and organelles such as mitochondria. I have argued that the fundamental mechanism of energy generation in cells made symbiosis necessary for the evolution of complexity: eukaryotic cells almost certainly could not have evolved by natural selection alone. Generating energy using mitochondria inside the cell made this leap possible. While symbiosis is commonplace in eukaryotic cells, however, endosymbiosis in bacteria (in which one bacterium lives inside another) is far less common. It seems that bacterial endosymbiosis gave rise to the complex eukaryotic cell on just one occasion, perhaps by way of the improbable train of events discussed in Part 1.

Yet once the first eukaryotes had evolved, we can legitimately talk about a ramp of ascending complexity: the progression from single cells to human beings certainly looks like a ramp, more than a little dizzying, even if we are deceived by appearances. Now a larger question looms: what drove the eukaryotes to acquire greater size and complexity? One answer that was popular in Darwin's day, and which enabled many biologists to reconcile evolution and religion, is that life innately becomes more complex. According to this line of reasoning, evolution leads to greater complexity in the same way that an embryo develops into an adult—it follows instructions, ordained by God, in which each step approaches closer to Heaven. Many of our turns of phrase, such as 'higher organisms' and the 'ascent of man', hark back to this philosophy, and are in common currency today despite the admonitions of evolutionists right back to Darwin himself. Such metaphors are powerful and poetic, but can be profoundly misleading. Another visually striking metaphor, that electrons orbit the nucleus of an atom in the same way that planets orbit the sun, long concealed the fantastic mysteries of quantum mechanics. The idea that evolution is akin to embryonic development conceals the fact that evolution has no foresight: it *cannot* operate as a program (whereas the development of an embryo is necessarily programmed by the genes). So complexity can't have evolved with the distant goal of approaching closer to God, but only as an immediate payback for an immediate advantage.

If the evolution of complexity was not programmed, are we to believe that it occurred merely by chance, or was it an inevitable outcome of the workings of natural selection? The fact that bacteria never showed the least tendency to become more complex (morphologically) argues against the possibility that natural selection inevitably favours complexity. Numerous other examples show that natural selection is as likely to favour simplicity as complexity. On the other hand, we have seen that bacteria are stymied by their respiration problem, but eukaryotes are not. Did complexity perhaps evolve in eukaryotes just because it could? Ridding himself of higher religious connotations, Stephen Jay Gould once compared complexity with the random meanderings of a drunkard: if a wall blocks his passage on one side of the pavement, then the drunkard is more likely to end up in the gutter, simply because there is nowhere else for him to go. In the case of complexity, the metaphorical wall is the base of life: it is not possible to be any simpler than a bacterium (at least as an independent organism), so life's random walk could only have been towards greater complexity. A related view is that life became more complex because evolutionary success was more likely to be found in the exploitation of new niches—an idea known as the 'pioneering' theory. Given that the simplest niches were already occupied by bacteria, the only direction in which life could evolve was towards greater complexity.

Both these arguments imply there was no intrinsic advantage to complexity—in other words, there was no trait inherent to the eukaryotes that encouraged the evolution of greater complexity—it was simply a response to the possibilities offered by the environment. I don't doubt for a moment that both of these theories account for certain trends in evolution, but I do find it hard to swallow that the entire edifice of complex life on Earth was erected by what amounts to evolutionary drift. The trouble with drift is its lack of direction, and I can't help but feel there is something inherently directed about eukaryotic evolution. The great chain of being may be an illusion, but it is a compelling one, one that held mankind in its sway for 2000 years (since the ancient Greeks). Just as we must account for the apparent evolution of 'purpose' in biology (the heart as a pump, etc), so too we must account for the apparent trajectory towards greater complexity. Can a random walk, stopping off at vacant niches on the way, really produce something that even *looks* like a ramp of complexity? To twist Stephen Jay Gould's analogy, how come so many meandering drunkards didn't end up in the gutter, but actually succeeded in crossing the road?

One possible solution, inherent to eukaryotic cells but not to bacteria, is sex. That there is a link between sex and complexity has been argued persuasively by Mark Ridley in *Mendel's Demon*. The trouble with asexual reproduction, says Ridley, is that it is not good at eliminating copying errors and harmful mutations in genes. The larger the genome, the greater the probability of a catastrophic error. The recombination of genes in sexual reproduction may lower this risk of error, and so raise the number of genes an organism can tolerate before undergoing a mutational meltdown (although this has never been proved). Clearly, however, the more genes an organism accumulates, the greater its possible complexity, so the invention of sex in eukaryotes might have opened the gates to complexity. While there is almost certainly some truth in this argument, there are also problems with the idea that sex stands at the gateway to complexity, as Ridley himself concedes. In particular, the number of genes in bacteria is well below the theoretical asexual limit, even if they relied on asexual reproduction alone, which they do not (lateral gene transfer in bacteria helps restore genetic integrity). Ridley acknowledges that the data are ambivalent, and the asexual limit to gene number may fall somewhere between fruit flies and human beings. If so the gates of complexity could hardly have been thrown open by the evolution of sex. Something else must have been the gate-keeper.

I do think there was an inherent tendency for eukaryotes to grow larger and more complex, but the reason relates to energy rather than sex. The efficiency of energy metabolism may have been the driving force behind the rampant ascent of eukaryotes to diversity and complexity. The same principles underpin energetic efficiency in all eukaryotic cells, giving an impetus to the evolution of larger size in both unicellular and multicellular organisms, whether plants, animals, or fungi. Rather than being a random walk through vacant niches, or a march driven by the imperative of sex, the trajectory of eukaryotic evolution is better explained as an inherent tendency to become larger, with an immediate payback for an immediate advantage—the economy of scale. As animals become larger, their metabolic rate falls, giving them a lower cost of living.

I am here conflating size with complexity. Even if it is true that greater size is favoured by a lower cost of living, is there really a connection between size and complexity? Complexity is not an easy term to define, and in attempting to do so we are inevitably biased towards ourselves: we tend to think of complex beings in terms of their intellect, behaviour, emotions, language, and so on, rather than, for example, a complex life cycle, as in an insect with its drastic morphological transitions, from caterpillar to butterfly. In particular, I am not alone in my bias towards larger size: for most of us, I suspect, a tree appears more complex than a blade of grass, even though, in terms of photosynthetic machinery, grasses might be said to be more highly evolved. We insist that multicellular creatures are more complex than bacteria, even though the biochemistry of bacteria (as a group) is far more sophisticated than anything we eukaryotes can muster. We are even inclined to see patterns in the fossil record implying an evolutionary trend towards greater size (and presumably complexity), known as Cope's Rule. While accepted with little question for a century, several systematic studies in the 1990s suggested that the trend is nought but an illusion: different species are equally likely to become smaller as they are larger. We are so mesmerized by our fellow large creatures that we easily overlook the smaller ones.

So do we conflate size with complexity, or is it fair to say that larger organisms are in general more complex? Any increment in size brings along a new set of problems, many of which are related to the troublesome ratio of surface area to volume that we discussed in the previous chapter. Some of these issues were highlighted by the great mathematical geneticist J. B. S. Haldane, in a delightful 1927 essay entitled *On Being the Right Size*. Haldane considered the example of a microscopic worm, which has a smooth skin across which oxygen can diffuse, a straight gut for absorbing food, and a simple kidney for excretion. If its size were increased tenfold in each dimension, its mass would rise by 10^3, or 1000-fold. If all the worm's cells retained the same metabolic rate it would need to take up a thousand times more oxygen and food, and excrete a thousand times more waste. The trouble is that if its shape didn't change, then its surface area (which is a two-dimensional sheet) would increase by a factor of 10^2, or 100-fold. To match the heightened requirements, each square millimetre of gut or skin would need to take up 10 times more food or oxygen every minute, while its kidneys would need to excrete 10 times as much waste.

At some point a limit must be reached, beyond which larger size can be attained only by way of specific adaptations. For example, specialized gills or lungs increase the surface area for taking up oxygen (a man has a hundred square metres of lung), while the absorptive area of the gut is increased by folding. All these refinements require greater morphological, and supporting genetic, complexity. Accordingly, larger organisms tend to have a larger number of specialized cell types (anything up to 200 in humans, depending on the definition we use), and more genes. As Haldane put it: 'The higher animals are not larger than the lower because they are more complicated. They are more complicated because they are larger. Comparative anatomy is largely the story of the struggle to increase surface area in proportion to volume.'

As if the purely geometric obstacles to large size were not intractable enough, there are other disadvantages to being big. Large animals struggle to fly, burrow, penetrate thick vegetation, or walk on boggy ground. The consequences of a fall for large animals can be catastrophic, as the air resistance during a fall is proportional to the surface area (which is smaller, relative to body mass, for large animals). If we drop a mouse down a mineshaft, it will be briefly stunned, before scampering away. If we drop a man, he will break; if we drop a horse, according to Haldane, it will 'splash' (though I'm not sure how he knew). Life looks bleak for giants; why bother getting bigger? Again, Haldane offers a few reasonable answers: larger size gives greater strength, which aids in the struggle for a mate, or in the battle between predator and prey; larger size can optimize the function of organs, such as the eyes, which are built from sensory cells of fixed size (so more cells means larger eyes and better vision); larger size reduces the problems of water tension, which can be lethal for insects (often forcing them to drink using a proboscis); and larger size retains heat (and for that matter water) better, which explains why small mammals and birds are rarely found anywhere near the poles.

These answers make good sense, but they betray a mammal-centric view of life: none begins to explain why something as large as a mammal should have evolved in the first place. The question I'm interested in answering is not whether large mammals are better adapted than small mammals, but why it was that small cells gave rise to large cells, then larger organisms, and finally to highly dynamic, energetic creatures like ourselves; in essence, why anything exists that we can see at all. If being larger demands greater complexity, which has an immediate cost—a need for new genes, better organization, more energy—was there any *immediate* payback, some advantage to being bigger for its own sake, which could counter-balance the costly new organization? In Part 4, we'll consider the possibility that the 'power laws' of biological scaling may have underpinned the apparent trajectory towards greater complexity that seems to have characterized the rise of the eukaryotes, while forever defying the bacteria.

9

The Power Laws of Biology

They say that in London everyone lives within 6 feet of a rat. Denizens of the night, these rats are presumably dozing the day away somewhere beneath the floorboards, or in the drains. Or perhaps you're reading this in bed, in which case they may be having a riot in the kitchen (in the house next door). Perhaps a few are decomposing in the drains too, as rats don't live much longer than three years. Once feared as carriers of the black plague, rats still symbolize squalor and filth, but we are also indebted to them: in the laboratory, their clean-living cousins have helped rewrite the medical texts, serving as models of human diseases and (in that archaic turn of phrase) as guinea pigs for many new treatments. Rats are useful laboratory animals because they are like us in many ways—they, too, are mammals, with the same organs, the same layout and basic functionality, the same senses, even sensibilities—they share a lively curiosity about their surroundings. Rats, too, suffer from the equivalent diseases of old age—cancer, atherosclerosis, diabetes, cataracts, and so on, but offer the tremendous advantage that we don't need to wait for seventy years to see whether a therapy is working—they suffer from such senile diseases within a couple of years. Like us, they are prone to overeat when bored, easily becoming obese. Anyone who owns a pet rat (commonly the researchers who work with them) knows they must guard against overfeeding and boredom. Hiding the raisins is a good idea.

We're so close to rats (in every sense) that it might come as a shock to appreciate how much faster their organs must work than ours: their heart, lungs, liver, kidneys, intestines (but not the skeletal muscle) must work on average *seven* times harder than ours. Let me specify what I mean by this. Let a modern-day Shylock take one gram of flesh from a rat, and another from a human being—perhaps a bit of liver. Both bits of liver contain roughly the same number of cells, which are about the same size in rats and humans. If we can keep the tissue alive for a while, and measure its activity, we'll see that the gram of rat liver consumes seven times as much oxygen and nutrients per minute as its human counterpart—even though we could hardly tell which piece was which down the microscope. I should stress that this is purely an empirical finding; *why* it happens is the subject of this chapter.

Even though the reasons behind this striking difference in metabolic rate are obscure, the consequences are certainly important. Because the cells in a rat and a human being are of a similar size, each individual rat cell must work seven times harder (nearly as fast as Haldane's geometrically challenged worm). The repercussions permeate all aspects of biology: each cell must copy its genes seven times faster, make seven times as many new proteins, pump seven times as much salt out of the cell, dispose of seven times as many dietary toxins, and so on. To sustain this rapid metabolism, the rat as a whole must eat seven times as much food relative to its size. Forget the appetite of a horse. If we had the appetite of a rat, instead of feeling full after a 12 ounce steak, we'd want to eat a five pounder! These are fundamental mathematical relationships, which have nothing to do with genes (or at least nothing directly), and go part way to explaining why rats live for three years, while we live out our three score years and ten.

Rats and humans sit on an extraordinary curve, which connects shrews, one of the smallest mammals, with elephants and even blue whales, the largest (see Figure 12). Large animals clearly consume more food and oxygen than small animals. However, given a doubling in mass, oxygen consumption does not rise by as much as one might predict. If the mass is doubled, so is the total number of cells. If each cell needs the same amount of energy to stay alive, then doubling the mass ought to double the quantity of food and oxygen required. This assumes an exact equivalence: for every rise in mass, there is an equivalent rise in metabolic rate. Yet this is not actually what happens. As animals become larger, their cells need *fewer* nutrients to stay alive. Effectively, large animals have a rather slower metabolic rate than they 'should' have. For every step in mass, there is a smaller step up in metabolic rate. We have seen there is a seven-fold difference between a rat and a man. The larger the animal gets, the less it needs to eat per gram weight. In the case of the elephant and the mouse, for example, if we work out the quantity of food needed to sustain each cell (or per gram weight), the elephant requires 20 times less food and oxygen every minute. Put the other way around, an elephant-sized pile of mice would consume 20 times more food and oxygen every minute than the elephant does itself. Clearly it's cost-effective to be an elephant; but can the cost savings of greater size explain the tendency of organisms to grow larger and more complex over evolution?

Metabolic rate is defined as the consumption of oxygen and nutrients. If the metabolic rate falls, then each cell consumes less food and oxygen. And if all the cells in the body consume less oxygen, then the breathing rate, heartbeat, and so forth, can all afford to slow down. This is why the heartbeat of an elephant is ponderous in comparison to the fluttering beat of a mouse—the individual cells of an elephant need less fuel and oxygen, so the elephant's heart doesn't need to beat as vigorously to provide them (this assumes that the heart is the same size, relative to the overall size of the animal). Another unexpected consequence is that the rate of ageing slows down. Mice live for 2 or 3 years, and elephants for about 60, yet both have a similar number of heartbeats in their lifetime, and over their lives their component cells consume around about the same quantity of oxygen and food (the elephant in 60 years, the mouse in 3). The cells seem able to burn a fixed amount of energy, but the elephant burns its quota more slowly than the mouse (its cells have a slower metabolic rate), and it does so, apparently, just because it is bigger. Such relationships have a profound effect on ecology and evolution. The size of animals influences their population density, the range of distances they travel in a day, the number of offspring, the time to reproductive maturity, the speed of population turnover, and the rate of evolution, such as the origin of new species. All of these traits can be predicted, with startling accuracy, from nothing more than the metabolic rate of individual animals.

Why metabolic rate should vary with size has perplexed biologists, and indeed physicists and mathematicians, for well over a century. The first person to study the relationship systematically was the German physiologist Max Rubner. In 1883, Rubner plotted the metabolic rates of 7 dogs, ranging in weight from 3.2 to 31.2 kilograms. The raw data trace a curve, but if instead the data are plotted out as a log–log plot, they fit a straight line. There are various reasons for using a log plot, but the most important is that this allows the multiplication factor to be seen clearly: instead of adding steps at a fixed distance along an axis (10 + 10 + 10, and so on), a log graph multiplies them (10 × 10 × 10, and so on). This shows how many multiplications of a parameter correspond to multiplications of another. Consider a simple cube. If we plot log surface area on one axis, and log volume on another, we can plot out how they change relative to one another as the size of the cube is increased. For every tenfold increase in the width of the cube, we see a 100-fold increase in the cube's surface area, and a 1000-fold increase in volume. On a log–log plot, an increase in surface area of 100-fold corresponds to two steps, and the increase in volume to three steps. This gives the slope of the line. In the case of the cube, the slope is 2/3, or 0.67— for every two steps in surface area there are three steps in volume. The slope of the line connecting the points is the *exponent*, which is usually written as a superscript after the number it applies to, so in this case the exponent would be written as $^{2/3}$. By definition, an exponent denotes how many times a number should be multiplied by itself (so $2^2 = 2 \times 2$, while $2^4 = 2 \times 2 \times 2 \times 2$) but when dealing with *fractional* exponents, like 2/3, it's much easier to think in terms of the slope of a line on a log-log graph. If the exponent is 1, this means that for every step along one axis, there is an equal step along the other axis: the two parameters are directly proportional. If the exponent is ¼, this means that for every step along one axis, there are four steps along the other: there is a consistent but disproportionate relationship.

Let's return to Max Rubner. When plotting log metabolic rate against log mass, Rubner discovered that the metabolic rate was proportional to the mass with an exponent of 2/3. In other words, for every two steps in log metabolic rate there were three steps in log mass. This is of course exactly the same as the relationship between the surface area and the volume of a cube, which we just discussed. For his dogs, Rubner explained the relationship in terms of heat loss. The amount of heat generated by metabolism depends on the number of cells, whereas the rate at which the heat is lost to the surroundings depends on the surface area (just as the amount of heat emitted by a radiator depends on its surface area). As animals get larger, their mass rises faster than their surface area. If all the cells continued to generate heat at the same rate, the overall rate of heat production would rise with body mass, but heat *loss* would depend on the surface area. Larger animals would retain more heat. If all the cells of an elephant generated heat at the same rate as those of a mouse, it would melt— literally. More constructively, if the point of a fast metabolic rate is to keep warm, and large animals retain heat better, then there is less need for an elephant to have a fast metabolic rate: it is as fast as it needs to be to maintain a stable body temperature of about 37°C. Thus, as animals increase in size their metabolic rate slows down by a factor that corresponds to the ratio of surface area to mass.

In considering dogs, of course, Rubner was considering only one species, even though the different breeds vary dramatically in size and appearance. Half a century later, the Swiss-American physiologist Max Kleiber plotted log metabolic rate against log mass of different species, and constructed his famous curve from mice to elephants. To his and everybody else's surprise, the exponent was not 2/3 as expected, but 3/4 (0.75; or in actual fact 0.73, rounded up; Figure 12). In other words, for every three steps in the log metabolic rate, there were four steps in log mass. Other researchers, notably the American Samuel Brody, came to a similar conclusion. Even more unexpectedly, the 0.75 exponent turned out to apply to more than just the mammals: birds, reptiles, fish, insects, trees, even single-celled organisms, have all been placed on the same curve: metabolic rate is claimed to vary with the 3/4 power of mass (or mass$^{3/4}$) across an extraordinary 21 orders of magnitude. Many other traits also vary with an exponent based on multiples of one quarter (such as 1/4 or 3/4), giving rise to the general term 'quarter-power scaling'. For example, the pulse rate, the diameter of the aorta (even the diameter of tree trunks), and lifespan, all conform roughly to 'quarter-power scaling'. A minority of researchers, most persuasively Alfred Heusner at UC Davis, have contested the universal validity of quarter-power scaling, but it has nonetheless entered virtually all standard

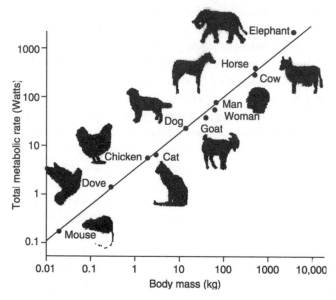

12 Graph showing the scaling of resting metabolic rate with body mass across mammals of widely differing mass, from mouse to elephant. The slope of the line on a log–log plot is ¾, or 0.75, which is to say that the line rises 3 steps up the vertical axis in the space of 4 steps on the horizontal axis. This slope gives the exponent. The metabolic rate is said to vary with the three-quarter power of mass, or $mass^{3/4}$.

biological texts as 'Kleiber's law'. It is often said to be one of the few universal laws in biology.[1]

Why on earth metabolic rate should vary with the 3/4 power of mass remained a mystery for another half century; and indeed the glimmerings of an answer are only now beginning to materialize, as we shall see. But one point was patent: while a 2/3 exponent, connecting the metabolic rate with the surface area to volume ratio, made sense for warm-blooded mammals and birds, there was no obvious reason why it should apply to cold-blooded animals, such as reptiles and insects: they don't generate heat internally (or at least not much), so the balance of heat generation and heat loss could hardly be the dominating factor. From this perspective a 3/4 exponent made as much, and as little, sense as a 2/3 exponent. But while various attempts have been made to rationalize the 3/4 exponent, none ever really convinced the whole field.

Then, in 1997, a high-energy particle physicist at Los Alamos National

---

[1] How do we reconcile Max Rubner's exponent of 2/3 with Max Kleiber's 3/4? The usual answer is that within species the metabolic rate does indeed vary with 2/3, and the 3/4 exponent only becomes apparent when we compare different species.

Laboratory, Geoffrey West, joined forces with the ecologists James Brown and Brian Enquist, at the University of New Mexico, Albuquerque (through the Santa Fe Institute, an organization that fosters cross-disciplinary collaborations). They came up with a radical explanation based on the fractal geometry of branching supply networks, such as the circulatory system of mammals, the respiratory tubes of insects (the trachea), and the plant vascular system. Their densely mathematical model was published in *Science* in 1997, and the ramifications (if not the maths) swiftly captured the imagination of many.

The fractal tree of life

Fractals (from the Latin *fractus*, broken) are geometric shapes that look similar at any scale. If a fractal is broken into its constituent parts, each part still looks more or less the same, because, as the pioneer of fractal geometry Benoit Mandelbrot put it, 'the shapes are made of parts similar to the whole in some way'. Fractals can be formed randomly by natural forces such as wind, rain, ice, erosion, and gravity, to generate natural fractals, like mountains, clouds, rivers, and coastlines. Indeed, Mandelbrot described fractals as 'the geometry of nature', and in his landmark paper, published in *Science*, in 1967, he applied this approach to the question advanced in its title: *How Long is the Coast of Britain?* Fractals can also be generated mathematically, often by using a reiterative geometric formula to specify the angle and density of branches (the 'fractal dimension').

Both types of fractal share a property known as scale invariance, which is to say they 'look' similar whatever the magnification. For example, the contours of a rock often resemble those of a cliff or even a mountain, and for this reason geologists like to leave a hammer lying around in photographs, to enable viewers to grasp the scale. Similarly, the pattern of river tributaries looks alike for a vast continental system, such as the Amazon basin when viewed from space, or small streams seen from the top of a hill, or even soil erosion in the back garden from the bathroom window. For mathematical 'iterative' fractals, a repeating geometric rule is used to generate an infinite number of similar shapes. Even the most complex and beautiful fractal images, seen adorning T-shirts and posters, are built from reiterations of geometric rules (often quite complicated ones), followed by plotting the points spatially. For many of us, this is as close as we'll ever get to the beauty of deep mathematics.

Most of nature's fractals are not really true fractals, in that their scale invariance does not extend to infinity. Even so, the pattern of twigs on a branch resembles the branching of the tree as a whole; and the branching pattern of blood vessels in a tissue or an organ resembles that of the body as a whole—it can be difficult to grasp the scale. Stepping up again in scale, the cardiovascular system of an elephant resembles that of a mouse, but the system as a whole is magnified by nearly six orders of magnitude (in other words, the cardiovascular system of an elephant is nearly a million times bigger than that of a mouse; a 1 followed by six zeros). When networks retain a similar appearance over such a scale, the natural language to describe them is fractal geometry; if nature's branching networks are not true fractals, they are still close enough to be modelled accurately using these mathematical principles.

West, Brown, and Enquist asked themselves whether the fractal geometry of nature's supply networks might account for the apparently universal scaling of metabolic rate with body size. This makes perfectly good sense, for the metabolic rate corresponds to the consumption of food and oxygen, and these don't arrive at the individual cells of an animal by diffusing across the body surface, but by way of the branching supply network—blood vessels in our own case. If the metabolic rate is constrained by the delivery of these nutrients, it's reasonable to assume that it should depend ultimately on the properties of the supply network. In their 1997 *Science* paper, West, Brown, and Enquist made three basic assumptions. First, they assumed that the network serves the entire organism—it must supply all cells, and so fill the entire volume of the organism. Second, they assumed that the smallest branch of the network, the capillary, is a size-invariant unit, which is to say that all capillaries are the same size in all animals, regardless of the size of the animal. And third, they supposed that the energy necessary to distribute resources through the network is minimized: over evolutionary time, natural selection has optimized the supply network to deliver nutrients with the minimum time and effort.

A number of other factors, relating to the elastic properties of the tubes themselves, also needed to be considered, but we needn't worry about these here. The upshot is this. To maintain a self-similar fractal network (one that 'looks' the same on any scale), while scaling up the body size over orders of magnitude, the total number of branches rises more slowly than the volume. This is observed to be true. For example, a whale is 10^7 (10 million) times heavier than a mouse, but has only 70 per cent more branches going from the aorta to the capillaries. According to the idealized calculations of fractal geometry, the supply network ought to take up relatively less space in a large animal, and so each capillary ought to serve a larger number of 'end-user' cells. Of course this means that cells must partition less food and oxygen between them; and if they receive less food to burn, then presumably they would be forced to have a slower metabolic rate. How much slower, exactly? The fractal model predicts that the metabolic rate should correspond to body mass to the power of ¾. Picture this as the slope of the line on a log-log plot: for every three steps in log metabolic rate, there are four steps in log mass. In other words, the fractal model calculates, from first principles, that the metabolic rate should scale with mass$^{3/4}$, thus explaining the universality of Kleiber's law, the quarter-power scaling rule. If this is true, then the entire living world is subject to the rules of fractal geometry. They determine body size, population density, lifespan, speed of evolution—everything.

As if this weren't enough, the fractal model goes still further and makes a radical general prediction. Because Kleiber's law apparently applies not only to large organisms that manifestly have a branching supply network, like mammals, insects, and trees, but also to simpler creatures that seem to lack a supply network, like single cells, then they too must have some sort of fractal supply network. This is radical as it implies that there exists a whole echelon of biological organization that we have not yet detected, and even the proponents feel obliged to talk about a 'virtual' network, whatever that might be. Even so, many biologists are receptive to the possibility, for the cytoplasm is now seen to be far more organized than the amorphous jelly that is passed off in text books. The nature of this organization is elusive, but it is clear that the cytoplasm 'streams' through the cell, and that many biochemical reactions are more carefully defined in space than had been assumed. Most cells have a complex internal architecture, including branching networks of cytoskeleton filaments and mitochondria; but is this really a fractal network, that obeys the same laws of fractal geometry? While it unquestionably branches, there is a very modest resemblance to the *tree-like* network of circulatory systems (Figure 13). If fractal geometry applies to self-similar systems, these do not look similar.

To address such intangibles, West, Brown, and Enquist recast their model to eliminate the need for explicit structures, like a branching anatomy, grounding it instead in the geometry of hierarchical networks (networks embedded within networks, like a set of Russian dolls). Other physicists, notably Jayanth Banavar at the University of Pennsylvania and his colleagues, have attempted to simplify the network model to eliminate the need for fractal geometry altogether; but they, too, specify a branching supply network. Various abstruse mathematical arguments have filled the pages of prestigious science journals every few months since the late 1990s, often delivering withering mathematical ripostes, such as 'this must be incorrect because it violates dimensional homogeneity. . .' The debate tends to polarize biologists, who are only too aware of exceptions to supposedly universal rules ('yeah, but what about the crayfish?'), and physicists, like West, who seek a single unifying explanation. West doesn't mince his words: 'If Galileo had been a biologist he would have written volumes cataloguing how objects of different shapes fall from the leaning tower of Pisa at slightly different velocities. He would not have seen through the distracting details to the underlying truth: if you ignore air resistance, all objects fall at the same rate regardless of their weight.'

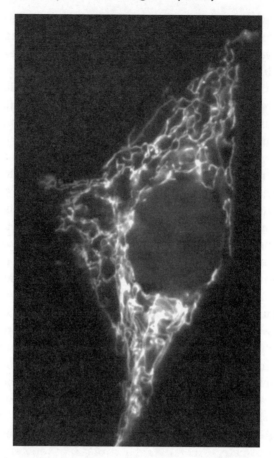

13 The mitochondrial network in a mammalian cell in a tissue culture, stained with the dye MitoTracker. Mitochondria often move around the cell, and can form into reticulated networks, as here, but these networks don't really resemble a fractal tree.

Supply and demand—or demand and supply?

West and Brown teamed up with the Los Alamos biochemist William Woodruff to report perhaps the most thought-provoking finding of all in 2002. They published data in the *Proceedings of the National Academy of Sciences* in which they extended their fractal model to the mitochondria. They showed that mitochondria, and even the thousands of miniscule respiratory complexes within individual mitochondria, could be plotted onto the same universal quarter-power scale. In other words, they said, the relationship between metabolic rate and body size extends from the level of individual respiratory complexes right up to the blue whale, spanning 'an astounding 27 orders of magnitude'. When I set out the proposal for this book, I had it in mind to discuss their paper. I had read it carefully enough to find the central argument compelling, but had not really come to grips with its implications. I've been struggling with them ever since— is it true that a straight line connects the metabolic rate of individual complexes within the mitochondria with the metabolic rate of a blue whale? And if it is true, what does it mean?

Because the metabolic rate is defined as the rate of oxygen consumption, which takes place mainly in the mitochondria, then ultimately the metabolic rate reflects the energetic turnover of the mitochondria themselves. The base-line rate of mitochondrial energy production is proportional to the size of the organism. According to West and his colleagues, the slope of this line is deter-mined by the properties of the supply network connecting to the cells, then to the mitochondria, and finally, deep inside, to the respiratory complexes themselves. This means that the telescopic series of networks *constrains* the metabolic rate, and *forces* a particular metabolic rate on individual mitochon-dria. West and his colleagues do refer to the network as a constraint, what they call 'network hierarchy hegemony'.[2] But if the supply networks do constrain the metabolic rate, then as animals become larger the metabolic rate of individ-ual mitochondria is *forced* to slow down, regardless of whether this is good or not. The maximum power they can possibly attain *must fall*. Why? Because as animals get bigger, the scaling of the network constrains each capillary to feed a larger number of cells (or the model doesn't work at all). Metabolic rate is *obliged* to fall in harness with capillary density. As West and colleagues admit, this is a *constraint* of larger size, not an opportunity, and has nothing to do with efficiency.

If this is true, then one of West's colloquial arguments must be wrong. He argues: 'As organisms grow in size, they become more efficient. That is why

---

[2] In fact they make a specific prediction based on this. The presence of a network obliges individual mitochondria to operate more slowly than they would if they were relieved from the constraints of the network. When grown in culture, cells have a lavish supply of nutrients delivered to them directly from the surrounding medium: there is no network, so cells can't be constrained by it. If unconstrained, the metabolic rate should rise. On this basis, West, Woodruff, and Brown calculate that cultured mammalian cells should become more meta-bolically active in culture, and they predict that cells should contain approximately 5000 mito-chondria after several generations in culture, each with about 3000 respiratory complexes. These numbers seem wrong. Mammalian cells tend to adapt to culture by *losing* mitochon-dria, becoming instead dependent on fermentation to provide energy, giving off the waste product lactate. Accumulation of lactate is known to impede the growth of mammalian cell cultures. As to the number of respiratory complexes in a single mitochondrion, most estimates are in the order of 30 000, not 3000. Far from 'agreeing with observation', West, Woodruff, and Brown's estimate appears to be an order of magnitude out.

nature has evolved large animals. It's a much better way of utilising energy.' If West's fractal argument is correct, then the truth must actually be the reverse. As animals become larger, their constituent cells are *forced* by the supply network to use less energy. Large animals must find a way of surviving with less power, at least in relation to their mass. This is not efficiency so much as rationing. If the network really does constrain metabolic rate, this adds up to another reason why the evolution of large size, and with it complexity, is so improbable.

So are organisms constrained by their network? The network is certainly important, and may well be fractal in its behaviour, but there are good reasons to question whether the network *constrains* the metabolic rate. In fact, the contrary may be true: there are certainly some instances in which the *demand* controls the network. The balance between supply and demand might seem more relevant to economists, but in this instance it makes the difference between an evolutionary trajectory towards greater complexity, and a world perpetually stuck in a bacterial rut, in which true complexity is unlikely to evolve. If cells and organisms become more efficient as they become bigger, then there really are rewards for larger size, an incentive to get bigger. And if size and complexity really do go hand in hand, then any rewards for larger size are equally rewards for greater complexity. There are good reasons for organisms to become larger and more complex in evolution. But if larger size is only rewarded by enforced frugality, the tight-fisted welcome of a miser, then why does life tend to get larger and more complex? Large size is already penalized by the requirement for more genes and better organization, but if the fractal model is right, size is also penalized by an everlasting vow of poverty—what's in it for giants?

Questioning the universal constant

There are various reasons to question whether the fractal model is really true, but one of the most important is the validity of the exponent itself—the slope of the line connecting the metabolic rate to the mass. The great merit of the fractal model is that it derives the relationship between metabolic rate and mass from first principles. By considering only the fractal geometry of branching supply networks in three-dimensional bodies, the model predicts that the metabolic rate of animals, plants, fungi, algae, and single celled organisms should all be proportional to their mass to the power of ¾, or mass$^{0.75}$. On the other hand, if the steady accumulation of empirical data shows that the exponent is *not* 0.75, then the fractal model has a problem. It comes up with an answer that is found empirically not to be true. The empirical failings of a theory may inculcate a fantastic new theory—the failings of the Newtonian universe ushered in relativity—but they also lead, of course, to the demise of the original model. In our case here, fractal geometry can only explain the power laws of biology if the power laws really exist—if the exponent really is a constant, the value of 0.75 genuinely universal.

I mentioned that Alfred Heusner and others have for decades contested the validity of the 3/4 exponent, arguing that Max Rubner's original 2/3 scaling was in fact more accurate. The matter came to a head in 2001 when the physicists Peter Dodds, Dan Rothman, and Joshua Weitz, then all at MIT in Cambridge, Massachusetts, re-examined 'the 3/4 law' of metabolism. They went back to the original data sets of Kleiber and Brody, as well as other seminal publications, to examine how robust the data really were.

As so often happens in science, the apparently solid foundations of a field turned to rubble on closer inspection. Although Kleiber's and Brody's data did indeed support an exponent of 3/4 (or in fact, of 0.73 and 0.72, respectively) their data sets were quite small, Kleiber's containing only 13 mammals. Later data sets, comprising several hundred species, generally failed to support the 3/4 exponent when re-analysed. Birds, for example, scale with an exponent close to 2/3, as do small mammals. Curiously, larger mammals seem to deviate upwards towards a higher exponent. This is in fact the basis of the 3/4 exponent. If a single straight line is drawn through the entire data set, spanning five or six orders of magnitude, then the slope is indeed approximately 3/4. But drawing a single line already makes an assumption that there *is* a universal scaling law. What if there is not? Then two separate lines, each with a different slope, better approximate the data, so large mammals are simply different from small mammals, for whatever reason.[3]

This may seem a little messy, but are there any strong empirical reasons to favour a nice crisp universal constant? Hardly. When plotted out reptiles have a steeper slope of about 0.88. Marsupials have a lower slope of 0.60. The frequently cited 1960 data set of A. M. Hemmingsen, which included single-celled organisms (making the 3/4 rule look truly universal) turned out to be a mirage, reforming itself around whichever group of organisms were selected, with slopes varying between 0.60 and 0.75. Dodds, Rothman, and Weitz concurred an earlier re-evaluation, that 'a 3/4 power scaling rule. . . for unicellular organisms generally is not at all persuasive.' They also found that aquatic invertebrates and algae scale with slopes of between 0.30 and 1.0. In short, a single universal constant cannot be supported within any individual phyla, and can only be perceived if we draw a single line through all phyla, incorporating many orders of magnitude. In this case, even though individual phyla don't support the universal constant, the slope of the line is about 0.75.

---

[3] Another re-analysis, published in 2003 by Craig White and Roger Seymour, at the University of Adelaide, came to a similar conclusion.

West and his collaborators argue that it is precisely this higher level of magnification that reveals the universal importance of fractal supply networks—the non-conformity of individual phyla is just irrelevant 'noise', like Galileo's air resistance. They may be right, but one must at least entertain the possibility that the 'universal' scaling law is a statistical artefact produced by drawing a single straight line through different groups, none of which conforms individually to the overall 'rule'. We might still favour a universal law if there was a good theoretical basis for believing it to exist—but it seems the fractal model is also questionable on theoretical grounds.

The limits of network limitation

There are some circumstances when it is clear that supply networks do constrain function. For example, the network of microtubules within individual cells are highly efficient at distributing molecules on a small scale, but probably set an upper limit to the size of the cell, beyond which a dedicated cardiovascular system is required to meet demand. Similarly, the system of blind-ending hollow tubes, known as trachea, which deliver oxygen to the individual cells of insects, impose quite a low limit to the maximum size that insects can attain, for which we can be eternally grateful. Interestingly, the high concentration of oxygen in the air during the Carboniferous period may have raised the bar, facilitating the evolution of dragonflies as big as seagulls, which I discussed in *Oxygen*. The supply system can also influence the lower limits to size. For example, the cardiovascular system of shrews almost certainly nears the lower size limit of mammals: if the aorta gets much smaller, the energy of the pulse is dissipated, and the drag caused by blood viscosity overcomes smooth flow.

Within such limits, does the supply network limit the rate of delivery of oxygen and nutrients, as specified by the fractal model? Not really. The trouble is that the fractal model links the *resting* metabolic rate with body size. The resting metabolic rate is defined as oxygen consumption at rest, while sitting quietly, well fed but not actively digesting a meal (the 'post-absorptive state'). It is therefore quite an artificial term—we don't spend much time resting in this state, still less do animals living in the wild. At rest, our metabolism cannot be limited by the delivery of oxygen and nutrients. If it was, we would not be able to break into a run, or indeed sustain any activity beyond sitting quietly. We wouldn't even have the reserves of stamina required to digest our food. In contrast, though, the *maximum* metabolic rate—defined as the limit of aerobic performance—is unquestionably limited by the rate of oxygen delivery. We are swiftly left gasping for breath, and accumulate lactic acid because our muscles must turn to fermentation to keep up with demand.

If the maximum metabolic rate also scaled with an exponent of 0.75, then the fractal model would hold, as the fractal geometry would predict the maximal aerobic scope, which is to say the range of aerobic capacity between resting and maximum exertion. This might happen if the maximal and resting metabolic rates were connected in some way, such that (evolutionarily) one could not rise unless the other did. This is not implausible. There certainly *is* a connection between the resting and maximal metabolic rates: in general, the higher the maximal metabolic rate, the higher the resting metabolic rate. For many years, the 'aerobic scope' (the increase in oxygen consumption from resting to maximal metabolic rate) was said to be fixed at 5 to 10-fold; in other words, all animals consume around 10 times more oxygen when at full stretch than at rest. If true, then both the resting and the maximal metabolic rates would scale with size to the power of 0.75. The entire respiratory apparatus would function as an indivisible unit, the scaling of which could be predicted by fractal geometry.

So does the maximum metabolic rate scale with an exponent of 0.75? It's hard to say for sure, as the scatter of data is confoundingly high. Some animals are more athletic than others, even within a species. Athletes have a greater aerobic scope than couch potatoes. While most of us can raise oxygen consumption 10-fold during exercise, some Olympic athletes have a 20-fold scope. Athletic dogs like greyhounds have a 30-fold scope, horses 50-fold; and the pronghorn antelope holds the mammalian record with a 65-fold scope. Athletic animals raise their aerobic scope by making specific adaptations to the respiratory and cardiovascular systems: relative to their body size, they have a greater lung volume, larger heart, more haemoglobin in red cells, a higher capillary density, and suchlike. These adaptations do not rule out the possibility that aerobic scope might be linked to size as well, but they do make it hard to disentangle size from the muddle of other factors.

Despite the scatter, there has long been a suspicion that the maximum metabolic rate *does* scale with size, albeit with an exponent that seemed greater than 0.75. Then in 1999, Charles Bishop, at the University of Wales in Bangor, developed a method of correcting for the athletic prowess of a species, to reveal the underlying influence of body size. Bishop noted that the average mammalian heart takes up about 1 per cent of body volume, while the average haemoglobin concentration is about 15 grams per 100 ml of blood. As we have seen, athletic mammals have larger hearts and a higher haemoglobin concentration. If these two factors are corrected against (to 'normalize' data to a standard), 95 per cent of the scatter is eliminated. Log maximum metabolic rate can then be plotted against log size to give a straight line. The slope of this line is 0.88—roughly, for every four steps in metabolic rate there are five steps in mass. Critically, this exponent of 0.88 is well above that for resting metabolic rate. What does that mean? It means that maximum metabolic rate and mass are closer to being directly proportional—we are closer to the expectation that for every step in mass we have a similar step in metabolic rate. If we double the body mass—double the number of cells—then we very nearly double the maximum metabolic rate. The discrepancy is less than we find for the resting metabolic rate. This means that the aerobic scope rises with body size—the larger the animal, the greater the difference between resting and maximum metabolic rate; in other words, larger animals can in general draw on greater reserves of stamina and power.

This is all fascinating, but the most important point, for our purposes, is that the slope of 0.88 for maximal metabolic rate does not tally with the prediction (0.75) of the fractal model—and the difference is statistically highly significant. On this count, too, it seems that the fractal model does not correspond to the data.

Just ask for more

So why does the maximal metabolic rate scale with a higher exponent? If doubling the number of cells doubles the metabolic rate, then each constituent cell consumes the same amount of food and oxygen as before. When the relationship is directly proportional, the exponent is 1. The closer an exponent is to 1, then the closer the animal is to retaining the same cellular metabolic power. In the case of maximum metabolic rate, this is vital. To understand why it is so important, let's think about muscle power: clearly we want to get stronger as we get bigger, not weaker. What actually happens?

The strength of any muscle depends on the number of fibres, just as the strength of a rope depends on the number of fibres. In both cases, the strength is proportional to the cross-sectional area; if you want to see how many fibres make up a rope, you had better cut the rope—it's strength depends on the diameter of the rope, not its length. On the other hand, the weight of the rope depends on its length as well as its diameter. A rope that is 1 cm in diameter and 20 metres long is the same strength, but half the weight, as a rope that is 1 cm in diameter and 40 metres long. Muscle strength is the same: it depends on the cross-sectional area, and so rises with the square of the dimensions, whereas the weight of the animal rises with the cube. This means that even if every muscle cell were to operate with the same power, the strength of the muscle as a whole could at best increase with mass to the power of 2/3 ($mass^{0.67}$). This is why ants lift twigs hundreds of times their own weight, and grasshoppers leap high into the air, whereas we can barely lift our own weight, or leap much higher than our own height. We are weak in relation to our mass, even though the muscle cells themselves are not weaker.

When the Superman cartoons first appeared in 1937, some captions used the scaling of muscle strength with body mass to give 'a scientific explanation of

Clark Kent's amazing strength.' On Superman's home planet of Krypton, the cartoon said, the inhabitants' physical structure was millions of years advanced of our own. Size and strength scaled on a one-to-one basis, which enabled Superman to perform feats equivalent, for his size, to those of an ant or a grasshopper. Ten years earlier, J. B. S. Haldane had demonstrated the fallacy of this idea, on earth or anywhere else: 'An angel whose muscles developed no more power, weight for weight, than those of an eagle or a pigeon would require a breast projecting for about four feet to house the muscles engaged in working its wings, while to economise in weight, its legs would have to be reduced to mere stilts.'

For biological fitness, it's plainly important to be strong in proportion to weight, as well as just having brute strength. Flight, and many gymnastic feats such as swinging from trees or climbing up rocks, all depend on the strength-to-weight ratio, not on brute strength alone. Numerous factors (including the lever-length and contraction speed) mean the forces generated by muscle can actually rise with weight. But all this is useless if the cells themselves grow weaker with size. This might sound nonsensical—why would they grow weaker? Well, they would grow weaker if they were limited by the supply of oxygen and nutrients, and this would happen if the muscle cells were constrained by a fractal network. Muscle would then have two disadvantages—the individual cell would be forced to become weaker, and at the same time the muscle as a whole would be obliged to bear greater weights. A double whammy. This is the last thing we would want: there is no way out of muscle having to bear greater weights with increasing size, but surely nature can prevent the muscle cells becoming weaker with size! Yes it can, but only because fractal geometry doesn't apply.

If muscle cells don't become weaker with larger size, their metabolic rate must be directly proportional to body mass: they should scale with an exponent of 1. For every step in mass there should be an equal step in metabolic rate, because if not the muscle cells can't sustain the same power. We can predict, then, that the metabolic power of individual muscle cells should not decline with size, but rather scale with mass to an exponent of 1 or more; they should not lose their metabolic power. This is indeed what happens. Unlike organs such as the liver (wherein the activity falls sevenfold from rat to man, as we've seen) the power and metabolic rate of the skeletal muscle is similar in all mammals *regardless of their size*. To sustain this similar metabolic rate, the individual muscle cells must draw on a comparable capillary density, such that each capillary serves about the same number of cells in mice and elephants. Far from scaling as a fractal, the capillary network in skeletal muscles hardly changes as body size rises.

The distinction between skeletal muscle and other organs is an extreme case of a general rule—the capillary density depends on the tissue *demand*, not on the limitations of a fractal supply network. If tissue demand rises, then the cells use up more oxygen. The tissue oxygen concentration falls and the cells become *hypoxic*—they don't have enough oxygen. What happens then? Such hypoxic cells send distress signals, chemical messengers like vascular-endothelial growth factor. The details needn't worry us, but the point is that these messengers induce the growth of new capillaries into the tissue. The process can be dangerous, as this is how tumours become infiltrated with blood vessels in cancer (the first step to metastasis, or the spreading of tumour outposts to other parts of the body). Other medical conditions involve the pathological growth of new blood vessels, such as macular degeneration of the retina, leading to one of the most common forms of adult blindness. But the growth of new vessels normally restores a physiological balance. If we start regular exercise, new capillaries start growing into the muscles to provide them with the extra oxygen they need. Likewise, when we acclimatize to high altitude in the mountains, the low atmospheric pressure of oxygen induces the growth of new capillaries. The brain may develop 50 per cent more capillaries over a few months, and lose them again on return to sea level. In all these cases—muscle, brain, and tumour—the capillary density depends on the tissue demand, and not on the fractal properties of the network. If a tissue needs more oxygen, it just asks for it—and the capillary network obliges by growing new feeder vessels.

One reason for capillary density to depend on tissue demand may be the toxicity of oxygen. Too much oxygen is dangerous, as we saw in the previous chapter, because it forms reactive free radicals. The best way to prevent such free radicals from forming is to keep tissue oxygen levels as low as possible. That this happens is nicely illustrated by the fact that tissue oxygen levels are maintained at a similar, surprisingly low level, across the entire animal king-dom, from aquatic invertebrates, such as crabs, to mammals. In all these cases, tissue oxygen levels average 3 or 4 kilopascals, which is to say about 3 to 4 per cent of atmospheric levels. If oxygen is consumed at a faster rate in energetic animals such as mammals, then it must be delivered faster: the through-flow, or *flux*, is faster, but the concentration of oxygen in the tissues need not, and does not, change. To sustain a faster flux, there must be a faster input, which is to say a stronger driving force. In the case of mammals, the driving force is pro-vided by extra red blood cells and haemoglobin, which supply far more oxygen than is available in crabs. Physically active animals therefore have a high red blood cell and haemoglobin count.

Now here is the crux. The toxicity of oxygen means that tissue delivery is restricted, to keep the oxygen concentration as low as possible. This is similar in all animals, and instead a higher demand is met by a faster flux. The tissue flux needs to keep up with maximum oxygen demand, and this sets the red blood cell count and haemoglobin levels for any species. However, different tissues have different oxygen demands. Because the haemoglobin content of blood is more or less fixed for any one species, it can't change if some tissues need more or less oxygen than others. But what *can* change is the capillary density. A low oxygen demand can be met by a low capillary density, so restricting excess oxygen delivery. Conversely, a high tissue oxygen demand *needs* more capillaries. If tissue demand fluctuates, as in skeletal muscle, then the only way to keep tissue oxygen levels at a constant low level is to divert the blood flow away from the muscle capillary beds when at rest. Accordingly, skeletal muscle contributes very little to resting metabolic rate, because blood is diverted to organs like the liver instead. In contrast, skeletal muscle accounts for a large part of oxygen consumption during vigorous exercise, to the point that some organs are obliged to partially shut down their circulation.

The diversion of blood to and from the skeletal muscle capillary beds explains the higher scaling exponent of 0.88 for maximal metabolic rate: a larger proportion of the overall metabolic rate comes from the muscle cells, which scale with mass to the power of 1—in other words, each muscle cell has the same power, regardless of the size of the animal. The metabolic rate is therefore somewhere in between the resting value of $mass^{2/3}$ or $mass^{3/4}$ (whichever value is correct) and the value for muscle, of mass to the power of 1. It doesn't quite reach an exponent of 1 because the organs still contribute to the metabolic rate, and their exponent is lower.

So the capillary density reflects tissue demand. Because the network as a whole adjusts to tissue demands, the capillary density *does* actually correlate with metabolic rate—tissues that don't need a lot of oxygen are supplied with relatively few blood vessels. Interestingly, if tissue *demand* scales with body size—in other words, if the organs of larger animals don't *need* to be supplied with as much food and oxygen as those of smaller animals—then the link between capillary network and demand would give an *impression* that the supply network scales with body size. This can only be an impression, because the network is always controlled by the demand, and not the other way around. It seems that West and colleagues may have confounded a correlation for causality.

Part and parcel of metabolism

The fact that resting metabolic rate scales with an exponent of less than 1 (it doesn't matter what the precise value is) implies that the energetic demand of cells falls with size—larger organisms do not need to spend as great a proportion of their resources on the business of staying alive. What's more, the fact that an exponent of less than 1 applies to all eukaryotic organisms, from single cells to blue whales (again, it doesn't matter if the exponent is not exactly the same in every case), implies that the energetic efficiencies are very pervasive. But that doesn't mean that the advantage of size is the same in every case. To see why energy demand falls, and what evolutionary opportunities this might offer, we need to understand the components of the metabolic rate, and how they change with size.

In fact, regardless of the network, we have yet to show that greater size actually yields efficiencies rather than constraints—from the exponent alone, it can be almost impossible to tell. For example, the metabolic rate of bacteria falls with size. As we saw in the previous chapter, this is because they rely on the cell membrane to generate energy. Their metabolic power therefore scales with the surface area to volume ratio, i.e. mass$^{2/3}$. This is a constraint, and helps to explain why bacteria are almost invariably small. Eukaryotic cells are not subject to this constraint because their energy is generated by mitochondria inside the cell. The fact that eukaryotic cells are much larger implies that their size is not constrained in this way. In the case of large animals, unless we can show *why* energy demand falls with size, we can't eliminate the possibility that scaling reflects a constraint rather than an opportunity.

We have noted that the large skeletal muscles contribute very little to the resting metabolic rate. This should alert us to the possibility that different organs contribute differently to the resting, and the maximal, metabolic rate. At rest, most oxygen consumption takes place in the bodily organs—the liver, the kidneys, the heart, and so on. The scale of their consumption depends on their size relative to the body as a whole (which may change with size), coupled to the metabolic rate of the cells that make up the organ (which depends on the demand). For example, the beating of the heart necessarily contributes to the resting metabolic rate of all animals. As animals get larger, their hearts beat more slowly. Because the proportion of the body filled by the heart remains roughly constant as size increases, but it beats more slowly, the contribution of the heart muscle to the overall metabolic rate must fall with size. Presumably something similar happens with other organs. The heart beats more slowly because it can *afford* to—and this must be because the oxygen demand of these other tissues has fallen. Conversely, if the tissue demand for oxygen rises, for example if we break into a run, then the heart must beat faster to provide it. The fact that the heart rate is slower in larger animals implies that there really are energetic efficiencies that can be gained from greater size.

Different organs and tissues respond differently to an increase in body size. A good example is bone. Like muscle, the strength of bone depends on the cross-sectional area, but unlike muscle the bone is metabolically almost inert. Both factors influence scaling. Imagine a 60-foot giant—ten times taller, ten times wider, and ten times thicker than an ordinary man. This is an example from Haldane again, who cites the giants Pope and Pagan from *The Pilgrim's Progress*—one of the few references that dates his essay, as I doubt that many science writers today would turn to Bunyan for an everyday analogy. Because bone strength depends on cross-sectional area, the giants' bones are 100 times the strength of ours, but the weight they must bear is 1000 times greater. Each square inch of giant bone must withstand ten times the weight of our own. Because the human thigh bone breaks under about ten times the human weight, Pope and Pagan would break their thighs every time they took a step. Haldane supposes this is why they were sitting down in his illustration.

The scaling of bone strength to weight explains why large, heavy animals need to be a different shape to smaller, lighter ones. Such a relationship was first described by Galileo in his *Dialogues Concerning Two New Sciences*, a delightful title that could hardly be matched these days. Galileo observed that the bones of larger animals grew more quickly in breadth than in length, compared with the slender bones of small animals. Sir Julian Huxley put Galileo's ideas on a firm mathematical footing in the 1930s. For a bone to retain the same strength relative to weight, its cross-sectional area must change at the same rate as body weight. Let's restrain ourselves to doubling the dimensions of our giant. His volume, and therefore weight, increases eightfold (2^3). To support this extra weight, his bones must grow eightfold in cross-sectional area. However, bones have length as well as cross-sectional area. If their cross-section is raised eightfold, and their length doubled, the skeleton is now sixteen (or 2^4) times heavier. In other words, the skeleton takes up a greater proportion of body mass. Theoretically, the scaling exponent is 4/3, or 1.33, although in reality it is less than this (about 1.08) because bone strength is not constant. Nonetheless, as Galileo realized in 1637, bone mass imposes an insurmountable limit on the size of any animal that must support its own weight—the point at which bone mass catches up with total mass. Whales can surpass the size limit of terrestrial animals because they are supported by the density of water.

The fact that bones necessarily take up a greater proportion of body mass as body size increases, coupled with their metabolic inertia, means that a greater proportion of a giant's body is metabolically inert. This lowers the total metabolic rate, and so contributes to the scaling of metabolic rate with size (the scaling exponent is 0.92). However, the difference in bone mass alone is not sufficient to account for the reduction in metabolic rate with size. But might other organs scale in a similar fashion? Might there be a threshold of liver or kidney function, beyond which there is little need to continue amassing ever more hepatic or renal cells? There are two reasons to think that there may indeed be a threshold of function in these organs. First, the relative size of many organs falls as body size rises. For example, the liver accounts for about 5.5 per cent of the body mass of a 20 gram mouse, about 4 per cent that of a rat, and just 0.5 per cent that of a 200 kg pony. Even if the metabolic rate of each liver cell remains the same, the proportionately lower mass of the liver would contribute to the lower metabolic rate of the pony. And second, the metabolic rate of each liver cell is *not* constant: oxygen consumption per cell falls about ninefold from the mouse to the horse. Presumably there is a limit to just how small an organ can be within the body cavity—it is better to maintain the size of the liver, so that it does not swing loose in the peritoneum, and instead restrict the metabolic rate of its component cells. The combination of both factors (a relatively small liver, along with a lower metabolic rate per cell) means that the contribution of the liver to metabolic rate falls quite dramatically with size.

By now we can begin to see that the resting metabolic rate of an animal is composed of many facets. To calculate the overall metabolic rate we need to know the contribution of each tissue, of each cell within that tissue and even of particular biochemical processes within cells. Such an approach can also explain how and why the metabolic rate changes from rest to aerobic exercise. This was the tack taken by Charles-Antoine Darveau and his colleagues at the University of British Columbia, Vancouver, in the lab of the Canadian guru of comparative biochemistry, Peter Hochachka, in work published in *Nature* in 2002. Darveau and colleagues attempted to sum up the contribution of each facet, and the influence of critical hormones (such as thyroid hormones and catecholamines) to derive an equation that could explain the scaling of metabolic rate with size, giving a flexible overall exponent of about 0.75 for the resting metabolic rate and 0.86 for maximal metabolic rate. Both West and colleagues, and Banavar and colleagues refuted their paper on mathematical grounds in the letters pages—and plainly Darveau's equations did need some refinement. Hochachka's group defended the soundness of their conceptual approach and did modify their equations, publishing a more detailed exposition in the journal *Comparative Biochemistry and Physiology* in 2003. Sadly, this was among the last works of Peter Hochachka, who died of prostate cancer at the age of sixty-five in September 2002. It is a measure of his unquenchable thirst for knowledge that his final paper was a study of the wayward metabolism of malignant prostate cells, published with his doctors as co-authors.

The strident mathematical dismissal of Hochachka's argument, and the concession of errors in its defence, may have caused some dispassionate observers (including me, initially) to suspect that if the maths was wrong, then so too, perhaps, was the whole approach. Not so: this might have been a flawed first approximation, but it was robustly grounded in biology, and I'm looking forward to more sophisticated revisions. But it already offers a quantitative demonstration that metabolic demand *does* fall with size, and that this controls the supply network, rather than the other way around. Even more importantly, it gives an insight into the evolution of complexity, and especially into a problem that has long eluded biologists—the evolution of warm-bloodedness in mammals and birds. There is no better illustration of the link between size and metabolic efficiency, and the way in which these attributes pave the way to greater complexity. For warm-bloodedness is about far more than just keeping warm in the cold: it adds a whole new energetic dimension to life.

10

The Warm-Blooded Revolution

Warm-bloodedness is a misleading term. It means that the temperature of the blood, and with it the body, is maintained at a stable temperature above that of the surroundings. But many so-called 'cold-blooded' creatures, such as lizards, are really warm-blooded in this sense, for they maintain a higher temperature than their surroundings through behaviour. They bask in the sun. While this sounds inherently inefficient, at least in England, many reptiles succeed in regulating their body temperature within tightly specified limits at a similar level to mammals—around 35 to 37°C (although it usually falls at night). The distinction between reptiles, such as lizards, and birds and mammals lies not in their ability to regulate temperature, but to generate heat internally. Reptiles are said to be 'ectothermic', in that they gain their body heat from the surroundings, whereas birds and mammals are 'endothermic'—they generate their heat internally.

Even the word endothermic needs some clarification. Many creatures, including some insects, snakes, crocodiles, sharks, tuna fish, even some plants, are endothermic: they generate heat internally, and can use this heat to regulate their body temperature above that of their surroundings. All of these groups evolved endothermy independently. Such animals generally use their muscles to generate heat during activity. The advantage of this is related directly to the temperature in the muscle. All biochemical reactions, including the metabolic rate, are dependent on temperature. The rate of metabolism doubles for each 10°C rise in temperature. Along with this, the aerobic capabilities of all species improve with higher body temperature (at least up to the point that the reactions become destructive). Speed and endurance are therefore enhanced at higher body temperature, and this clearly offers many advantages, whether in the competition for mates or in the battle for survival between predators and prey.[1]

---

[1] Lizards are sluggish when cold (as are torpid mammals or birds), and so vulnerable to predators. The earless lizard finds a way around this problem by using a blood sinus on top of its head. In the mornings, it pokes its head out of its burrow, and remains there, keeping a wary eye out for predators, ready to duck back inside if necessary. It can warm its whole body via the

Birds and mammals stand apart in that their endothermy is not dependent on muscle activity, but on the activity of their organs, such as liver and heart. In mammals, muscles contribute to heat generation only during shivering in intense cold, or during vigorous exercise. When at rest, the body temperature of all other groups falls (unless they maintain it by basking in the sun) whereas the mammals and birds maintain a constant and high temperature even at rest. The difference in resource use is profligate and shocking. If an equally sized reptile and mammal maintain the same temperature, through behavioural and metabolic means, respectively, the mammal needs to burn six to ten times as much fuel to maintain this temperature. If the surrounding temperature falls, the distinction becomes even greater, because the temperature of the reptile will fall, whereas the mammal strives to maintain a constant temperature of 37°C, by increasing its metabolic rate. At 20°C, a reptile uses only about 2 or 3 per cent of the energy needed by a mammal, and at 10°C barely 1 per cent. On 'average', in the wild, a mammal uses about thirty times more energy to stay alive than an equivalent reptile. In practical terms, this means that a mammal must eat *in one day* the amount of food that would sustain a reptile for a whole month.

The evolutionary costs of such a profligate lifestyle are profound. Instead of merely keeping warm, a mammal could divert thirty times more energy towards growth and reproduction. I shudder to think of the consequences on teenage angst; but given that natural selection is all about surviving to maturity and reproducing, the costs are grave indeed. The benefits must at least equal these costs, or natural selection would favour the reptilian lifestyle, and the evolution of mammals and birds would have been snuffed out at the beginning. Most attempts to explain the evolution of warm-bloodedness for its own sake fall prey to this difficulty.

For example, the benefits of endothermy include the ability to operate at night, and to expand ecological niches into temperate and even polar climates. A high body temperature, as we have seen, also speeds up the metabolic rate, with potential benefits on speed, stamina, and reaction time. The drawback is the cost-to-benefit ratio, and in particular the large amount of energy needed to raise the body temperature by a trifling degree. Revealingly, digesting a very large meal can raise the resting metabolic rate of lizards by as much as fourfold for a period of several days, but only raises the body temperature by 0.5°C. To sustain such a rise in body temperature would require the reptile to eat on average four times as much food—and this is no easy matter, as it inevitably

---

blood sinus on its head, and only when warm and up to speed does it venture out. Natural selection never misses a trick: some lizards have a connection from this sinus to the eyelids, through which they can squirt blood at predators, especially dogs, who find the taste repugnant.

demands extra hours of foraging, with a concurrent exposure to danger. The advantage in speed and endurance is also trivial: a 0.5°C rise in temperature speeds the rate of chemical reactions by about 4 per cent—well within the inter-individual variability of athleticism for most species. The problem is not merely one of heat loss, which could be offset by a fur coat or feathers. One amusing experiment, in which a lizard was dressed in a specially tailored fur coat, showed that far from warming the body by improving heat retention, the fur had the opposite effect: it interfered with the lizard's ability to absorb heat from its surroundings. Insulation, of course, keeps the heat out as well as in. In short, there are serious and immediate costs to raising body temperature, which more than offset the trifling advantages. How, then, do we explain the rise of endothermy in mammals and birds?

Much the most coherent and plausible (albeit still unproven) explanation for the evolution of endothermy was put forward in an illuminating and un-surpassed paper in *Science* in 1979 by Albert Bennett and John Ruben, then (and indeed still) at UC Irvine and Oregon State University, respectively. Their theory, known as the 'aerobic capacity' hypothesis, makes two assumptions. First, it postulates that the initial advantage was not related to temperature at all, but to the aerobic capacity of animals. In other words, selection was primarily directed at speed and endurance—at the maximum metabolic rate and muscular performance, not at the resting metabolic rate and body temperature. Second, the hypothesis postulates that there is a direct connection between the resting and the maximal metabolic rate, such that it is not possible (evolutionarily) to raise one without raising the other. Thus, selection for a faster maximal metabolic rate (a higher aerobic capacity) necessarily entails raising the resting metabolic rate. This is plausible: we've already noted that there *is* a link between the resting and the maximal metabolic rate, and that the aerobic scope (the factorial difference between the two) rises with body size. So there certainly is a link; but is it causal? If one rises or falls, *must* the other?

Bennett and Ruben argued that the resting metabolic rate was eventually ele-vated to the point that internal heat production could raise body temperature permanently. At this point, the advantages of endothermy—niche expansion, and so on—were selected for their own benefit. Selection was now directed at maintaining internally generated heat, favouring the evolution of insulatory layers such as subcutaneous fat, fur, down, and feathers.

Sizing up to complexity

For the aerobic capacity hypothesis to work, both the maximal and the resting metabolic rate of mammals and birds need to be substantially higher than those of lizards. This is well known to be the case.[2] Lizards become exhausted quickly and have a low capacity for aerobic exercise. While they can move very fleetly (when warmed up) their muscles are mostly powered by anaerobic respiration to produce lactate (see Part 2). They can sustain a burst of speed for little more than 30 seconds, enabling them to dart for the nearest hole and hide, whereupon they often need several hours to recover. In contrast, the aerobic performance of similarly sized mammals and birds is at least six to tenfold greater. While not quicker off the mark or fleeter of foot, they can sustain the pace for far longer. As Bennett and Ruben put it in their original *Science* paper: 'The selective advantages of increased activity are not subtle but rather are central to survival and reproduction. An animal with greater stamina has an advantage that is readily comprehensible in selective terms. It can sustain greater levels of pursuit or flight in gathering food or avoiding becoming food. It will be superior in territorial defense or invasion. It will be more successful in courtship or mating.'

What must an animal do to improve its stamina and speed? Above all else, it has to augment the aerobic power of its skeletal muscles. To do so requires more mitochondria, more capillaries and more muscle fibres. We immediately run into a difficulty with space allocation. If the entire tissue is taken up with muscle fibres, there is no room left over for mitochondria to power muscle contraction, or for capillaries to deliver oxygen. There must be an optimal tissue distribution. To a point, aerobic power can be improved by a tighter packing of these components, but beyond that improvements can only be made by greater efficiency. This is indeed what happens. According to Australian researchers Tony Hulbert and Paul Else, at the University of Wollongong, New South Wales, mammalian skeletal muscles have twice as many mitochondria as the equivalent lizard muscles, and these are in turn more densely packed with membranes and respiratory complexes. The activity of respiratory enzymes in rat skeletal muscle is also about twice that of the lizard. In total, the aerobic performance of rat muscle is nearly eight times that of the lizard—a difference that wholly accounts for its greater maximum metabolic rate and aerobic capacity.

This deals with the first part of the aerobic capacity hypothesis: selection for

---

[2] The scaling equation is given as metabolic rate $= aM^b$ in which a is a species-specific constant, M is the mass, and b is the scaling exponent. The constant a is fivefold greater in mammals than in reptiles, but both groups still scale with size (the lines are parallel). The fractal model can't explain why different species should have different a constants, in other words why the resting metabolic rate and capillary density in various organs is different in mammals and reptiles; nor can it explain the rise of endothermy. The explanation again lies in the tissue demand for more oxygen to power greater aerobic performance: this is the driving force that leads to the remodelling of muscle and organ architecture, and with it the fractal supply network.

endurance raises the mitochondrial power of muscle cells, leading to a faster maximum metabolic rate; but what about the second part? Why is there a link between maximal and resting metabolic rate? The reason is not clear, insofar as none of the possible explanations has been proved. Even so, there is a good intuitive reason to expect a connection. I mentioned that lizards may often take several hours to recover from exhaustion, even after a few minutes of vigorous exertion. Such a slow recovery is less dependent on muscles than on organs, such as the liver and kidneys, which process the metabolic waste and other breakdown products of vigorous exercise. The rate at which these organs operate depends on their own metabolic power, which in turn depends on their mitochondrial power—the more mitochondria, the faster the recovery. Presumably the advantages of endurance also apply to recovery time: given the eightfold rise in aerobic power of mammalian muscles, if there were no compensating changes in organ function it would take a whole day, rather than a few hours to recover from exercise.

Unlike muscles, organs are not faced with a dilemma of space allocation—while the density of mitochondria doesn't change with size in muscles, it does in the organs. As animals get bigger, the power laws that we discussed in the last chapter mean that their organs become more sparsely populated with mitochondria. This is an opportunity in waiting. For the organs of a large animal to gain power, the tissue architecture doesn't need to be restructured as it does in muscle: it can simply be repopulated with mitochondria. This opportunity seems to have given rise to endothermy. In their classic comparative studies, Hulbert and Else showed that the organs of mammals contain five times as many mitochondria as an equivalent lizard, but in all other respects the mitochondria are no different. For example, the efficiency of their respiratory enzymes is exactly the same. In other words, for every hard-won increment in muscle power, it's relatively simple to counterbalance the new power by filling up the half-empty organs with a few more mitochondria, so as to ensure speedy recovery from aerobic exertion. The important point is that the function of organs like the liver is coupled to muscle demand, and not with the need to keep warm.

Proton leak

But there is a diabolical catch. We have seen that the muscles contribute little to the resting metabolic rate: the danger of oxygen toxicity means that blood is diverted away from the muscles and into the organs, where there are relatively few mitochondria to cause damage. So what might have happened in the first mammals? They had extra mitochondria in their organs to compensate for their higher aerobic capacity, but nowhere to divert the blood, which had to pass through either the organs or the muscles.

Once our prototype mammal has digested his food, caught so easily with his newfound aerobic prowess, he goes to sleep. Beyond replenishing his reserves of glycogen and fat, there is little call to expend energy. His mitochondria fill up with electrons extracted from food. This is a dangerous situation. The respiratory chains in the mitochondria become packed with electrons, because there is only a sluggish electron flow. At the same time, there is plenty of oxygen around, as the blood flow can't be diverted. In these conditions, electrons easily escape from the respiratory chains to form reactive free radicals, which can damage the cell. What might be done?

According to Martin Brand in Cambridge, one answer might be to *waste energy* by keeping the whole system ticking over. The danger from free radicals is at its greatest when there is no electron flow down the chain. Electrons pass most readily on to the next complex in the chain, and so tend to react with oxygen only when that complex is choked up with electrons, blocking normal flow. Restarting electron flow usually requires the consumption of ATP.[3] If there is no demand for ATP, the whole system clogs up and becomes reactive. This is the situation when resting after a large meal. One possible escape is to uncouple the proton gradient, so electron flow is not tied to ATP production. In Part 2, we compared this to a hydroelectric dam, in which an overflow channel prevents flooding in times of low demand. In the case of the respiratory chains, instead of passing through the ATPase to generate ATP (the main dam gates), some protons pass back through other pores in the membrane (the overflow channels), so that part of the energy stored in the gradient is dissipated as heat. By uncoupling the proton gradient in this way, slow electron flow is maintained, and this restricts free-radical damage (just as the overflow channel prevents flooding). The fact that such a mechanism *does* protect against free-radical damage was verified in a fascinating study of mice by John Speakman and his colleagues in Aberdeen, working with Martin Brand. Their title said it all: 'Uncoupled and surviving: individual mice with high metabolism have greater mitochondrial uncoupling and live longer.' We'll look into this further in Part 7, but in short they live longer because they accrue less free-radical damage.

In resting mammals, perhaps a quarter of the proton gradient is dissipated as heat. The same is true of reptiles, but they have barely a fifth the mitochondria in each cell and so generate five times less heat per gram. Their organs are also relatively small, so reptiles have fewer mitochondria altogether, adding up to a

---

[3] In respiration, ATP is formed from ADP and phosphate, and during cellular work it is converted back to them. If all the cell's ADP and phosphate has already been converted to ATP, then there is a shortage of raw materials, which means that respiration must come to a halt. Once the cell has consumed some ATP, more ADP and phosphate are formed, and respiration starts again. Thus the speed of respiration is tied to the demand for ATP.

tenfold difference in heat production. In the first large mammals, proton leak may well have generated enough heat to raise body temperature by a significant degree, simply as a side-effect of aerobic health. Once heat is generated in this way, selection can take place for endothermy for its own sake—for keeping warm. In contrast, small animals can only generate enough heat to maintain their body temperature if they insulate themselves better, and even step up the rate of heat production. These properties probably evolved in the descendents of animals that were already endothermic—otherwise we're back to the problems of raising body temperature for its own sake. In other words, it's likely that endothermy evolved in animals that were large enough to balance heat production with heat loss, while the smaller descendents of the first endothermic mammals had to make further adjustments to offset their heat-retention problem. Small mammals like rats need to supplement their normal heat production with brown fat, rich in mitochondria, which are dedicated to heat production—here, all the protons leak back across the mitochondrial membranes to give off heat. This in turn means that the resting metabolic activity of small mammals doesn't correlate with muscular capacity for hard work, but rather with the rate of heat loss.

These ideas explain several long-standing puzzles, and scotch any fond lingering notions of a universal constant (metabolic rate scales with mass$^{3/4}$ across the entire living world) once and for all. It is plain why small mammals and birds (none of which approach large mammals in size) scale with an exponent of about 2/3: the greater part of their metabolic rate is linked not to muscle function, but to maintaining body heat instead. In contrast, for larger mammals and reptiles, heat generation is not a top priority—quite the contrary, overheating is much more of a problem—so the metabolic power of their organs only needs to balance muscular demand, and not heat production. Because the maximal metabolic rate scales with an exponent of 0.88, so too does the resting metabolic rate.

How closely animals meet these expectations depends on other factors, such as diet, environment, and species. So, for example, marsupials have a lower resting metabolic rate than most other mammals, as do desert-dwellers and all ant-eaters: we can predict that they should take longer to recover from vigorous physical exertion, or will be less likely to partake in it at all; and generally this is the case.[4] It seems that the scaling of energetic efficiency presents an opportu-

---

[4] Some marsupials, such as kangaroos, are capable of moving at great speeds despite a low resting metabolic rate. They can do this because hopping differs from running, in that oxygen consumption tapers off with speed—they can hop faster and faster without consuming more and more oxygen. Hopping is more efficient because it makes use of the elastic rebound, which can be dissociated from aerobic muscle contraction to some extent.

nity that can be met in different ways, from new pinnacles of aerobic power in energetic birds and mammals, to varying degrees of sloth in well-protected but less energetic animals, whether armadillos or tortoises.

First steps up the ramp

Generating energy with mitochondria enables eukaryotic cells to be much larger than bacteria—on 'average', perhaps 10 000 to 100 000 times the size. Large size brings with it the gift of energetic efficiency. Within limits that are probably defined by the efficiency of the supply network, the bigger the better. This is an immediate payback for an immediate advantage, and is likely to counterbalance the immediate disadvantages of larger size—the requirement for more genes, more energy, and better organization. The immediate reward of energetic efficiency may have helped push eukaryotes up the ramp of ascending complexity.

Still a couple of conundrums tease me, but I think they can be explained. First, energetic efficiency is often dismissed as a target for natural selection, on the grounds that large animals still have to eat more food than small animals: the energetic savings are only apparent on a cell-by-cell, or gram-weight basis. The critic is quick to point out that natural selection works at the level of individuals, usually, and certainly not on a gram-weight basis. This is obviously true, but the environment and the needs of an organism still relate to its size. We saw that a rat is seven times as hungry as a human being: in relation to its body size it must find and eat seven times as much food as we do. But a rat is no stronger or faster in relation to its surroundings than we are. The term *relative* here is real. Clearly, a rat can't hunt buffalo, but we can, or anything else down to the size of a rat and beyond. The world that animals inhabit is shaped by their size, and in our own world we need to eat seven times less food than does a rat, day in day out. On the same basis, we can survive seven times as long without food or water. The scale of the advantage can be seen even more clearly if we think in terms of how much we need to eat relative to our body weight. A mouse, for example, must eat *half* its body weight every day to avoid starvation, whereas we need only consume about 2 per cent of our own body weight. Surely this is a genuine advantage. That is not to say size is invariably a dominant advantage—in many circumstances, small size may offer strong advantages, leading to different evolutionary trends; but the energetic efficiency of greater size does seem to have had a deep influence on the direction of eukaryotic evolution.

The second conundrum that teases me relates to the very pervasiveness of energetic advantage. In Part 4, we've considered mostly the mammals and reptiles. We have broken down the energy savings into their components, to conclude that they offer genuine opportunities, rather than merely the constraints of a fractal network. On the other hand, I have also pointed out that bacteria are limited by their surface area-to-volume ratio, and that this is a constraint, not an opportunity. Do single eukaryotic cells, such as amoeba, really have an advantage with larger size? Do trees, or shrimps? Have we, in rejecting a universal constant, also relinquished any right to generalize beyond the example of mammals?

I don't think so. I have left other examples aside until now because the answers are less certain—they have received far less attention than mammals and reptiles. Nonetheless, I suspect that most organisms, including single cells, gain the same benefits. In larger organisms, these benefits are the familiar economies of scale: it's cheaper by the dozen. As in society, such benefits depend on the set-up costs, operational costs, and distribution costs, and these impose outer limits on the economies of scale. But within these limits, the benefits ought to apply widely. This is because living organisms are highly conservative in their operational principles. In particular, their organization is invariably modular. Both single cells and multicellular organisms are made up from a mosaic of functional parts. In multicellular organisms, the organs perform particular functions, such as breathing or detoxification; within cells, discrete functions are carried out by organelles like mitochondria. Modular functions within single cells include genetic transcription, protein synthesis, packaging, membrane synthesis, pumping salts, digesting food, detecting and responding to signals, generating energy, moving around, trafficking of molecules, and so on. I imagine that the economies of scale apply as much to these modular aspects of single cells as they do to multicellular organisms.

This idea brings us back to the question of gene number, which I touched on at the start of the chapter. We noted that complex organisms need more genes, and thought about Mark Ridley's argument that the invention of sex enabled the accumulation of genes, opening the gates to complexity. But sex, as we saw, may not have been the gatekeeper, and certainly didn't limit gene number in bacteria or single eukaryotic cells. I wonder whether the accumulation of genes in eukaryotes is better explained in terms of the energetic efficiency of larger cells. Larger cells usually have a larger nucleus. It seems that balanced growth during the cell cycle requires the ratio of nuclear volume to cell volume to be basically constant—another power law! This means that, over evolution, the nuclear size, and with it the DNA content, adjusts to changes in cell volume for optimal function. So as cells grow larger they adjust by developing a larger nucleus with more DNA, even if this extra DNA does not necessarily code for more genes. This explains the C-value paradox discussed in Chapter 1, and is why cells like *Amoeba dubia* have 200 times more DNA than a human being, albeit coding for fewer genes.

The extra DNA is often dismissed as junk, and may be purely structural, but it can also be called upon to serve useful purposes, from forming the structural scaffolding of chromosomes, to providing binding sites that regulate the activity of many genes. This extra DNA also forms raw material for new genes, building the foundations of complexity. The sequences of many genes betray their ancestry as junk DNA. Might it be that the origin of complexity was as simple as a scale? As soon as eukaryotic cells became powered with mitochondria, there was a selective advantage to them being bigger. Bigger cells need more DNA, and with this they had the raw material needed for more genes and greater complexity. Notice that this is the reverse of bacteria: whereas a heavy selection pressure to lose genes oppresses bacteria, eukaryotes are under pressure to gain them. If Ridley is correct that sex postpones mutational meltdown, it might be that the requirement for more DNA with larger size was an underlying selection pressure that gave rise to sex itself.

For eukaryotic cells, the possession of mitochondria raised the ceiling on the possibilities of life. Mitochondria made larger size probable, rather than staggeringly unlikely, inverting the constrained world of bacteria. With larger size came greater complexity. But there were disadvantages too, arising from a conflict between the mitochondria and the host cells. The consequences of this long battle were equally pervasive, marking life forever with deep scars; yet even these scars had the power to create and destroy. Without mitochondria we would have had no cell suicide, but no multicellular 'individuals'; no ageing, but no sexes. The dark side of mitochondria had even more power to rewrite the script of life.

PART 5

Murder or Suicide
The Troubled Birth of the Individual

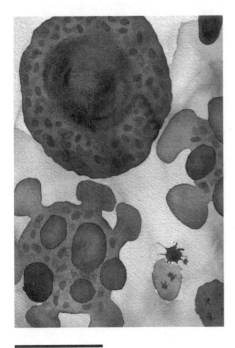

Death by apoptosis—mitochondria determine whether a cell lives or dies by enforced suicide

When cells in the body become worn out or damaged, they die by enforced suicide, or apoptosis. The cell blebs, is packaged up, and reabsorbed. If the mechanisms controlling apoptosis fail, the result is cancer, a conflict of interest between cells and the body as a whole. Apoptosis seems to be necessary for the integrity and cohesion of multicellular individuals, but how did once-independent cells come to accept death for the greater good? Today apoptosis is policed by mitochondria, and the machinery of death was inherited from their bacterial ancestors, suggesting a history of murder. So was the cohesion of the individual forged in deadly conflict?

'I think therefore I am' said Descartes, begging the rejoinder 'But what exactly am I?' The nature of the individual has long eluded philosophers and scientists, and is only now coming into focus. We can say that an individual is an organism composed of genetically identical cells, which are specialized to perform diverse tasks for the good of the organism as a whole. From an evolutionary point of view, the question is: why did these cells subordinate their selfish interests to collaborate so altruistically in the body? Inevitably there were conflicts between the various levels of organization in the body, between genes, organelles, and cells, but paradoxically without these internecine battles the strong bonds that forge the individual might never have evolved. Such conflicts spurred the evolution of a molecular 'police force', which curbs selfish interests much as the legal system enforces acceptable behaviour in society. In the body, programmed cell death, *apoptosis*, is central to the policing of conflicts. Today, apoptosis is enforced by mitochondria, raising the possibility that they may have been key to the evolution of individuals. In this Part we shall see that, back in the mists of evolutionary time, mitochondria were indeed intimately linked with the rise of multicellular individuals.

There has been more spleen vented about selfish genes, altruism, and the limits of natural selection than is seemly in polite scientific society. Underpinning many of the arguments was a simple question: what does natural selection act upon—genes, individuals, groups of individuals (such as a kin group), or the species as a whole? In 1962, Vero Wynne-Edwards' eloquent treatise on animal behaviour, *Animal Dispersion in Relation to Social Behaviour*, concentrated minds. He ascribed many aspects of social behaviour to selection not at the level of the individual, as had been assumed by Darwin, but at the level of the species. Behaviour was just the tip of the iceberg. Many other traits seemed easier to explain by thinking of the species rather than the individual. For example, ageing doesn't seem to benefit the individual in any way (what do we gain from getting old and dying?) but does look like a useful service to the species, for it leads to population turnover, preventing overcrowding and overconsumption of lean resources. Similarly, sex seemed pointless for individuals, so much so that we must be bribed by intense erotic pleasure; presumably, mild pleasure is not enough. Rather than simply dividing in two like a bacterium, such that one parent produces two daughter cells, sex takes two parents to produce a single offspring, making it twice as costly as clonal reproduction—

the twofold cost of sex—to say nothing of the trouble of finding a mate. Worse, sex randomizes the very genes that had ensured the success of the parents, making it a potential liability. Its most obvious value is the fast dissemination of variation, and beneficial adaptations, throughout a population: a benefit to the species.

The reaction to these ideas is often dismissed as ultra-Darwinism, a term of disparagement meaning little. How, one must ask, does species-level selection work? There are ways in which it might. For example, a fast population turnover may lead to a fast rate of evolution, which might benefit one species over another if conditions change quickly (for example during rapid global warming, or after a meteorite impact). Another possibility, which Richard Dawkins refers to as the 'evolution of evolvability' relates to the genetic 'flexibility' of a species—some species have more scope for further evolution in their form and behaviour than others. In most instances, however, the blindness of evolution means that such species-level selection just can't develop. Sex is complicated and didn't evolve overnight. If the only benefits are at the species level, and are deferred until sex has actually evolved, what happens in the meantime? Any individuals in a population that take a tentative step towards sex will lose out, and eventually be eliminated by natural selection, because they suffer from the twofold cost of sex and the randomization of any beneficial traits, before any advantages can take over. Similarly, individuals who don't age will leave their anti-ageing genes behind, which will come to dominate the population simply because the carriers have more time to have more children, who can pass on the same anti-ageing genes. Thus, on one hand there seemed few ways that selection could work at the level of the species, and on the other, some noble self-effacing traits could only be explained (at the time) by recourse to selection at the level of the species.

From the 1960s onwards, William Hamilton, George C. Williams, John Maynard Smith, and others, sought to explain apparently altruistic traits by means of selection at the level of the individual, the kin group, or the gene. The new approach boiled down to a mathematical exploration of *inclusive fitness*—the idea famously expressed in a pub conversation by J. B. S. Haldane: 'Would I lay down my life to save my brother? No, but I would to save two brothers, four nephews, or eight cousins' (on the grounds that he shared 50 per cent of his genes with his brothers, 25 per cent with his nephews and 12.5 per cent with his cousins, so his genes at least would break even). Much of the ensuing acrimony centred on the use of such loaded terms as 'selfish'—terms that have a specific definition in biology, but emotive overtones in general usage. In particular, Richard Dawkins' *The Selfish Gene* either inspired or raised the hackles of an entire generation, at least partly because it was so well written that everyone could feel the icy blast of its conclusion—living organisms are the throwaway survival machines of their genes, temporary puppets controlled by virtually immortal puppet-master genes. The only logical way to think about evolution, said Dawkins, is to stop gazing at our own belly buttons, and take a genes'-eye view of population dynamics.

The idea that the gene is the 'unit of selection' has been attacked from many quarters. The most common line of attack is the claim that genes are invisible to natural selection: they are inert stretches of ticker tape that do no more than code for proteins or RNA. What's more, there is an ambiguous relationship between a gene and the protein it encodes: the same gene may be split up in different ways, so that it codes for several different proteins; and we now realize that many proteins fulfil more than one function. Genes can also have very different effects, depending on the body they find themselves in. For example, it's often pointed out that a variant of the haemoglobin gene protects against malaria when present in half dose (heterozygous), but causes sickle cell an-aemia when present in full dose (homozygous). All this is true, but none of it undermines the power of a gene-centred approach to explain the currents of evolution: the individual may be the object of selection, but only the genes are passed on to the next generation. The key to the selfish gene is that, in sexual reproduction, the individual does not persist from one generation to the next; no more do any of the individual cells, nor even chromosomes. Bodies dissolve and reform like wisps of cloud, each one fleeting and different. According to Dawkins, only the genes persist, resistant to being scrambled, old as the moun-tains. From the perspective of a population over evolutionary time, the changes in gene frequencies are the best means of quantifying evolution. To an extent this is a mathematical crutch to a complex problem, but it is also a reality, how-ever unpalatable it may be.

From the point of view of selfish genes, the evolution of an individual is not a problem. If the conglomeration of cells that we call a body happens to be successful at passing on its genes to the next generation, then these genes will thrive to the detriment of the genes that don't collaborate in this way. A body is the product of genes collaborating together to serve their own selfish end of being copied in ever greater numbers. Dawkins is explicit on the point: 'Some people use the metaphor of the colony, describing a body as a colony of cells. I prefer to think of the body as a colony of *genes*, and of the cell as a convenient working unit for the chemical industries of the genes.'

The crux of the selfish gene is that only the gene passes from one generation to the next, so the gene is the most stable evolutionary unit: it is the 'replicator'. Dawkins makes it clear that this perspective is restricted to sexually reproduc-ing organisms, like most (but not all) eukaryotes. It doesn't apply to bacteria with the same force, because they replicate clonally. In this case, the individual cell *can* be said to persist from one generation to the next, whereas accumulating mutations mean that the genes themselves *do* change. In fact, in physically stressful circumstances, bacteria can even speed up the mutation rate in their genes. So there is a dilemma in bacteria about whether selection is 'for' the genes or the cell as a whole. In many respects the cell is the replicator.

Mutations in a gene don't necessarily change the phenotype (the function or appearance of the organism) but by definition they must change the gene itself, perhaps even scrambling its sequence out of recognition over aeons. Mutations accumulate because many of them have little or no effect on function, and so go unnoticed by natural selection—they are said to be 'neutral'. Most of the genetic differences between people, on average one in every 1000 DNA letters, millions of letters in total, are likely to result from neutral mutations. When we consider very different species, two sequences can be so dissimilar that it is not possible to discern any relationship between them, unless we take into consideration the spectrum of intermediary forms in more closely related species. Then we can see that two apparently unrelated genes are indeed related. The physical structure and function of proteins encoded by utterly dissimilar genes is often strikingly well conserved, even though the amino acid components are now mostly different. Plainly, the structure and function of the protein has been selected 'for', whereas the sequence of the gene is relatively plastic. It's like returning to a company that you once worked for, to discover that none of your former colleagues still works there, but that the type of business, ethos, and management structures are exactly as you remembered them, a ghostly echo of the past.

Because genes can change, while the cell and its constituents remain essentially unchanged, the bacterial cell might be considered more stable an evolutionary unit than its genes. For example, cyanobacteria (the bacteria that 'invented' photosynthesis) have certainly changed their gene sequences over evolution, but if the fossil evidence can be believed, the phenotype has barely changed over billions of years. If, as Dawkins has argued, the worst enemy of the selfish gene is a competitive (polymorphic, or altered) form of the same gene, then neutral mutations are the selfish gene scrambler *par excellence*: gene sequences diverge over time as neutral mutations accumulate. There may be millions of different forms of the same gene in different species, all scrambled to varying degrees; this is the basis of any gene tree. So evolution pits the selfish interests of genes (which 'want' to produce exact copies of themselves) against the randomizing power of mutation, which forever scrambles the sequence of genes, turning the selfish gene into its own worst enemy, the gene it used to hate.

Several other considerations militate against the gene as the 'unit of selection' in bacteria. It is said that in clonal replication all the genes are passed on together, so there is no distinction between the fate of the genes and the fate of the cell. This isn't quite true. Bacteria swap genes, and are prey to viruses called bacteriophages which load up cassettes of selfish DNA. Yet whereas eukaryotes are stuffed with selfishly replicating 'parasitic' DNA (DNA sequences that replicate for their own benefit, rather than that of the organism), bacteria have small genomes and next to no parasitic DNA. As we saw in Part 3, bacteria lose excess DNA, including functional genes, because this speeds up their replication. If these genes are 'selfish', they are punished for it by being regularly thrust out into the hostile world. Perhaps it's reasonable to think of lateral gene transfer in bacteria as a selfish rearguard action on the part of the genes themselves, but in general such lateral gene transfers only last as long as the cell needs the extra genes, and then they are lost again, along with any other genes that are not needed. I don't doubt we could interpret all of this in terms of selfish genes, but I find such behaviour much easier to grasp in terms of the costs and benefits to the cells themselves, not the genes.

There is another sense in which it might be better to see the cell as the selfish unit, rather than its genes, at least in bacteria. This is that genes do not code for cells: they code for the machinery that makes up cells, the proteins and RNA that in turn build everything that is needed. This may seem a trivial distinction, but it is not. All cells have a highly elaborate structure, even bacterial cells, and the more we learn about them, the more we appreciate that cellular function depends on this structure; as we saw in Part 2, cells are emphatically *not* just a bag of enzymes. Intriguingly, there seems to be nothing in the genes that codes for the *structure* of cells. For example, membrane proteins are directed to their particular membranes by means of well-known coding sequences, but nothing stipulates how to create such a membrane from scratch, or determines where it should be built: lipids and proteins are added to existing membranes. Similarly, new mitochondria are always formed from old mitochondria—they cannot be made from scratch. The same goes for other components of the cell like centrioles (the bodies that organize the cytoskeleton).

At the fundamental level of the cell, then, nature *depends* on nurture, and vice versa. In other words, the power of the genes depends absolutely on the pre-existence of the cell itself, while the cell can only be perpetuated through the action of the genes. Accordingly, the genes are *always* passed on within a cell, such as an egg or a bacterium, never as a discrete packet. Viruses, which are a discrete packet, only come alive when they gain access to the machinery of an existing cell. The microbiologist Franklin Harold, whom we met in Part 2, has pondered long and deep about these matters; he put it thus some twenty years ago, and little has changed:

The genome is the sole repository of hereditary information and must ultimately determine form, subject only to limited modulation by the environment. But the inquiry into just how the genome does this leads through another set of Chinese boxes, to show the innermost one empty. . ... Gene products come into a pre-existing organized matrix consisting of previous gene products, and their functional expression is channelled by the places into which they come, and by the signals they receive. Form is not explicitly spelled out in any message but is implicit in its combination with a particular structural context. At the end of the day, only cells make cells.

On balance, then, there are many reasons to see the bacterial cell as the selfish unit of evolution, rather than its genes. Perhaps, as Dawkins said, the invention of sex in the eukaryotes changed all that; but if we wish to understand the deeper currents of evolution we must look to the bacteria, which alone held dominion over the world for two billion years.

These differences in perspective help to explain why microbiologists, such as Lynn Margulis, are among the most prominent critics of the selfish gene. In fact, Margulis has become an outspoken critic of mathematical neo-Darwinism in general, going so far as to dismiss it as being reminiscent of phrenology, that Victorian obsession with cranial shape and criminality, and likely to suffer the same ignominious fate.

While one senses that Margulis is repelled by the concept of the selfish gene, it is also true that bacteria are rather more likely to behave in a civil manner, forming communities that live together in harmony rather than 'eating' each other: the idea of bacteria as merely pathogenic is persistent but false. For Margulis, evolution is largely a bacterial affair, and can be explained in terms of mutual collaborations between consortia of bacteria, including endosymbioses, such as those which founded the eukaryotic cell. These consortia work well in bacteria because predatory behaviour doesn't pay: as we saw in Part 3, the mechanism of respiration across the cell membrane means that large, energy-rich bacterial cells capable of physically engulfing other cells (phagocytosis) are virtually precluded by natural selection. Bacteria are obliged to compete with each other by the speed of their growth, rather than the size of their mouth. Given the reality of food shortage in bacterial ecosystems, bacteria gain more by living from each others' excrement than they do by fighting over the same raw materials. If one bacterium lives by fermenting glucose to form lactic acid, then there is scope for another to live by oxidizing the waste lactic acid to carbon dioxide; and for another to convert the carbon dioxide into methane; and another to oxidize the methane; and so on. Bacteria live by endless recycling, which is best achieved via cooperative networks.

Perhaps it's worth remembering that even cooperative partnerships can only persist if the partners do better within the partnership than without. Whether we measure 'success' by the survival of cells or the survival of their genes, we still see only the survivors—the cells or genes that *did* copy themselves most successfully. Those cells whose altruism is so extreme that they die for another are doomed to disappear without trace, just as many young war heroes fought and died for their country, leaving behind a mourning family but no children of their own. My point is that collaboration is not necessarily altruistic. Even so, a world of mutual collaboration seems a far cry from the conventional idea, expressed by Tennyson, of 'nature, red in tooth and claw'. Collaboration might not be altruistic, but neither is it 'aggressive'—it doesn't make us think of jaws dripping in blood.

This discrepancy is partly responsible for the schism that has opened between Margulis and neo-Darwinists like Dawkins. As we have seen, Dawkins' ideas about selfish genes are equivocal when applied to bacteria (which he does not try to do). For Margulis, however, the whole tapestry of evolution is woven by the collaborations of bacteria, which form not just colonies but the very fabric of individual bodies and minds, responsible even for our consciousness, via the threadlike networks of microtubules in the brain. Indeed, Margulis pictures the entire biosphere as the construct of collaborating bacteria—Gaia, the concept that she pioneered with James Lovelock. In her most recent book, *Acquiring Genomes: A Theory of the Origins of Species*, written with her son Dorion Sagan, Margulis argues that even among plants and animals, new species are formed by means of a bacterial-style merging of genomes, rather than the gradual divergence pictured by Darwin, and accepted by virtually every other biologist. Such a theory of merging genomes might be true in some instances, but in most cases it flies in the face of a century of careful evolutionary analysis. In dismissing neo-Darwinism, Margulis deliberately provokes the majority of mainstream evolutionists.[1] Few have the patience displayed by the late Ernst Mayr, who contributed a wise foreword to the book, in which he commended Margulis's vision of bacterial evolution, while cautioning the reader that her ideas don't apply to the overwhelming majority of multicellular organisms, including all 9000 species of bird, Mayr's own particular field of expertise. The reality of sexual reproduction means that genes must compete for space on the chromosomes; and the rise of predation in the eukaryotes means that nature, at this level, really is red in tooth and claw, however much we may wish it otherwise.

Given their different perspectives, it's ironic that the views of Dawkins and Margulis do not diverge as far as one might think when it comes to the individual. As we have seen, Dawkins wrote of the individual as a colony of collaborating genes, while Margulis thinks of an individual as a colony of collaborating

---

[1] These ideas attract passionate devotees, some of whom will doubtless feel that because Lynn Margulis is a visionary who has proved an entire field wrong before, then she is necessarily right. Another of my heroes, Peter Mitchell, revolutionized biochemistry, but towards the end of his life was proved completely wrong about several aspects of his own theory. Likewise, I fear Margulis is plain wrong to condemn neo-Darwinism as irrelevant.

bacteria, which might be construed as a colony of collaborating bacterial genes. Both see the individual as a fundamentally collaborative entity. Here is Dawkins, for example, in his splendid book *The Ancestor's Tale*: 'My first book, *The Selfish Gene*, could equally have been called *The Cooperative Gene* without a word of the book itself needing to be changed… Selfishness and cooperation are two sides of a Darwinian coin. Each gene promotes its own selfish welfare, by cooperating with other genes in the sexually stirred gene pool which is the gene's environment, to build shared bodies.'

But the ideal of collaboration does not give proper weight to the conflict between the various selfish entities that make up an individual, and in particular to the cells and mitochondria within the cells. While conflict between various selfish entities is entirely in keeping with Dawkins's philosophy, he did not develop the idea in *The Selfish Gene*—these ideas awaited his own later book *The Extended Phenotype*, and in the 1980s and 1990s the important work of Yale biologist Leo Buss and others. Thanks to the exploration of such conflicts and their resolutions, evolutionary biologists now appreciate that colonies of cells (or genes, if you like) do not constitute true individuals, but rather form a looser association, in which individual cells may still act independently. For example, multicellular colonies like sponges often fragment into bits, each of which is able to establish a new colony. Any commonality of purpose is transitory, for the fate of individual cells is not tied to the fate of the multicellular colony.

Such cavalier behaviour is ruthlessly suppressed in true individuals, in whom all selfish interests are subordinated to a common purpose. Various means are employed to guarantee a common purpose, including the early sequestration of a dedicated germ-cell line, so that the great majority of cells in the body (so-called somatic cells) never pass on their own genes directly, and can only participate in the next generation voyeuristically, as it were. Such voyeurism could not possibly work if the individual cells within the body did not share identical genetic bonds—all derive from a single parent cell, the fertilized egg (the zygote), by asexual, or clonal, replication. Although their own genes are not passed on directly to the next generation, the germ-line cells do pass on exact copies of them, which is the next best thing, and ultimately little different. Even so, carrot measures are not enough: stick measures are also needed. The resolution of selfish conflicts between the cells themselves, even though they are genetically identical, can only be achieved by the imposition of a police state reminiscent of Stalinist Russia. Offenders are not prosecuted but eliminated.

The consequence of this draconian system is that natural selection ceases to pick and choose between the independent entities that make up an individual, and begins to operate at a new and higher level, now choosing between the competing individuals themselves. Yet even within apparently robust individuals, we can still detect echoes of dissent, a reminder that the unity of an individual was hard won, and all too easily lost. One such echo of the past is cancer, and it is to this, and the lessons we can learn from it, that we turn in the next chapter.

11

Conflict in the Body

Cancer is a chilling ghost of conflict within an individual. A single cell opts out of the body's centralized control and proliferates like a bacterium. At a molecular level, the sequence of events is one of the most graphic illustrations of natural selection at work. Let's consider briefly what happens.

Cancer is usually, but not always, the result of genetic mutations. A single mutation is rarely enough. Typically, a cell must accumulate eight to ten mutations in rather specific genes before it can transform into a malignant cell, whereupon the transformed cell puts its own interests before those of the body. Genetic mutations tend to accumulate at random as we grow older, but it takes a particular combination to cause cancer: mostly the mutations must be in two sets of genes known as oncogenes and tumour-suppressor genes. Both sets code for proteins that control the normal 'cell cycle'—the way in which cells proliferate or die in response to signals from elsewhere in the body. The products of oncogenes normally signal a cell to divide in response to a particular stimulus (for example, to replace dead cells after an infection) but in cancer they get stuck in the 'on' position. Conversely, the products of tumour suppressor genes normally act as a brake on uncontrolled cell division: they countermand the signals for proliferation, making cells quiescent, or forcing them to commit suicide instead. In cancer, they tend to get stuck in the 'off' position. There are numerous checks and balances in cells, which is why it takes an average of eight to ten particular mutations before a cell transforms into a cancer cell. People with a genetic predisposition to cancer may inherit some of these mutations from their parents, leaving them with a lower threshold of 'new' mutations that must accumulate before the onset of cancer.

Transformed cells no longer respond normally to the body's instructions. As they proliferate, they form into a tumour. Yet there is still a big distinction between a benign growth and a malignant tumour: many other changes still have to take place for a cancer to spread. First of all, to grow larger than a couple of millimetres across, the tumour requires sustenance. Slow absorption of nutrients across the surface of the tumour is no longer enough—the tumour cells need an internal blood supply. To acquire a blood supply, they need to produce the right chemical messengers (or growth factors) in appropriate quantities to stimulate the growth of new blood vessels into the tumour. Further growth requires digestion of the surrounding tissues, giving the tumour space to invade: the cells need to spray potent enzymes that break down the tissue structure. Perhaps the most feared step is the leap to remote sites elsewhere in the body—*metastasis.* The properties required are opposing and specific. Cells must be slippery enough to escape the clutches of the tumour, and yet sticky enough to bind to the walls of blood vessels elsewhere in the body. They must be able to evade the attentions of the immune system during their passage through the blood or lymph system, often by 'sheltering' in a clump of cells that bind together despite their slipperiness. On arrival, the cells must be able to bore their way through the vessel walls, into the safe haven of the tissue behind—but then stop there. And throughout this hazardous solo journey they must retain their ability to proliferate, to found a cancerous outpost in the new continent of a different organ.

Luckily very few cells come equipped with the dialectical qualities needed to cause metastatic cancer. Yet few of us are untouched by cancer, if not ourselves, then our family, relatives, and friends. How, then, do cells acquire all the properties needed? The answer is that cancer cells *evolve* by natural selection. In the course of our lifetime, cells acquire hundreds of mutations, some of which may just happen to affect the oncogenes and tumour-suppressor genes that control the cell cycle. If a single cell is freed from the shackles that normally prohibit its proliferation, it proliferates. Soon it is not a single cell but a colony of cells, all of which are busily picking up new mutations. Many of these mutations are neutral, others are detrimental to the cells, but in time a few will cause a single cell to take the next step down the road to malignancy, then the next, and the next. Each time, the descendents proliferate: what had been a singular mutant becomes a heaving population, until this, too, is displaced by another single cell adapted to the next step. In the space of a few years, even a few months, the body becomes riddled with cancer. The cancer cells have no prospects—they are doomed to die as surely as we are. They simply do what they must, grow and change, a progression dictated by the inexorable blind logic of variation and selection.

What is the unit of selection in cancer, the gene or the cell? As we saw with bacteria, it makes more sense to think of the cells themselves as the selfish unit. The cells do not replicate by sex, but in the manner of bacteria, by asexual replication. The genes may change faster than the phenotype of the cell, which at least for a period retains many aspects of its provenance, including its appearance down the microscope. Even metastatic cancers betray their origins: if we scrutinize a tumour in the lung, it is usually possible to tell whether it is a 'primary' tumour, derived from the lung cells, or a 'secondary' tumour, a metastatic outpost of cells from a distant tissue such as the breast. We know because they still retain some atavistic traits of 'breast' cells, such as hormone production. At the same time, cancer cells are notorious for their genetic instability: chromosomes are lost, or broken, or cobbled together in wild rearrangements. So while the cells retain a semblance of their former appearance, their genes are scrambled out of recognition by mutations and rearrangements. If there is a 'selfish' evolutionary unit, surely it is the cell, which leaps all hurdles in its way until finally killing its master, a course as heavily laden with fate as that of Macbeth.

In cancer, the word 'selfish' rings hollow. There is no sense in which a malignant tumour is making a bid for freedom—it is simply a ghost in the machine, a pointless reversion to an earlier type, which ruled before the evolution of the 'individual'—that of cells doing their own thing. In this sense, cancer gives a dull and empty sense of the sheer meaninglessness of evolution. Cells replicate, and the cells that replicate best leave the most descendants. That's it. It's hard to think of any deeper meaning for cancer: it is mindless mechanics and no more. This contrasts with that other revealing view of evolution in microcosm, bacterial infection, where for all the grinding levers of bacterial replication there is still a strong whiff of purpose: we may find infections abhorrent, but we do accept that bacteria have a point—a life cycle, a future, an 'objective'. They're not doomed, but go on to infect another individual. (Of course, this distinction is in itself imaginary—neither bacteria nor cancer cells have any 'purpose'. However, cancer is a useful example, for it is plain that cancer cells are not equipped to outlive the body, and so the futility of their short-term success in self-replication is transparent.)

If cancer has no meaning, it does at least illustrate the obstacles that must be overcome to forge an individual. If today we still succumb to the lawlessness of cancer, what hope had the first individuals? In those days of looser associations, deserters had the same chance as bacteria of making it alone: desertion was not futile. How did the first individuals quell the strong tendency of their own cells to rebel? It seems they did so in the same way that we do today: they killed the transgressors via a mechanism known as programmed cell death, or apoptosis—they forced the dissident cells to commit suicide. Apoptosis exists even in cells that spend part of the time as independent free-living cells, and part of the time in colonies, begging the question: how and why did apoptosis evolve in single-celled organisms? Why would a potentially independent cell 'agree' to kill itself?

Much of our understanding of apoptosis comes from the study of its role in cancer. The more we learn, the more we appreciate that mitochondria play the title role in apoptosis. And as we trace our way back through evolutionary time, it emerges that apoptosis evolved out of the manipulative campaigns between mitochondria and their host cells in the first eukaryotes—at a time when colonies were far from the rule.

Chronicle of a death foretold

There are two main forms of cell death: the violent, unexpected, swift demise known as necrosis, in which the carpet is left stained with blood and gore; and the silent, premeditated swallow of a cyanide pill, apoptosis, in which all evidence of the deed is spirited away. This is the spooks' end, and it seems appropriate in the Stalinist state of the body. In contrast, death by necrosis incites an unruly inflammatory reaction, equivalent to an incendiary police investigation, in which more bodies turn up, and the ructions take a long time to fade.

Historically, there has been a curious reluctance among biologists to cede full significance to apoptosis. Biology, after all, is the study of life and there is a sense in which death, the absence of life, is beyond the remit of biology. Many of the early observations of programmed cell death were treated as curiosities without wider meaning. One of the earliest observations was in 1842, from the German revolutionary, savant, and materialist philosopher, Karl Vogt, whose politics had forced him to flee to Geneva, and whose dealings with Napoleon III later made him the target of Karl Marx's brilliant polemical pamphlet, *Herr Vogt* (1860). Perhaps it's more edifying to remember Vogt for his careful studies of the metamorphosis of the midwife toad, from the tadpole into the adult. In particular, Vogt used a microscope to follow the fate of the flexible, primitive backbone of the tadpole, the notochord: did the cells of the notochord transform into the spinal column of the adult toad, or did they disappear, making way for new cells which formed the spinal column? The answer turned out to be the latter: the cells of the notochord die off, by apoptosis as we now know, and are replaced by new cells.

Other nineteenth-century observations also concerned metamorphosis. The great German pioneer of evolutionary biology, August Weismann, noted in the 1860s that many cells die off quietly during the transformation of the caterpillar into the moth, but curiously he did not discuss his findings in relation to ageing and death, subjects that later made him famous. Most subsequent descriptions of orderly cell death also came from embryology—the changes that take place during development. Most strikingly, whole populations of neurons (nerve cells) were found to die off in fish and chick embryos. The same applies to us. Neurons disappear in successive waves during embryonic development. In some regions of the brain, more than 80 per cent of the neurons formed during the early phases of development disappear before birth! Cell death allows the brain to be 'wired' with great precision: functional connections are made between specific neurons, enabling the formation of neuronal networks. But the same general theme of sculpting pervades all of embryology. Just as the sculptor chips away at a block of marble to create a work of art, so too the sculpting of the body is achieved by subtraction rather than addition. Our fingers and toes, for example, are formed by orderly cell death between the digits, not by forming discrete extensions to a 'stump'. In web-footed animals such as ducks, some of the cells do not die, so the feet remain webbed.

Despite its importance in embryology, the role of apoptosis in adults was not appreciated until much later. The name itself was coined in 1972 by John Kerr, Andrew Wyllie, and Alastair Currie, all then at Aberdeen University, following the suggestion of James Cormack, professor of Greek at that university. It means 'falling off', and was introduced in the title of their paper in the *British Journal of Cancer*: 'Apoptosis: a basic biological phenomenon with wide-ranging implications in tissue kinetics.' Being Greek, the second 'p' is silent, so the term should be pronounced 'ape-oh-toe-sis'. The word had been used by the ancient Greeks, originally Hippocrates, to mean 'the falling off of the bones', an opaque phrase that referred to the erosion of fractured bone beneath gangrenous bandages; while Galen later extended its meaning to 'the dropping off of scabs'.

In modern times, John Kerr noticed that in rats the size of the liver was not fixed, but changed dynamically with fluctuations in blood flow. If blood flow was impaired to certain lobes of the liver, the affected lobes compensated by becoming gradually smaller over a period of weeks, as cells were lost by apoptosis. Conversely, if blood flow was restored, the corresponding lobes gained in weight, again over weeks, as cells proliferated in response. This balancing act is generally applicable. Every day in the human body, some 10 *billion* cells die and are replaced by new cells. The cells that die do not meet a violent unpremeditated end, but are removed silently and unnoticed by apoptosis, all evidence of their demise *eaten* by neighbouring cells. This means that apoptosis balances cell division in the body. It follows that apoptosis is just as important as cell division in normal physiology.

In their 1972 paper, Kerr, Wyllie, and Currie presented evidence that the form of cell death is basically similar in numerous disparate circumstances—in normal embryonic development as well as teratogenesis (malformation of the embryo); in healthy adult tissues, cancers, and tumour regression; and in the shrinkage of tissues with disuse and ageing. Apoptosis is also critical to immune function: immune cells that react against our own body tissues commit apoptosis during development, enabling the immune system to distinguish between 'self' and 'non-self'. Thereafter, immune cells exert many of their own effects by inducing damaged or infected cells to undergo apoptosis themselves. This kind of screening by immune cells eliminates incipient cancer cells before they get a chance to proliferate.

The sequence of events in apoptosis is precisely choreographed. The cell shrinks and begins to develop bubble-like blebs on its surface. The DNA and proteins in the nucleus (the chromatin) condense in the vicinity of the nuclear membrane. Finally, the cell breaks up into small membrane-wrapped structures called apoptotic bodies, which are taken up by immune cells. Effectively, the cell packages itself into bite-sized chunks, which are then cannibalized without fanfare. Consistent with such a choreography, apoptosis requires a source of energy in the form of ATP—if deprived of ATP, a cell cannot undergo apoptosis. So the process is very different from the swelling and rupture characteristic of necrosis, the violent unpremeditated form of cell death. Also unlike necrosis, there is no aftermath to apoptosis, in particular no inflammation: nothing to mark the passing of a cell but its absence. It is a death foretold, but unremembered.

The executioners

For more than a decade, Andrew Wyllie and a handful of others persevered, evangelists of apoptosis, in the face of indifference in the wider biological community. Wyllie began to convert the unbelievers through his discovery that, in apoptosis, the chromosomes break up into segments that exhibit a characteristic laddering pattern on biochemical analysis. This finding enabled apoptosis to be diagnosed in the lab, overcoming the cynical biochemists' perpetual suspicion of electron-microscope artefact. But the real turning point came in the mid 1980s, when Bob Horvitz, at MIT in Boston, set about identifying the genes responsible for apoptosis in the nematode worm *Caenorhabditis elegans*, research for which he shared the Nobel Prize in 2002. *C. elegans* is a tiny, microscopic worm which offered several big advantages—first, it is transparent, so researchers could actually make out the fate of individual cells down the microscope; second, a small, predictable group of cells, 131 of the 1090 somatic cells (body cells, as opposed to germ-line cells) comprising the nematode, die by apoptosis during embryonic development; and third, the mean lifespan of *C. elegans* is barely 20 days, so its swift development is easily tracked in the lab.

Horvitz and his colleagues discovered several genes that coded for the effectors of cell death in nematodes—the death genes. Their findings were fascinating in their own right, but by far the most unexpected and important discovery was that there were exact equivalents of the death genes in flies, mammals, and even plants. Cancer researchers had already identified some of these genes at the time, but why or how they were involved in cancer was still unknown. The link with nematodes made their function clear, while giving another demonstration of the fundamental unity of life. Not only were the human genes unambiguously related to the nematode genes, but also they could even be genetically engineered to replace the nematode genes in the worms themselves, where they worked equally well! Mutations that disabled any of the death genes prevented the nematodes from losing their 131 cells by apoptosis as usual. The implications for cancer were plain: if the same mutations had a similar effect in people, then incipient cancer cells would likewise fail to commit suicide, and would instead continue to proliferate to form a tumour.

By the early 1990s, researchers realized that a number of oncogenes and tumour suppressor genes, which we discussed earlier as the causes of cancer, did indeed control the fate of the cell through their effect on apoptosis. In other words, cancers arise from cells that have lost the ability to kill themselves by apoptosis, after mutations in the death genes. The death genes are any genes that normally cause a cell to commit apoptosis, and so can potentially include both oncogenes and tumour-suppressor genes, both of which can overrule a cell's commitment to die in the interest of the body as a whole. As Wyllie put it at the time: 'The ticket to cancer comes with a ticket to apoptosis built in; the apoptosis ticket has to be cancelled before reaching cancer.'

The executioners responsible for carrying out the cell death program are proteins known as *caspases* (a rather more evocative name than the biochemists' original 'cysteine-dependent aspartate-specific proteases'). More than a dozen different caspases have now been discovered in animals, 11 of which also operate in people. All work in essentially the same way: they slice up proteins into bits and pieces, some of which are activated in turn, such that they go on to degrade other components of the cell, like DNA. Interestingly, the caspases are not made to order when needed, but are produced continuously, whereupon they wait in an inactive state for the call to arms: they hang poised over the cell like the sword of Damocles, suspended by a thread above the man who would be king. It is a sobering thought that almost all eukaryotic cells contain within themselves, at all times, this silent apparatus of death.

We can be grateful, we who sit beneath the suspended sword, that the thread is strong. Once the caspases have been activated, there is little hope of turning back the clock; but many checks and balances must be triggered before the ancient machinery grinds into operation. These controls are the subject of nearly two decades of intensive research, and the welter of names and acronyms is enough to confuse all but the most dedicated student. The situation is not helped by the retention of historical names for the same gene in different organisms. I am reminded of Celtic music, in which the same tune goes by several names, and the same title refers to several different tunes: an endless stream of lovely variation, but scarcely conducive to a straightforward understanding. Just to give a genetic example, the gene *ced-3* in nematode worms is known as *nedd-2* in mice, *dcp-1* in *Drosophila*, and *ICE*, or interleukin-1 beta-converting enzyme, in humans (as at the time it was known to be involved in the production of the immune messenger, interleukin 1-beta). After discovering its importance in nematode worms, ICE turned out to be the prototype caspase in humans too, and it is now known as *caspase-1*, although it seems to play a lesser role in human apoptosis. Similar caspase enzymes, and related ones called para-caspases and meta-caspases, have been found in fungi, green plants, algae, protozoa, even sponges: they are virtually universal among the eukaryotes, and so presumably their forerunners were present in some of the earliest eukaryotes, perhaps 1.5 to 2 billion years ago.

There is no need for us to get bogged down in the detail. Suffice to say that the regulation of apoptosis is complex, involving a number of steps in which one caspase sets off the next in a cascade, leading finally to the activation of a small army of executioners, which slice up the cell.[1] Virtually all these steps are opposed by other proteins, which are responsible for counteracting the cascade, thereby preventing a false alarm turning into an orgy of death.

Mitochondria, angels of death

This was the state of knowledge a decade ago, in the mid 1990s. None of it has been contradicted. Yet since then, there has been a change of emphasis that amounts to a revolution, overturning the nascent paradigm. The paradigm was that the nucleus is the operations centre of the cell, and controls its fate. In many respects this is of course true, but in the case of apoptosis it is not. Remarkably, cells lacking a nucleus can still commit apoptosis. The radical discovery was that the mitochondria control the fate of the cell: they determine whether a cell shall live or die.

There are two ways in which the apparatus of death is sprung. These originally seemed quite distinct, but more recent work shows that they share some common features. The first mechanism is known as the extrinsic pathway, because death is signalled from the outside, via 'death' receptors on the outer membrane of cell. For example, activated immune cells produce chemical signals (such as tumour necrosis factor) that bind to the death receptors on incipient cancer cells. The death receptors pass on the message, activating the caspases within the cell, to induce apoptosis. While many details clearly needed filling in, the broad outline seemed plain enough. Not a bit of it!

The second route to apoptotic death is called the intrinsic pathway. As the name suggests, the impetus to commit suicide comes from within, usually from

---

[1] The caspase cascade amplifies a signal through the action of enzymes. An enzyme is a catalyst, which acts on a substrate but is unchanged itself, enabling it to act on many substrates. If these substrates are themselves enzymes, activated by the first enzyme, then each step amplifies the response. If the first enzyme activates 100 secondary enzymes, and each of these activates 100 executioners, then we would have an army of executioners 10 000 strong—each of which is also an enzyme that strikes repeatedly. Add in another intermediary step and we have a caspase army a million strong.

cell damage. For example, DNA damage from ultraviolet radiation activates the intrinsic pathway, leading to apoptosis of the cell, without any external signal. Hundreds of triggers have now been discovered that activate the intrinsic pathway of apoptosis—they do not operate through 'death receptors', but damage the cell directly. The sheer variety of these is breathtaking. Many toxins and pollutants can cause apoptosis, as do some chemotherapeutic drugs used for treating cancer. Viruses and bacteria can provoke it directly, most notoriously in the case of AIDS, where the immune cells themselves die. Many physical stresses cause apoptosis, including heat and cold, inflammation, and oxidative stress. And cells may commit apoptosis in waves of death following a heart attack or stroke, or in a transplanted organ. All these diverse triggers independently bring about the same response, the activation of the caspase cascade, and so produce a similar pattern of cell death by apoptosis in each case. Presumably, the signals had somehow to converge on the same 'switch'. All somehow had to convert an inactive form of a caspase enzyme into the active form, a biochemical task that is as specific as the turn of a key in a lock. But what on earth could recognize such a diverse array of signals, calibrate their strength, and then integrate them into a single common pathway by turning the caspase key in its lock?

The first part of the answer was supplied in 1995 by Naoufal Zamzami and his colleagues, in Guido Kroemer's research team at the Centre National de la Recherche Scientifique, in Villejuif, France. Their results were published in two papers in the *Journal of Experimental Medicine*, which came to be among the most widely cited papers in medical research. A number of factors had already suggested that mitochondria might play a role in apoptosis, but Kroemer's team proved that mitochondria are in fact key to the process. In particular, they showed that a loss of the membrane potential across the inner mitochondrial membrane—the proton gradient generated by respiration (see Part 2)—was one main trigger for apoptosis. If the inner membrane depolarized for a period, then the cells *invariably* went on to commit apoptosis. In their second paper, Kroemer's team showed that the process takes place in two steps. The initial membrane depolarization was followed by a burst of oxygen free-radical generation, which seemed to be required for apoptosis to progress to the next stage.

This mitochondrial two-step—membrane depolarization and free-radical release—takes place in response to virtually all intrinsic triggers. In other words, the mitochondria act as both sensors and transducers of a wide variety of cell damage. Transferring apoptotic mitochondria into a normal cell is enough to cause the nucleus to fragment and the cell to die by apoptosis. Conversely, blocking the mitochondrial two-step can delay or even prevent cells from committing apoptosis. But a question remained: how do apoptotic mitochondria communicate with the rest of the cell? In particular, how do they activate the caspase enzymes?

The answer came from Xiaodong Wang's group at Emory University in Atlanta, Georgia, in 1996, and was greeted with 'general stupefaction', as one expert put it. It was *cytochrome c*. We met cytochrome c, if you recall, in Part 2. It is a protein component of the respiratory chain, originally discovered by David Keilin in 1930, and is responsible for shuttling electrons from complex III to complex IV of the chain. It is normally tethered to the outside of the inner mitochondrial membrane, adjacent to the inter-membrane space (see Figure 5, page 77). Wang's group discovered that, in apoptosis, cytochrome c is released from the mitochondria. Once free in the cell, it binds to several other molecules to form a complex (the *apoptosome*), which in turn activates one of the final executioners, caspase 3. Release of cytochrome c from the mitochondria commits the cell inexorably to die—as indeed does just injecting it into a healthy cell. In other words, an integral component of the respiratory chain (which generates the energy needed for the life of the cell) turns out to be an integral component of apoptosis, responsible for the death of the cell. The link between life and death hinges on the subcellular location of a single molecule. Nothing in biology quite compares with this two-faced Janus: life, looking one way, death the other, the difference between the two but a few millionths of a millimetre.

Cytochrome c is not the only protein released from the mitochondria in this way. A number of other proteins are also released, which play a role in apoptosis too—sometimes a more prominent role than cytochrome c. Some of these additional proteins activate caspase enzymes, while others (such as the apoptosis-inducing factor, or AIF) attack other molecules, like DNA, without the involvement of the caspase enzymes. Like so much in biochemistry, the details often seem endlessly involved, but the underlying principles are simple enough: depolarization of the mitochondrial inner membrane and free-radical generation releases cytochrome c and other proteins into the cytosol, which set in motion the enzymes that slice up the cell.

Battle of life and death

If the life or death of the cell depends on the location of cytochrome c and its companions of doom, it's not surprising that medical research has focused on the specific mechanism that releases these molecules from the mitochondria. Again the answer is complicated, but helps clarify the link between the intrinsic and extrinsic pathways of apoptosis. Overlooking a few exceptions, likely to be refinements, these findings place the mitochondria at the centre of both forms of cell death. In almost all cases, the basic apparatus of death is controlled by the mitochondria. When enough mitochondria in a cell spill out their death proteins—probably beyond a threshold point of no return—then the cell inexorably goes on to kill itself.

The release of cytochrome c takes place in two steps, according to recent research from Sten Orrenius and his colleagues at the Karolinska Institutet in Stockholm. In the first step, the protein is mobilized from the membrane itself. Cytochrome c is normally bound loosely to lipids (especially cardiolipin) in the inner mitochondrial membrane, and is released only upon the oxidation of these lipids. This explains the requirement for free radicals in apoptosis: they oxidize the lipids in the inner membrane to release cytochrome c from its shackles. But this is still only half the story. Cytochrome c is mobilized into the inter-membrane space, and it can't escape from the mitochondria altogether until the outer membrane has become more permeable. This is because cytochrome c is a protein, and so is too large to cross the membrane in normal circumstances. If it is to escape the mitochondrion, some sort of pore must breach the outer membrane.

The nature of the pore that opens in the outer mitochondrial membrane has foxed researchers for a decade or more. It seems probable that several distinct mechanisms can operate under different circumstances, giving rise to at least two different types of pore. One mechanism apparently involves metabolic stress to the mitochondria themselves, which leads to excess free-radical generation. The rising stress opens up a pore in the outer membrane, known as the *permeability transition pore*, leading to swelling and rupture of the outer membrane, coupled with the release of proteins.

Another pore, which may be of more general significance, involves a large family of proteins known rather dryly as the *bcl-2* family. The name is now largely irrelevant, and stands for 'B cell lymphoma/leukaemia-2', which refers to the oncogene discovered by cancer researchers in the 1980s. At least 21 related genes have since been discovered, which code for proteins in the family. These proteins fall into two broad groups, which battle among themselves in ways that are complex and still largely obscure. One group protects against apoptosis. They are found in the outer mitochondrial membrane and seem to prevent the formation of pores, thus blocking the release of proteins like cytochrome c into the cytosol. The other group are diametrically opposed: they act to form pores, apparently large enough to allow the escape of proteins from the mitochondria directly. This group thereby promotes apoptosis. They are normally found elsewhere in the cell, and migrate to the mitochondria upon receiving some sort of signal. The final outcome—whether or not the cell commits apoptosis—depends on the numerical balance of the feuding family members in the mitochondrial membrane, and the number of mitochondria embroiled in the battle. If the pro-apoptosis members outnumber their protective cousins in a sufficient number of mitochondria, the pores open, the death proteins are spilled out from the mitochondria, and the cell goes on to kill itself.

The existence of the feuding *bcl-2* family helps explain the links between the two different forms of apoptosis, the intrinsic and extrinsic pathways. Many different signals alter the balance of the feud in the mitochondria, either in favour of or against apoptosis. For example, both the 'death' signals from outside the cell (the extrinsic pathway) and the 'damage' signals from inside the cell (the intrinsic pathway) alter the feuding family balance in favour of apoptosis.[2] Thus the *bcl-2* proteins integrate a diverse array of signals from both outside and inside the cell, and calibrate their strength in the mitochondria. If the balance favours death, pores form in the outer membrane, cytochrome c and other proteins spill out, and the caspase cascade is activated. Thus the final events are the same in most cases.

The centrality of mitochondria to both the main forms of apoptosis raises the possibility that it was ever thus. We have discussed the fact that bacteria and cancer cells act independently in their own interests, and so can be seen as 'units of selection'. At one and the same time, selection can operate at the level of the cell and that of the individual. Mitochondria were once free-living bacteria, and at that time operated independently. Once incorporated into the eukaryotic cell, they presumably retained the power to operate as independent cells, at least for a while: they were independent cells living within a larger organism, and could rebel in the same way as cancer cells (also independent cells living within a larger organism).

If, today, mitochondria bring about the demise of their host cell, might it be that from the very beginning, the mitochondria killed their host cells in their own interests? In other words, the origin of apoptosis was not an altruistic act on behalf of the individual, but a selfish act on behalf of the tenants themselves. If this view is correct, then apoptosis is better seen as murder than suicide. And if so, the reason why single cells should apparently commit suicide is clear: they are sabotaged from within. So is there any evidence that the mitochondria brought with them to the eukaryotic merger the apparatus of death? Indeed there is.

Parasite wars?

We know that the gene for cytochrome c was brought to the eukaryotic merger by the ancestors of the mitochondria, rather than the host cell, and was later

---

[2] Some forms of the extrinsic pathway of apoptosis, mediated by the death receptors, do bypass the mitochondria altogether, but these are likely to be refinements to the original pathway, which probably did involve mitochondria; otherwise it is hard to explain why most forms of the extrinsic pathways do involve mitochondria.

transferred to the nucleus (see Part 3). We know this because the gene sequence has almost identical counterparts in the α-proteobacteria, and is part of the respiratory chain that was their most important contribution to the partnership. Just how important cytochrome c was in the early evolution of apoptosis is less certain. While it seems to play a pivotal role in mammals, and perhaps in plants, it is not needed for apoptosis in fruit flies or nematode worms: certainly it is not a universal player. But it might once have played a central role in apoptosis, before being superseded in a few species, or it might have assumed its decisive role more recently, independently, in plants and mammals. We won't know which is closer to the truth until we know more about how apoptosis works in the most primitive eukaryotes.

But cytochrome c, as we have seen, is only one of quite a number of proteins released from the mitochondria during apoptosis—proteins with strange names, like Smac/DIABLO, Omi/HtrA2, endonuclease G, and the AIF (and in fruit flies, more evocatively, Reaper, Grim, and Sickle). Their names need not concern us, but we should note that some of these proteins can at times play a more important role than cytochrome c itself. Most of them have only been identified since the turn of the millennium, but already, from the prolific genome sequencing projects around the world, we know something of their provenance. The pattern is striking. With the sole exception of AIF (apoptosis-inducing factor), all known apoptotic proteins released from the mitochondria are *bacterial* in origin, and are absent from the archaea. (Recall from Part 1, page 48, that the host cell was almost certainly an archaeon, whereas the mitochondria are bacterial in origin.) The bacterial origin of these proteins means that they were *not* contributed by the host cell, which must have had little in the way of death machinery. Not all of these proteins were necessarily brought to the merger by the mitochondria themselves—a few seem to have gained access to eukaryotic cells more recently, by lateral gene transfers from other bacteria—but it looks as if only AIF came from the archaeal host cell, and even this does not act to kill in archaea.

These mitochondrial proteins are not the only ones to have come from bacteria. If their gene sequences are to be believed, the caspase enzymes, too, almost certainly came from the bacteria, probably by way of the mitochondrial merger. It's worth noting, though, that the bacterial caspases are tame—they slice up proteins but do not cause cell death. More intriguing is the ancestry of the *bcl-2* family. Here, the gene sequences have little in common with either the bacteria or archaea—but the 3-dimensional structure of the proteins does betray a possible link with bacterial proteins, in particular a group of toxins found in infectious bacteria such as diphtheria. Like the pro-apoptosis proteins of the *bcl-2* family, the bacterial toxins form pores in the host cell membranes, sometimes even inducing apoptosis, suggesting a plausible functional link.

Taken together, these findings imply that most of the machinery of death was brought to the eukaryotic merger by the ancestors of the mitochondria. It really does look like murder from within, rather than suicide, a thankless act of ingratitude on the part of the tenant. This idea was developed into a forceful hypothesis by José Frade and Theologos Michaelidis, of the Max-Planck Institute for Psychiatry in Martinsried in Germany, in 1997. Much of the evidence discussed above, accumulated since 1997, seems to lend support to their case.

Frade and Michaelidis drew a parallel between the behaviour of the modern bacterium *Neisseria gonorrhoeae*, the cause of the sexually transmitted disease gonorrhoea, and the possible behaviour of the proto-mitochondria in the early days of the eukaryotic merger. *N. gonorrhoeae* infects the cells of the urethra and cervix, along with white blood cells. Once inside these cells, *N. gonorrhoeae* brings to bear a diabolical cunning. The bacteria produce a pore-forming protein known as PorB (which is similar to the mitochondrial *bcl-2* proteins). The PorB protein is inserted into the host cell membrane, as well as the vacuole membranes that enwrap the bacteria inside the cell. These pores are kept firmly closed by their interaction with the host cell's ATP—again, some of the *bcl-2* proteins behave in a similar way—but when the host's ATP is depleted, the pores open. Opening of the pores triggers the host cell's apoptosis machinery, leading to cell death. The *gonorrhoeae* themselves survive the experience. They take the opportunity to escape from the freshly disintegrated host cell, making use of the neatly packaged remnants of the cell for fuel. Thus, the bacteria subsist within their host cell for as long as the cell is healthy, by monitoring its ability to maintain good stocks of ATP (implying plentiful fuel); but as soon as the host cell begins to run down, outlasting its usefulness, it is summarily executed and the bacteria move on to pastures new. Bastards!

Frade and Michaelidis note that *N. gonorrhoea* is not the only bacterium to make use of such an insidious trick—the deadly bacterial predator *Bdellovibrio*, which we met in Part 1, employs similar tactics when inside other bacteria. It, too, monitors their metabolic health for a period before devouring its prey from within. *Bdellovibrio* has been cited by Lynn Margulis as a possible ancestor of the mitochondria themselves. Another contender, which we discussed in Parts 1 and 3, is *Rickettsia prowazekii*, also a parasite that lives inside other cells. These examples have in common a parasitical relationship with the host. Reconstructing such biochemical archaeology suggests that, in the first eukaryotes, the relationship between the mitochondria and their host cells was parasitical. The proto-mitochondria presumably got into an archaeon and monitored its health for a period, before triggering the death of this host, devouring its packaged remains, and moving on to the next.

If apoptosis grew out of an armed struggle between the cells that were later to be united as the eukaryotic cell, then the eukaryotic merger grew out of a relationship in which the parasite initially killed its host, and moved on to another. This is of course exactly what Lynn Margulis and others propose. The relationship ultimately bequeathed the eukaryotic cell with the machinery of death, which was only later employed for the more 'altruistic' purpose of programmed cell suicide in multicellular organisms. But a parasite war is not the story that we told in Part 1, when we considered the origin of the eukaryotic cell; there we talked about a collaboration between two peaceful prokaryotes which lived side by side, in what amounted to metabolic wedlock. When we considered the evidence, we dismissed the possibility that the relationship between the two cells was parasitic. But now, from a different perspective, there is a challenge to that view. Nothing is certain in this kind of science—it is all about weighting the bits and pieces of evidence that have a bearing on the matter; and this evidence most certainly bears on the matter. So does it overturn our already unstable craft? Should I, horror of horrors, go back and start rewriting Part 1?

12

Foundations of the Individual

The multicellular individual is made up of cells that collaborate together for the greater good. Nonetheless, this collaboration is not a cellular love fest—it is enforced by the death penalty for any cells that try to abscond and return to their ancestral way of life. Occasionally, selfish cells escape detection and evade the death penalty, and when they do the result is cancer. Cancer cells replicate furiously, in their own interests rather than those of the body, undermining the integrity of the body. Ultimately, having evaded death temporarily themselves, they bring about the death of their erstwhile master, and with it their own.

Cancer can persist because it is rare in younger individuals: if the body were to tear itself asunder by internal squabbles before the community of cells had engineered their own reproduction, though the germ-line, then the individual as a whole would fail to pass on its genes, and the selfish genes would be lost from the population. In the early days of multicellular organisms, however, the selfish cells that make up a multicellular body had a far better chance of independent survival—unlike cancer cells they could survive alone, and they had retained the potential to found a new colony of cells. This kind of independence still occurs in sponges and other simple animals today, but their laissez-faire laws of coexistence prevent them from scaling the heights of multicellular complexity. True commitment to the multicellular way of life demands the ultimate sacrifice—death for the greater good. But if cells could survive alone, how was the death penalty imposed upon them in the first place?

Today, the cellular death penalty, known as *apoptosis*, is executed by the mitochondria. The mitochondria integrate signals coming from different sources, and if the balance of signals indicates that the cell is damaged, and so prone to act in its own interests, then they activate the cell's silent machinery of death. Swift and smooth, almost unnoticed, some 10 *billion* cells die by apoptosis every day in the human body, to be replaced by fresh, undamaged cells. The death apparatus consists of a number of proteins that are released from the mitochondria into the cell, which then activate the latent death enzymes, the caspases. These enzymes dismember the cell from within, and package its contents for reuse later by other cells. Nothing goes to waste.

Virtually all the death proteins that are released from the mitochondria, along with the caspase enzymes themselves, were brought to the eukaryotic cell by the bacterial ancestors of the mitochondria, back in the mists of evolutionary time. They still have close analogues among free-living, and especially parasitic, bacteria today. In modern bacteria, many of the death proteins are used for other purposes and are relatively 'tame'—they don't bring about the death of bacteria or anything else. On the other hand, one family of proteins, the bacterial *porins*, are actually targeted at other cells, instruments of war and murder rather than fruitful collaboration. This raises the prospect that it was ever thus: once upon a time, the bacterial ancestors of the mitochondria were parasites, and used proteins like the porins to attack and dismember their host cell from within, feeding on its remains before moving on to another cell.

Whether or not this case is reasonable hinges on the true identity of the bacterial porins. In modern parasites, they are plugged into the membranes of the host cell, and ruthlessly execute its demise as soon as it shows signs of flagging, of being unable to keep pace with the metabolic demands of its parasite. These bacterial porins bear an uncanny physical—but not genetic—resemblance to the mitochondrial porins: the *bcl-2* proteins that activate the cell's machinery of death by forming pores in the mitochondrial membranes. The larger implication is that the eukaryotic cell itself was forged in the crucible of war between an intracellular parasite, which was later tamed and went on to become the mitochondria, and the host cells, which learnt to survive the infection.

This sounds simple enough, but it presents a conundrum. In Part 1, we looked into some of the theories of the origin of the eukaryotic cell, in particular the 'parasite model', in which the mitochondria are derived from a *Rickettsia*-like bacterium, and the hydrogen hypothesis, which contends that the initial alliance grew out of mutual metabolic benefits—both partners lived from the metabolic waste products of the other. There I argued that the evidence at present supports the hydrogen hypothesis, rather than the parasite model. But the parasite story developed above doesn't sit comfortably with the hydrogen hypothesis, which involves a peaceful metabolic union. A parasite may well benefit from killing its host and moving on to another, but a chemical addict can gain little from killing its supplier, and especially if it has no means of finding another one. So either the parasite story undermines the validity of the hydrogen hypothesis, or else it can't be correct itself, despite its apparent explanatory power. I don't see how both of the theories can be right. So which view is correct?

To make a stab at answering this question, we first need to distinguish between the evidence that is known to be true, or at least is not disputed, and ingenious surmise. This is not hard. It's plain that the mitochondria supplied most of the death machinery: they are central to apoptosis today, and almost certainly were instrumental in its evolution. But the link between the *bcl-2* proteins, and the bacterial porins, such as those of *Neisseria gonorrhoeae*, which we discussed in the final pages of the last chapter, must be classed as ingenious surmise. Certainly there are intriguing structural similarities, but this does not constitute proof of an evolutionary relationship.

There are three possible relationships between the *bcl-2* proteins and the bacterial porins, on the basis of what is known today. First, the similarities may result from convergent evolution, such that the mitochondria and *N. gonorrhoeae* both independently invented similar-looking proteins with similar purposes. Nothing in the known gene sequences rules out this possibility—and anyone who doubts the power of convergent evolution at a molecular level should read *Life's Solution* by Simon Conway Morris. If this possibility were true then we wouldn't expect to find any genetic relationship between the *bcl-2* proteins and bacterial porins, since they evolved from different starting places; but we would expect to find structural similarities, as their functional purpose is similar. As there are only a few possible ways to form a large pore in a lipid membrane, these must place functional constraints on possible 3-dimensional structures. If two different cells both need large pores, they are obliged to come up with something similar.

The second possibility is that the mitochondria really did inherit the *bcl-2* proteins from their bacterial ancestors, as suggested by Frade and Michaelidis, and discussed in the previous chapter. This can only be proved by similarities in the gene sequences, which have not been found so far. What's more, such similarities would need to be found among representatives of the α-proteobacteria, the known ancestors of the mitochondria, or else lateral gene transfer at a later stage could not be ruled out. Clearly, if lateral gene transfers took place later on, that would say nothing about the initial relationship between the mitochondria and the host cells. So a more systematic sampling of genes across the α-proteobacteria might lend support to this hypothesis, but in the meantime the structural similarities are suggestive at best.

Finally, it's feasible that *N. gonorrhoea* and other parasitic bacteria acquired their porins from the mitochondria, rather than the other way around. Such transfers of genes from host to parasite are common. If this were the case, we might expect to find similarities in the gene sequences between the mitochondria and the parasites. The lack of such genetic similarities might only be for want of trying—they'll turn up when we have sequenced more genes—or it might be that sequence similarities have simply been lost over time, smothering any evidence of common ancestry. This is not unlikely, as the unceasing evolutionary war waged between parasite and host means that parasite genes are notoriously volatile. Furthermore, the bacterial porins do not themselves bring about the whole of apoptosis—they merely plug into the host's existing death apparatus. Effectively, they bring with them a portable 'on' switch, which triggers the host's own machinery. The behaviour of parasites that cause cell death today is therefore not comparable with the inferred role of the proto-mitochondria, for they would have had to bring the entire death apparatus with them, and implement it in the host cell without killing themselves in the process. (Today, of course, the mitochondria die along with their host cell.)

On the basis of evidence to date, it isn't possible to resolve between these three possibilities. Nonetheless, the picture of parasite wars painted by Frade and Michaelidis does seem at least coherent and plausible. Or does it? There are a few other, rather knotty, problems with the story. First and foremost, mitochondria are no longer independently replicating cells, and in all probability they would have lost their independence soon after their genes began to be transferred to the host cell nucleus. Once a few critical genes were held hostage in the nucleus, then the mitochondria could gain nothing from killing their host, as they could no longer survive independently out in the wild. Their future was tied to that of their host. That's not to say they could gain nothing from *manipulating* their host, but surely they could not gain from actually killing it. In contrast, none of the parasites that we've discussed, even tiny *Rickettsia*, has ever lost its independence. They all maintain complete control over their life cycle and resources. They can get away with murder in a way that mitochondria cannot.

Exactly when mitochondria lost power over their own future is unknown, but it is likely to have happened quite early in the evolution of the eukaryotic cell. Consider, for example, the evolution of the ATP carrier, the membrane pump that exports ATP from the mitochondria (see page 145). For the first time, this enabled eukaryotic cells to extract energy in the form of ATP from mitochondria (which could hardly even be called mitochondria until then). It was a symbolic moment, for the symbionts no longer had control of their own energy resources—they had suffered a loss of sovereignty. For the mitochondria, it marked the transition from a symbiotic relationship to a captive state. We can date the transition reasonably accurately by comparing the sequences of the ATP-carrier gene in the various different groups of eukaryotes. In particular, the fact that the carrier is found in all groups of eukaryotes, including plants, animals, fungi, algae, and protozoa, implies that it evolved before the divergence of these groups, placing it very early in the history of the eukaryotic cell. I need hardly say that this places it well before the evolution of multicellular organisms; from fossil evidence, probably by a few hundred million years.

So we have a gap. It seems very likely that the mitochondria lost their autonomy well before the evolution of true multicellular organisms. During this period, the mitochondria could gain nothing from killing their hosts, for they could not survive independently. Nor could their hosts gain anything from being killed, for they were not yet part of a multicellular organism. Thus the current advantages of apoptosis, the ruthless maintenance of a police state in multicellular organisms, could not apply.

This is a paradox. The persistence of a dedicated machinery of death must have been actively detrimental to both host and mitochondria. We might expect it to have been jettisoned by natural selection, yet we know that it was maintained. We also know that much of the death apparatus was inherited from the mitochondria, rather than from the host (or evolving more recently). And to cap it all, I have argued in favour of the hydrogen hypothesis, which contends that the eukaryotic cell originated in a metabolic union between two peacefully cohabiting cells, neither of which could gain from killing the other. I seem to have argued us into a blind alley, to wit: a collaborative cell brought with it to a peaceful union a fully developed death apparatus, detrimental to both parties, which persisted against all the odds for a few hundred million years before it happened to find a use. Can this crazy scenario be rationalized? Yes, but only if we are prepared to make a concession—the death apparatus did not always cause death. Once upon a time, it caused sex.

Sex and the origin of death

Let's consider the first eukaryotes from the point of view of the peaceful cohabitation proposed by the hydrogen hypothesis. In the introduction to Part 5, we discussed the different levels at which natural selection operates—the level of the individual as a whole, or its constituent cells, or the mitochondria within the cells, or of course the genes themselves. We saw that it is not necessarily helpful, when considering cells that replicate asexually like bacteria, to think about natural selection operating at the level of the genes. Instead, selection works mostly at the level of the individual cells, which in this case are the true replicating units. This background will now prove invaluable to us, for we must consider the interests of the mitochondria and their host cells separately, in the early days of the eukaryotic merger. In those days, both the mitochondria and the host cells could be thought of as separate cells (and we shall see in the next few chapters that in many ways it still helps to consider them in this way).

So what were the private interests of the proto-mitochondria and their host cells? Given their combination of autonomy and uneasy mutual dependency, how could they have acted out their own interests? A compelling answer was put forward in 1999, by one of the most fertile thinkers in evolutionary biochemistry, Neil Blackstone, at Northern Illinois University, along with Douglas Green, one of the pioneers of cytochrome c release in apoptosis, at UCSD, La Jolla.

Like all cells, it is in the interest of mitochondria to proliferate. As soon as their own future has been tied to that of their host, they can gain nothing from killing this host and moving on to another—they could not survive the interim in the wild. There's also a limit to how far the mitochondria can proliferate within a single host cell: a mitochondrial 'cancer' within the host would be detrimental to the cell as a whole, which would perish, along with all of its mitochondria. So the only way that the mitochondria can successfully proliferate is in line with the host cell. Each time the host cell divides, the mitochondrial population must double, to provide a contingent for each daughter cell. Of course, there's nothing the host cell likes better than dividing either, so the interests of host and mitochondria are in common. If they were not, it is quite doubtful that the arrangement could have persisted as a stable relationship for two billion years. It would surely have torn itself asunder early on, and we would not have been here to be any the wiser.

But the interests of the mitochondria and the host cell are not always in common. What might happen if, for some reason, the host cell refused to divide? Clearly neither the host cell nor its mitochondria could then proliferate (well, the mitochondria *could* proliferate, but only to a certain point: it would be detrimental to the host, and so to the mitochondria themselves, if they continued proliferating until they produced a mitochondrial 'cancer' inside the cell). The consequences might differ depending on the reason the host cell refused to divide. The most likely reason is lack of food. In Part 3 we noted that most bacteria spend most of their lives in stasis, despite their enormous capacity to replicate. The same must have applied to the early eukaryotes. If so, there was nothing to do but wait out the lean times, and resume proliferating again as soon as food became available. In this case, the interests of the mitochondria and the host cell are again in common: if the mitochondria pressed the host to divide without sufficient resources, both would perish. Better to devote the remaining resources to bolstering resistance to any physical stress likely to be encountered during the period of deprivation, such as heat, cold, and ultraviolet radiation. Under these conditions, many cells form a resistant spore, which survives the wait in a dormant state before springing back to life in times of plenty.

Another reason that might prevent the host cell from dividing is damage, in particular to the DNA of the cell nucleus. Now the interests of the host and the mitochondria begin to diverge. Let's assume that food is plentiful, but the host cell is nonetheless unable to divide. You can almost picture the trapped mitochondria, faces pressed against the bars, yelling 'Let me out! Unfairly imprisoned!' In the meantime, their neighbouring cells grin and divide away, their mitochondria proliferating happily. What are the trapped mitochondria to do? They don't gain anything from killing their host, as they'd soon be dead themselves. But they *would* gain if the host cell *fused* with another, and recombined its DNA with that of the partner. Recombination of DNA is common in bacteria, and is the very basis of sex in eukaryotes. The fused cell gains a new lease of life—and the mitochondria a new playground.

Why sex evolved in eukaryotes is still fiercely contested, given its twofold cost (see page 191). It seems likely that several different factors contribute. Sex tends to mask damaged DNA, as the damaged gene is likely to be paired with an undamaged copy of the same gene; and the variety generated by recombination probably gives cells a competitive edge over parasites—a theory championed by Bill Hamilton. Recent data imply that neither reason alone is sufficiently strong in all circumstances to account for the evolution of sex; but they don't conflict with each other, and it seems likely that the benefits of sex are many pronged. On the other hand, its origin is a mystery. Bacteria recombine genes, but never *fuse* cells. In contrast, sexual reproduction in most eukaryotes involves the fusion of two cells, then the fusion of their nuclei, and finally the recombination of their genes, an altogether more committing act. What made eukaryotic cells fuse in the first place? Losing the unwieldy cell wall of bacteria no doubt made the physical act of fusion far more practicable, but this still does not account for the actual *urge* to fuse. Cells don't fuse all the time, so there is nothing about the wall-less state in itself that promotes fusion. Might it be that early eukaryotic cells were manipulated by their mitochondria to fuse together? If so, could mitochondrial sabotage explain the origin of sexual fusion? Tom Cavalier-Smith, whom we met in Part 1, has reasoned that cell fusion would have been common in the early eukaryotes: he argues that the form of cell division in sex (meiosis), in which the chromosomal number is first doubled, and then halved, evolved via a few simple steps, as a means of restoring the original number of genes and nuclei after cellular fusions. In this case, mitochondria might have agitated for a fusion that was likely to happen in due course anyway.

The question of whether the mitochondria can manipulate the host cell is a serious one. We know they do today: they cause apoptosis. But might they have done so in the early days of the eukaryotic cell too? Neil Blackstone has suggested an ingenious way in which they could have done, and it explains both the urge to fuse and ultimately the evolution of apoptosis.

Free-radical signal

Think about the respiratory chain. We discussed the leakage of free radicals from the chain in Part 3. Paradoxically, the rate of free-radical leakage does not correspond to the rate of respiration, as one might think intuitively, but rather depends on the availability of electrons (ultimately derived from food) and oxygen. Because these factors vary continuously, free-radical production shifts according to circumstances. Sudden bursts of free-radical production can affect the behaviour of the cell.

If a cell is growing and dividing quickly, and so has a high demand for fuel (and plenty to meet this demand), there is a fast flux of electrons down the respiratory chain to oxygen. In these circumstances, relatively few free radicals leak from the chain. This is because they are more likely to pass down the line of least resistance, from one electron acceptor to the next in the chain, and finally to oxygen. Blackstone describes the chain in these circumstances as a well-insulated wire, through which electricity flows as a current of electrons. So, fast growth with plentiful fuel equates to a low leakage of free radicals.

What about times of starvation? Now there is little fuel, and practically no electrons passing down the respiratory chain. There may be plenty of oxygen around, but no spare electrons to stray off and form free radicals. If we think of the respiratory chains as little electrical wires, then starvation equates to a grid power failure: it's impossible to suffer an electric shock if the mains supply is dead. Free-radical leakage is low because there is no electron flow at all.

But now think about what happens if a cell is damaged, leaving it with plenty of fuel, but no longer able to divide. The mitochondria are trapped in their prison. Because there is no cell division, there is a very low demand for ATP, and the cellular stocks remain high. The rate of electron flow down the chain depends on the rate of consumption of ATP. If ATP consumption is fast, then the electrons flow swiftly to keep up, as if sucked on by a vacuum; but if there is no demand, then the respiratory chain becomes choked up with spare electrons, which have nowhere to go. Now there is plenty of oxygen as well as spare electrons. The rate of free-radical leakage is far higher. The respiratory chain behaves like a badly insulated wire, easily giving an electric shock. So if the host cells are damaged, and don't grow or divide despite plentiful fuel, their mitochondria give them an electric shock from within: a sudden burst of free radicals.[1]

Any burst of free radicals tends to oxidize the lipids in the mitochondrial membranes, and release cytochrome c from its shackles into the intermembrane space; this in turn *completely* blocks electron flow down the chain, as cytochrome c is an integral part of the respiratory chain. Losing cytochrome c from the chain is like clipping a live wire. The earlier part of the chain chokes up with electrons, and continues to leak free radicals, just as the live part of a clipped live wire still gives a shock. But cessation of electron flow eventually dissipates the membrane potential (for proton leak is no longer balanced by proton pumping), and as stress mounts the pores open in the outer mitochon-

---

[1] As we noted in Part 4, there is a way out of high free-radical production when the chain is blocked, and that is to uncouple electron flow from ATP production (see also Part 2, page 92). The proton gradient is dissipated as heat, which reduces free-radical production and may have contributed to the evolution of endothermy.

drial membrane, spewing apoptotic proteins, including cytochrome c, into the rest of the cell. In other words, these circumstances simulate the first steps of apoptosis.

Where does all this leave us? It means the interests of the mitochondria and the host cell are aligned at most times. If both proliferate, all is well and good. The cell is in a reduced (as opposed to oxidized) state, but free-radical leakage is minimal. Conversely, if resources are scarce, then neither can proliferate, and the cell will do best to bolster its resistance, to wait out the lean times ahead. The cell is now in an oxidized state, and again free-radical leakage is minimal. When the host cell is damaged, however, and can't divide despite having plenty of fuel, then the mitochondria signal their displeasure by producing an angry burst of free radicals. This is significant, says Blackstone, for the free-radicals attack the DNA in the cell nucleus (and the presence of cytochrome c in the cytosol would actually promote free-radical formation there). In yeasts and other simple eukaryotes, DNA damage constitutes a signal for sexual recombination. Even more strikingly, in the primitive multicellular alga *Volvox carteri*, a luminously beautiful hollow green ball, a twofold rise in free-radical production activates the sex genes, leading to the formation of new sex cells (gametes). Importantly, this effect can be induced by a blockage of the respiratory chain. So Blackstone's theory can be furnished with some concrete examples. The long and short of it is this. The first few steps of apoptosis in single cells might once have stimulated sex, not death.

First steps to the individual

This view is entirely compatible with the hydrogen hypothesis, for it implies that the cells involved in the initial eukaryotic merger lived peacefully together, but nonetheless retained their own interests. These interests stretched to manipulating the host for sex, in the case of mitochondria, but not to murder, from which neither side could gain. Moreover, such a gently manipulative relationship, in which most interests are aligned at most times, explains why the machinery of apoptosis might have survived in single cells for possibly hundreds of millions of years—sex benefits both the damaged host and the mitochondria, and so would not be penalized by natural selection.

But the question remains: how did sex turn into death? We know that the mitochondria brought most of the death machinery with them, and they certainly use it to kill their hosts by apoptosis today. If we accept that the original purpose of the death machinery was sex, not death, what led to such a portentous change in purpose? When did the drive for sex become punishable by death, and why?

Sex and death are entwined. To an extent, both serve the same purpose.

Consider why yeasts and *Volvox* are driven to recombine their genes when their DNA is damaged: recombination of genes probably enables the damaged copy to be replaced, or masked, by an undamaged copy of the same gene. Similarly, free radicals promote lateral gene transfer in bacteria (the uptake of genes from other cells or the environment). Again, the damaged genes are replaced or masked. What of programmed cell death, then? In multicellular organisms, apoptosis is also a means of repairing damage. Rather than the costly option of fixing a broken cell, apoptosis takes the cost-effective approach of eliminating it from the body, making space for an undamaged replacement—the first steps towards our modern 'throwaway' culture. So sex helps to eliminate damaged genes, while apoptosis eliminates damaged cells. Seen from the point of view of the 'higher' organism, sex repairs damaged cells and apoptosis repairs damaged bodies.

In Blackstone's view, the machinery of apoptosis originally signalled cells to fuse, instigating recombination and repair of damage. At a later stage, in multicellular organisms, the machinery was rededicated to death. In principle, all that was required was the insertion of a new step—the caspase cascade. We noted earlier that the caspase enzymes were inherited from α-proteobacteria (probably by way of the mitochondrial merger), but that they serve a different purpose in bacteria—they slice up some proteins, but do not bring about the death of the cell. In this respect, it's interesting that different groups of eukaryotes appear to have integrated the caspase enzymes into programmed cell death quite independently. Plants, for example, bring about cell death using a group of related proteins known as meta-caspases, whereas mammals use the familiar caspase cascade. Both, however, trigger cell death through the release of cytochrome c and other proteins from the mitochondria. This implies that the death machinery of apoptosis arose independently more than once in the eukaryotes, in response to a common signal (free radicals, and the release of proteins from stressed mitochondria) and a common selection pressure—the need to eliminate damaged cells from a multicellular organism.

If apoptosis is linked with the need to police the multicellular state, rather than a parasite war, and multicellular organisms evolved independently more than once, which they certainly did, then it is not surprising that the detailed execution of apoptosis differs in different groups. On the face of it, it is more surprising that there is so much in common—that somewhat similar machinery was pressed into service more than once. Why was this?

Again Blackstone suggests an answer. He has spent many years studying some of the most primitive animals, such as marine colonial hydroids (colonies of cells that are capable of reproducing sexually or asexually, by fragmentation). He argues that a multicellular colony offers various advantages over individual cells, but as soon as the cells within a colony begin to differentiate—

so that some are obliged to fulfil menial tasks, like paddling (moving the colony around), while others form fruiting bodies that pass on their genes—then a tension must develop. What stops the menial 'slave' cells from revolting?

Although they are all genetically identical (for a period at least), the cells in a colony don't have equal opportunities—a 'caste' system develops in which some cells reap privilege at the expense of others. Blackstone argues that redox gradients are set up by the food and oxygen supply, which varies with currents, other local fluctuations, and the position of cells within the colony (at the surface or buried under other cells). Some cells have plenty of oxygen and food while others are deprived of one or the other, and so find themselves in a different redox state. The differentiation of cells is controlled by their redox state, by way of signals from the mitochondria. We have already noted, for example, that a lack of respiratory electrons, due to starvation, generates a signal for stress resistance.

The urge for independent sex—welling up as a burst of free radicals from the mitochondria—is also a redox signal. In a colony, damaged cells that attempt to have sex with other cells are likely to jeopardize the survival of the colony as a whole—only chaos can ensue. The very signal for sex is a confession of damage to the cell. It is as much as to say that the cell can no longer perform its normal tasks. In somatic (body) cells, there must have been a strong selection pressure to transmute a redox signal for sex into a signal for death. And in time, the selective removal of damaged cells for the greater good paved the way for the evolution of the individual, in whom common purpose is policed by apoptosis. So the cries for freedom of captive mitochondria, which may once have urged for sex in single cells, were met with death in a multicellular body—their own, along with their damaged host cells.

This answer gives a beautiful insight into the vested interests of different cells, and how these can ebb and flow over time. The final outcome may depend on the environment that the cells find themselves in. In the first eukaryotic cells, the host cells and their mitochondria each had their own selfish interests. For the most part, these interests were aligned, but that was not always the case. In particular, if a host cell became genetically damaged, in a way that prevented it from dividing, the mitochondria were effectively imprisoned, for they no longer had the autonomy to survive outside the host. Their only escape was through the act of sexual fusion, for in this way they can be passed on to another cell directly. One signal for sexual fusion in simple single-celled organisms is a burst of free radicals emanating from the mitochondria, so mitochondria can indeed manipulate their host cells in this way.

When the host cells formed into colonies, however, times changed. There are many advantages to living in primitive colonies, without the constituent cells having to give up the possibility of a return to free living. But for this reason, the path from a colony to a genuine multicellular individual is tricky. The fact that all multicellular individuals make use of apoptosis suggests that cells must accept the death penalty if they step out of line. But why did they do so? Perhaps because the damaged cells were betrayed by their own mitochondria. The free-radical signals, welling up from the mitochondria, amount to a confession of damage to the host cell. In a colony, the future of the other cells is jeopardized: the majority gain if the damaged cell is eliminated. So the battleground shifts from the mitochondria and their host cells, to the cells of the colony, and finally to the more familiar setting of competing multicellular individuals.

One question that emerges from this scenario is, how did the colony as a whole reproduce? If any cells in a colony that 'want' to have sex are eliminated, then the colony as a whole is under pressure to find a common, agreed method of reproduction. Today, individuals produce dedicated sex cells from a seques-trated germ-line that is hived off well before birth. How and why such seques-tration got started is a conundrum, but if the punishment for sex was generally death, then it must have been much easier to make a single exception rather than many. Surely this must have been a strong selective pressure to seques-trate a germ-line. Such an executive decision might have had a startling out-come. Once a sequestrated germ-line had been established, then multicellular individuals could only replicate by way of sex. The individual no longer per-sisted from one generation to the next; no more did any of the individual cells, nor even chromosomes. Bodies dissolved and reformed like wisps of cloud, each one fleeting and different. Does this sound at all familiar? I'm repeating myself from the beginning of this Part: these conditions codified the selfish gene. Ironically, the long battles between individual cells that ultimately gave rise to the multicellular individual may in the end have crowned a different victor, who slipped in through the back door: the gene.

Primitive multicellular colonies stand at the gates of sex and death, of selfish cells and selfish genes, and it will be revealing to learn more about their behaviour. It will be revealing, too, to learn more of the mitochondrial signals for sex in single cells. For while sex looks like a good idea from the point of view of mitochondria, the fusion of two cells leads to another conflict—between the two populations of mitochondria derived from the two fusing cells. These populations are not the same, and so can compete among themselves to the detriment of the newly fused host cell. Today, sexual organisms go to extra-ordinary lengths to block the entry of mitochondria from one of the two parents. Indeed, at a cellular level, the inheritance of mitochondria from only one of the two parents is among the defining attributes of gender. Mito-chondria might once have pushed for sex, but they left us everlastingly with two sexes.

PART 6

Battle of the Sexes
Human Pre-History and the Nature of Gender

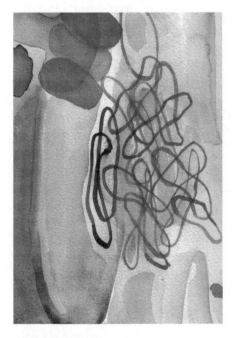

Males have sperm and females have eggs. Both pass on the genes in their nucleus, but under normal circumstances only the egg passes on mitochondria to the next generation—along with their tiny but critical genomes. The maternal inheritance of mitochondrial DNA has been used to trace the ancestry of all human races back to 'Mitochondrial Eve', in Africa 170 000 years ago. Recent data challenge this paradigm, but give a fresh insight into why it is normally the mother who passes on mitochondria. The new findings help explain why it was ever necessary for two sexes to evolve at all.

Mitochondrial DNA—a tiny circular genome in the mitochondria, inherited from the mother

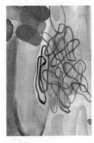

What is the deepest biological difference between the sexes? Most of us, I imagine, would venture the Y chromosome, but this isn't actually the case. The Y chromosome is allegedly pivotal to our sexual development, yet its presence is far from categorical, even for us. About 1 in 60 000 women are known to carry a Y chromosome, giving them the typical masculine chromosome combination of XY, yet they are nonetheless female. One unfortunate example was the Spanish 60-metre hurdles champion Maria Patino, who was publicly humiliated and stripped of her medals in 1985 after failing a mandatory sex test, despite the fact she was plainly not a man, nor a drugs cheat. She was in fact 'androgen resistant'—her body could not respond to the natural presence of testosterone, and so she developed by default as a woman. She had no 'unfair' hormonal or muscular advantage. After a legal battle she was reinstated by the International Amateur Athletics Federation nearly three years later. The IAAF abolished the tests altogether in 1992, and in May 2004, in time for the Athens Olympics, the International Olympic Committee ruled that even transsexuals would be permitted to compete in the Games, as they, too, do not gain a hormonal advantage.

Interestingly, 1 in 500 female Olympic athletes carry a Y chromosome, substantially more than the general population, implying there might be some kind of physical advantage, albeit not hormonal. A relatively high proportion of models and actresses also carry a single Y chromosome. It seems to promote a long, leggy physique, ironically attractive to heterosexual men. Conversely, some men carry two X chromosomes but no Y chromosome; in their case, one X chromosome usually incorporates a tiny fragment of the Y chromosome, bearing a critical sex-determining gene, which stimulates development as a man, but this is not always the case: it's possible to develop as a man without any Y chromosome genes at all. Rather more common (about 1 in 500 male births) is the XXY combination, known as Klinefelter's syndrome. Strangely enough, men with this combination would once have qualified for the women's Olympic Games by the same test that disqualified Maria Patino—the second X chromosome marks them histologically as women, even though they are not. Various other unusual combinations are also possible, some giving rise to hermaphroditism, in which the organs of both sexes are present, for example both ovaries and testes.

The superficiality of the Y chromosome is exposed if we consider sex determination more widely across species. Essentially all mammals share the familiar X/Y chromosome system, but there are some exceptions. As regularly publicized in the media, the Y chromosome is in perpetual decline. With no possibility of recombination between Y chromosomes (men usually only have one copy), it is difficult to correct mutations, as there is no 'clean' copy that can act as a template, so mutations accumulate over many generations, potentially leading to a mutational melt-down. The Y chromosome has duly degenerated completely in some species, such as the Asiatic 'mole voles' *Ellobius tancrei* and *E. lutescens*. In *E. tancrei*, both sexes have unpaired X chromosomes; in *E. lutescens*, the females and males both carry two X chromosomes. Exactly how their sexes are determined remains an enigma, but it is reassuring to know that the decay of the Y chromosome does not inevitably herald the demise of men.

If we venture further afield the X and Y chromosomes soon begin to look parochial. The sex chromosomes of birds, for example, contain a different set of genes to the mammalian chromosomes, implying that they evolved independently; accordingly, they are denoted W and Z. Their inheritance reverses the mammalian pattern: males carry two Z chromosomes making them chromosomally equivalent to female mammals, whereas females carry a single copy of the W and Z chromosomes. Interestingly, in reptiles, the evolutionary ancestors of birds and mammals, both chromosomal systems exist, along with other variations. Most startlingly, sex determination in the cold-blooded reptiles often depends not on sex chromosomes at all, but on the temperature at which the eggs are incubated. In alligators, for example, males are produced from eggs incubated above about 34°C, and females from eggs cooler than about 30°C; if the temperature is intermediate, a mixture is produced. This relationship is reversed in other reptiles; in sea turtles, the females develop from eggs incubated at higher temperatures.

Even reptiles fail to exhaust the cornucopia of sex determinators. In *Hymenoptera*, such as ants, wasps, and bees, the males often develop from an unfertilized egg, whereas the females develop from fertilized eggs. So if a queen bee mates with a drone, the daughters share three quarters of their genes, rather than half, as in the X/Y or W/Z systems. Such genetic similarities might have favoured selection at the level of the colony over that of individuals, facilitating the evolution of eusocial structures (in which reproduction is carried out by a specialized caste in a colony of non-reproductive individuals).

In some crustaceans, sex is not fixed but plastic: individuals can undergo a sex-change. Perhaps the most peculiar example is furnished by the diverse range of arthropods that become infected with the reproductive bacteria *Wolbachia*, which converts males into females, thus ensuring its own transmission in the egg (they are not passed on in sperm). In other words, sex is determined by infection. Other examples of sexual plasticity are unrelated to infection. For example, many tropical fish change sex, notably the colourful teleosts (the most common type of bony fish) that dwell in coral reefs—a source of confusion that could have added a whole new dimension to *Finding Nemo*. In fact, most reef fish switch sex at some point in their lives, and the few shrinking violets that don't are contemptuously labelled gonochoristic. The rest are ardently transsexual: males change to females and vice versa; some change sex in both directions, and others manage to be both sexes at the same time (hermaphrodites).

If any order emerges from this sexual cacophony, it's certainly not the Y chromosome. From an evolutionary point of view, sex seems as accidental and shifting as a kaleidoscope. One of the few enduring pillars is the occurrence of two sexes. With the exception of some fungi (which we'll come to later) there are few unequivocal examples of more than two sexes. What is rather more curious, though, is the need for any sexes at all. The trouble is that having two sexes halves the number of possible mates. This begs the question what's wrong with having only one sex, which amounts to having no sexes at all? That would give everyone twice the choice of partners, and indeed would spirit away the distinction between homosexuals and heterosexuals; couldn't everyone be happy? Unfortunately not. In Part 6, we shall see that, for better or worse, we are generally doomed to two sexes. The culprits, need I say it, are mitochondria.

13

The Asymmetry of Sex

There are two fundamental aspects to sex: the first is the need for a mate at all, and the second is the need for specialized mating types, which is to say, for having two sexes rather than just any old partner. We touched on the need to mate in Part 5. Sex is often said to be the ultimate existential absurdity, as there is a twofold cost to overcome—two partners are required to produce one offspring—while in clonal, or parthenogenic, reproduction (in which an organism produces an exact replica of itself), only one parent is required to produce two identical copies. Radical feminists and evolutionists agree that males are a serious cost to society.

Most evolutionists believe that the advantage of sex lies in the recombination of DNA from distinct sources, which may help to eliminate broken genes and to foster variety, keeping one step ahead of inventive parasites, or rapid changes in environmental conditions (although any of this has yet to be proved by experiment). Of course it takes two to recombine, hence the need for at least two parents; but even if we accept the need to recombine genes, and so to mate, why can't we be free to mate with anyone? Why do we need specialized sexes, rather than all being the same sex, or, given the mechanical constraints of fertilization, hermaphrodites, uniting both sexual functions in the same body?

A quick look at the hermaphrodite lifestyle answers this question: they don't have it easy, by any means. The misogynist German philosopher Arthur Schopenhauer once asked why men seemed to get along with each other quite amicably, whereas women were rather bitchy. His answer was that all women were occupied in the same profession—the business, presumably, of winning men—while men had their own professions and so didn't need to compete so ruthlessly with each other. I hasten to say I couldn't disagree more, but his words do help to explain why so few species are hermaphrodites (plants excepted)—they must all compete with each other, using the same tools of the trade.

Just how awkward this can be is illustrated by the marine flatworm *Pseudobiceros bedfordi*, which engages in a sperm battle when mating. Each is equipped with two penises, with which they fence, attempting to smear sperm onto the other without being fertilized themselves. The ejaculate burns a hole in the skin of the recipient, which is sometimes cavernous enough to cause the loser to tear in half. The problem is that the flatworms all want to be male. The female, almost by definition, invests more of her resources in the offspring, which means that individuals pass on more of their genes if they succeed in fertilizing others, while avoiding being fertilized themselves. This equates to spraying sperm around liberally without becoming pregnant. Penis envy is more than mere psychology. According to the Belgian evolutionary biologist Nico Michiels, the male mating strategy—of spraying sperm—is often adopted by the entire species, leading to such bizarre mating conflicts as penis fencing flatworms. Having two specialized sexes offers a way out of this trap. Females and males have their own ideas about when to mate, and with whom; males tend to be keener, females choosier. The outcome is an evolutionary arms race in which each sex exerts an influence over the adaptations of the other, countering some of the more preposterous mating strategies. As a rule of thumb, the hermaphrodite lifestyle works well if the prospects of finding a mate are slim, for example in low-density or immobile populations (explaining why many plants are hermaphrodites), while separate sexes develop in species with higher population densities or greater mobility.

This is all very well, but it conceals a deeper mystery: the origin of the asymmetry between the male and female roles. I mentioned that females 'almost by definition' invest more of their resources in the offspring. To some this may sound a rather chauvinist remark, implying the father is somehow free to walk away. I do not mean it in that sense. In many sexually reproducing creatures, there is little difference in the input of parental care. Amphibians and fish, for example, produce eggs that are fertilized outside the body, which often develop without any further parental input; and in some crustaceans, only the fathers guard the young. In sea horses, the father nurses the fertilized eggs in his brood pouch, effectively undergoing pregnancy, and giving birth to as many as 150 offspring. Nonetheless there is still a basic inequality of input that is evident at the level of the sex cells, or gametes, themselves—the difference between the sperm and the eggs. Sperm are tiny and disposable. Men, and males in general, produce them by the bucket-load. In contrast women, and females in general, produce far fewer, much larger eggs. Unlike the slippery distinction between the sexes on the basis of their sex chromosomes, this distinction is definitive. Females produce large, immobile eggs, whereas males produce small, motile sperm.

What is the basis of this asymmetry? Various explanations have been put forward. One of most convincing is the destabilizing tug between quality and quantity—between a small number of large gametes, and a large number of small gametes. This is because the fertilized egg provides not just the genes, but also all the nutrients and cytoplasm (including all the mitochondria) required for the new organism to grow. Inevitably there is a tension between the needs of the offspring and the parents. For a good start in life, the offspring 'wants' to be lavishly supplied with nutrients and cytoplasm, whereas the parents 'want' to sacrifice as little as possible, and to fertilize as much as possible. Parental sacrifice is all the more costly if the parents are microscopically small, as was the case early in the evolution of sex, more than a billion years ago.

If the success of a fertilized egg depends at least in part on being properly furnished with resources, one might naively imagine that natural selection would favour an equal input from both parents, minimizing the cost to the parents, while maximizing the benefit to offspring. By this measure, sperm add next to nothing, beyond their genes, to the next generation, and so 'ought' to be selected against. In fact, they behave like parasites, giving nothing and gaining everything. While parasitic behaviour might plausibly develop in many cases, why are sperm *always* parasitic? In the case of amphibians and fish, whose eggs are free-floating, the answer might be that millions of tiny sperm can fertilize more eggs, by virtue of their 'blanket' coverage. It seems peculiar, though, that the sperm and eggs have retained their extreme size discrepancy even when fertilization is internal. Now, millions of tiny sperm are targeted at one or two eggs closeted in a fallopian tube, rather than thousands of eggs dispersed in the ocean. Is it just too late to change, or not worth bothering; or is there a more fundamental reason for an extreme size discrepancy? There's a strong argument to say there is.

Uniparental inheritance

The search for a fundamental difference between the sexes takes us back to primitive eukaryotic organisms, such as algae and fungi, some of which have two sexes despite there being no obvious distinctions between their gametes (sex cells). They are said to be *isogamous*, meaning that their sex cells are equally sized. In fact, the two sexes appear to be identical in every way. Because they are basically the same, it makes more sense to refer to them as mating types rather than sexes. But the very lack of differences between the two mating types highlights the fact that there are still two of them. Individuals are restricted to mating with only half the population. As the pioneers of this field, Laurence Hurst and William Hamilton, pointed out, if finding a mate presents a problem then halving the population size ought to be a serious constraint. Imagine that a mutant mating type appeared in the population, which was able to mate with both the existing mating types. This third type ought to spread swiftly, as it has twice the choice of mates. Any subsequent mutants that could mate with all three types would have a similar advantage. The number of mating types should therefore tend towards infinity; and indeed the widespread 'split gill' mushroom *Schizophyllum commune* has 28 000. Short of having no sexes at all (all are the same sex) then it makes sense to have as many as possible. Two is the worst of all possible worlds.

So why do many isogamous species still have two mating types? If there really is a deep asymmetry between the sexes, a grain of inequality from which all other inequalities grow, then the algae and fungi are the place to look.

The answer betrays a fundamental intolerance that makes our own battle of the sexes look like a love-in. Take the primitive alga *Vulva*, for example, also known as the sea lettuce. It is a multicellular alga, which forms into sheets of cells only two cells thick but up to a metre long, giving the appearance of leaves. Sea lettuces produce identical gametes, or isogametes, which contain both chloroplasts and mitochondria. The two gametes, and their nuclei, fuse together in a perfectly normal fashion, but after cell fusion the organelles attack each other with savage ferocity. Within a couple of hours of fusion, the chloroplasts and mitochondria deriving from one of the gametes have been pulped to a swollen mass, and soon afterwards disintegrate altogether.

This is an extreme example of a general trend. The common denominator is an intolerance of the organelles from one of the two parents, but the method of extermination varies widely. Perhaps the most illuminating example is the single-celled alga *Chlamydomonas rheinhardtii*, which at first sight appears to buck the trend. Instead of destroying half its chloroplasts in a wanton display of violence, the chloroplasts fuse together peaceably. But biochemical scrutiny shows this alga is no more tolerant than its cousins; rather, it is more refined in its intolerance, like a cultivated Nazi. To use the correct, rather chilling euphemism, *Chlamydomonas* practices 'selective silencing': it eliminates the DNA in the organelles rather than the whole organelles, leaving the infrastructure intact. The organelle DNA from each parent attacks the other with lethal DNA-digesting enzymes. According to some reports, 95 per cent of all the organelle DNA is dissolved, but the speed of destruction is slightly faster on one side than the other. The surviving DNA, by definition, derives from the 'maternal' parent.

The upshot is that nuclear fusion and recombination are fine, but the organelles—the chloroplasts and mitochondria—are almost invariably inherited from only one parent. The problem is not with the organelles, but with their DNA. There's something about this DNA that fate abhors. Two cells fuse, but only one of them passes on the organelle DNA.

Here lies the deepest difference between the sexes: the female sex passes on organelles, the male sex does not. The result is *uniparental inheritance*, which means that organelles, such as mitochondria, are normally inherited only down the maternal line, like Judaism. The realization that mitochondria are inherited only from the mother is not age-old knowledge: it was first reported in 1974, in horse-donkey hybrids, by the geneticist and jazz pianist Clyde Hutchison III and his colleagues at the University of North Carolina.

Is this really the deepest difference between sexes? The best place to look for a reality check is at any apparent exceptions to the rule. We already noted, for example, that the mushroom *S. commune* has 28 000 mating types. These are encoded by two 'incompatibility' genes on different chromosomes, each of which comes in many possible versions (alleles). An individual inherits one out of more than 300 possible alleles on one chromosome, and one out of more than 90 on the other, giving a total of 28 000 possible combinations. If two cells share the same allele on either chromosome, they cannot mate. This is likely to be the case among siblings, which encourages out-breeding. However, if the gametes have different alleles at both loci, they are free to mate, and this allows them to mate with more than 99 per cent of the population, rather than a feeble 50 per cent like the rest of us.

But with all these sexes, how on earth do the fungi keep track of their organelles? Can they, too, ensure that organelles are inherited from only one of two parents? If so, with 28 000 sexes, how do they know which is the 'mother'? In fact they solve the problem by engaging in a fastidious form of sex, a loveless fungal missionary position, and are zealous never to mix bodily fluids. Sex, for *S. commune*, is all about getting two nuclei into the same cell, and the cytoplasm never conjoins in blissful union: cell fusion does not take place. In other words, these fungi get around the sex problem altogether by evading the issue. While it can be said that they have 28 000 different sexes, it is better to say they don't have any at all: they have incompatibility types instead.

Intriguingly, incompatibility types can coexist with sexes in the same individual, implying that these adaptations really do serve different functions. The best examples come from the flowering plants, or angiosperms, many of which, as we have seen, are hermaphrodites (the individuals are both male and female). In principle, this means that plants can fertilize themselves or their closest relations—and in practice, given the difficulties of dispersal faced by the sessile plants, this would be the most likely scenario. The trouble is that local fertilization favours in-breeding, thereby losing the benefits of sex altogether. Many angiosperms get around the problem by having incompatibility types as well as two sexes, ensuring that out-breeding takes place.

In principle, it's possible to have more than two sexes, while maintaining uniparental inheritance. There are examples of this among primitive eukaryotes, notably the slime moulds, which fuse cells together into a matrix with numerous nuclei sharing the same vast cell. Slime moulds look similar to fungi, overgrowing woody mulch or grass as an amorphous mass; some bright yellow moulds have been likened to dog vomit. From our point of view, the most important aspect is that some of them have more than two sexes, even though the whole gametes fuse together, not just the nuclei. The best known example is *Physarum polycephalum*, which has at least 13 sexes, encoded by different alleles of a gene known as *matA*. While looking the same, however, these sexes are not equal—their mitochondrial DNA is ranked in a pecking order. Upon fusion of the gametes, the mitochondrial DNA of the more dominant strain persists, while that of the subordinate strain is digested, and disappears completely within a couple of hours; the vacant sheaths are eliminated within three days of fusion. So uniparental inheritance is preserved despite the occurrence of multiple sexes. Presumably there is a limit to how high a pecking order can rise; it's hard to imagine a hierarchy accommodating all 28 000 sexes of *S. commune*, for example. And in practice, more than two sexes is rare.

To draw a general conclusion, we can say that the act of *sex* involves nuclear fusion (and out-breeding can be enforced by having incompatibility types), but proper *sexes* can only be distinguished when cytoplasm is shared. In other words, *sexes* develop when the cells as well as their nuclei fuse. Then the female passes on some of her organelles, and the male must accept the untimely demise of all of his. Even when there are multiple sexes, uniparental inheritance of the mitochondria is the rule.

Selfish competition

Why is uniparental inheritance so important? And why are multiple sexes so uncommon, given that they expand the mating opportunities and are technically feasible? The most widely accepted reason was developed as a forceful hypothesis by Leda Cosmides and John Tooby, at Harvard, in 1981. They argued that mixing the cytoplasm from two different cells creates an opportunity for conflict between different cytoplasmic genomes. These include both mitochondrial and chloroplast genomes, but also any other cytoplasmic 'passengers' such as viruses, bacteria, endosymbionts, and so on. If these passengers are genetically identical, there can be no competition between them; but as soon as they begin to differ, there is scope for competition to gain entrance to the gametes.

Consider, for example, two different populations of mitochondria, one of which replicates faster than the other. If one population becomes more numerous, it gains preferential entry to the gametes. The other population will be eliminated unless it speeds up its own replication, and to do so almost certainly means that it will fail to do its proper job, generating energy, as effectively. This is because the easiest way to speed up replication is to jettison 'unnecessary' genes, as we saw in Part 3; and the genes that are unnecessary for mitochondrial replication are of course exactly the genes that the cell as a whole needs for energy production. So competition between mitochondrial genomes leads to an evolutionary arms race, in which selfish interests take precedence over the interests of the host cell.

The host cell inevitably suffers from competition between mitochondrial genomes, and this in turn generates a strong selective pressure on the genes in the nucleus to ensure that all the mitochondria are identical, thereby preventing such conflict. This can be achieved by the 'selective silencing' of one population, as in *Chlamydomonas*, but in general it is safest to preclude their entrance altogether; this simultaneously precludes competition between other cytoplasmic elements, such as bacteria and viruses. Thus, in this selfish theory, the reason that two sexes develop is because this is the most effective means of preventing conflict between selfish cytoplasmic genomes.

The male mitochondria don't take their purging lying down. Any attempt to exclude them is met with stiff resistance. The angiosperms attest eloquently to the reality of selfish mitochondrial behaviour. In these hermaphrodite flowering plants the mitochondria strive to avoid being caged in the male part of the plant, a dead end for them because they are not passed on in pollen. They avoid ending up in pollen by sterilizing the male sex organs, usually by bringing about the abortion of pollen development. Not surprisingly, this is an important trait in agriculture, discussed at length by Darwin himself, and known rather forbiddingly as *male cytoplasmic sterility*. By sterilizing the male sex organs, the mitochondria convert a hermaphrodite into a female, thereby helping safeguard their own transmission. However, because this upsets the sex balance of the population as a whole, which now comprises females and hermaphrodites, various nuclear genes that counteract the selfish mitochondrial actions have been selected over evolution, restoring full fertility. The battle is still being waged. A trail of selfish mitochondrial mutants and nuclear suppressor genes shows that female conversion took place repeatedly, only to be suppressed each time. In Europe today, 7.5 per cent of angiosperm species are gynodioecious, to use the term Darwin coined—their population comprises both females and hermaphrodites.

Hermaphrodites are particularly vulnerable to male sterilization, because the female organs leave open the possibility of mitochondrial transmission in the same individual. But even when the male and female sex organs are housed in separate individuals, there are indications that mitochondria attempt to distort the sex balance by harming males. Some diseases, notably Leigh's hereditary optic neuropathy, are caused by mutations in the mitochondrial DNA, and are more prevalent in men than women. This situation is similar to the action of *Wolbachia* in arthropods, which we noted earlier. In crustaceans, infection with *Wolbachia* converts males into females, but in many insects the effect is even more drastic: males are simply killed. The 'objective' of the bacteria, which are passed on from one generation to the next only in the egg, is to convert the entire population into females, thereby improving their own chances of transmission. Mitochondria, too, can help safeguard their own transmission in the egg by eliminating males. Unlike *Wolbachia*, however, their success seems to be very limited. Presumably, this is because there has been a stronger counter-selection against selfish mitochondria. Fully functional mitochondria are essential to our survival and health, and selfish mutants are less likely to be effective at respiring. They are therefore likely to be heavily selected against. *Wolbachia*, in contrast, distorts the sex ratio but doesn't necessarily cause much damage otherwise; accordingly, there is a lower selective pressure against it.

All these various attempts to subvert the sex ratio come about because mitochondria, along with other cytoplasmic elements such as chloroplasts and *Wolbachia*, are passed on only in the egg. The pressure to bend the rules has almost certainly polarized the existing differences between the sperm and eggs even further. For example, the pressure exerted by selfish mitochondria probably contributed to the extreme size difference between sperm and eggs. The simplest way to tip the scales against the selfish mitochondria is to stack the odds against them. There are some 100 000 mitochondria in human egg cells, but fewer than 100 in sperm. If the male mitochondria get into the egg at all (they do in many species, including ourselves), they are simply diluted out. But even dilution is not enough. Numerous tricks have evolved to exclude male mitochondria from the fertilized egg altogether, or to ensure the permanent silencing of the few that do get in. In mice and humans, for example, the male mitochondria are tagged with a protein called ubiquitin, which marks them up for destruction in the egg. In most cases, the male mitochondria are degraded within a few days of entry to the egg. In other species the male mitochondria are excluded from the egg altogether, or even from the sperm, as in crayfish and some plants.

Perhaps the most bizarre method of excluding the male mitochondria is found in the giant sperm of some species of fruit fly (*Drosophila*), which can be more than ten times longer than the total male body length when uncoiled. The testes required to produce such mammoth sperm comprise more than 10 per cent of the total adult body mass, and retard male development markedly. Their evolutionary purpose is unknown. Such extraordinary sperm add far more cytoplasm to the egg than normal. What's more, the sperm tail persists in the egg, raising the question of its fate. According to Scott Pitnick and Timothy Karr, at Syracuse University, New York, and the University of Chicago, respectively, the mitochondria fuse together during sperm development to form two enormous mitochondria, which extend the entire length of the tail. These two vast mitochondria fill 50 to 90 per cent of the total cell volume. They are not digested in the egg, but rather are sequestered throughout embryonic development, ultimately in the midgut. The sperm tail can still be observed in the midgut of the larvae after hatching, and is defaecated out soon after birth, conforming to the spirit of uniparental inheritance, if in an oddly convoluted manner.

The fact that there are so many radically different methods of excluding the male mitochondria implies that uniparental inheritance evolved repeatedly, in response to similar selection pressures. This in turn suggests that uniparental inheritance was also lost repeatedly, and later regained by way of whatever trick was most readily pressed into service at the time. I suspect this means that losing uniparental inheritance was weakening but rarely fatal, and indeed there are some examples of mitochondrial mixing, or *heteroplasmy*, notably among the fungi and angiosperms. For example, in one large study of 295 angiosperm species, nearly 20 per cent of all the species examined showed some degree of bi-parental inheritance. Interestingly, bats, too, are often heteroplasmic. Bats are long-lived, vigorously active mammals, so it is curious that they are not compromised by heteroplasmy. Little is known about the circumstances or selection pressures involved, but there is a hint that some sort of selection for the fittest mitochondria might take place in the flight muscles themselves.

We have brought mitochondrial heteroplasmy upon ourselves in some assisted reproductive technologies, especially ooplasmic transfer. This technique involves the injection of cytoplasm, along with its mitochondria, from a healthy donor egg into the egg cell of an infertile woman, thereby mixing mitochondria from two different women. We touched on this technique in the Introduction, for it found fame in a newspaper under the headline 'Babies born with two mothers and one father.' More than thirty apparently healthy babies have been born by this method, despite the pungent criticism that it 'may be akin to trying to improve a bottle of spoiled milk by adding a cup of fresh.' The profound disquiet felt about mixing two mitochondrial populations, which nature strives so hard to avoid, combined with the suspiciously high rate of developmental abnormalities leading to miscarriage, has led to the technique being placed on hold in the United States. Even so, to an open-minded sceptic, perhaps the most surprising finding is that it works at all. Certainly heteroplasmy is worrying, probably weakening, but not out of the question.

If, as we have seen, the deepest distinction between the sexes relates to restricting the germ-line passage of mitochondria, then the barrier between the sexes seems curiously shaky. The language one tends to read in journals or books speaks of outright conflict, along the lines of: 'organelles from more than one parent are not tolerated in the offspring.' In more mundane reality, the condition that forces nature to differentiate two sexes, obliging us to mate with only half the population, is constantly collapsing and reforming. Mitochondrial heteroplasmy seems to be tolerated with surprisingly few detrimental effects in many cases—there is little sign of conflict. So while the evidence suggests that mitochondria really are central to the evolution of two sexes, genomic conflict may not be all there is to it. Recent research suggests that there are other, more subtle, but probably more pervasive and fundamental reasons, too.

The area of research forcing this new thinking, ironically, is another field altogether: the study of human prehistory and population movements, by tracking human mitochondrial genes. Some of the most arresting insights into prehistory, such as our relationship with the Neanderthals, have derived from such studies of mitochondrial DNA. All these studies rest on the assumption that the inheritance of mitochondrial DNA is strictly maternal, that any mixing is simply not tolerated. In this hotbed of research, controversial data have recently raised questions about the validity of this assumption in our own case. But if some of the once iron-cast conclusions now look a bit more rickety, they do give a new insight not just into the origin of the sexes, but also into previously unexplained aspects of infertility. In the next two chapters, we'll find out why.

14

What Human Pre-History Says About the Sexes

In 1987, Rebecca Cann, Mark Stoneking, and Allan Wilson, at Berkeley, published a celebrated paper in *Nature*, which (although it built on earlier work) was to revolutionize our understanding of our own past. Instead of looking to the fossil record, or to genes in the nucleus, they studied the mitochondrial DNA of 147 living people, drawn from five geographical populations. They concluded that the samples were closely related, and ultimately inherited from a single woman, who lived in Africa about 200 000 years ago. She became known as 'African Eve', or 'Mitochondrial Eve', and to the best of our knowledge everyone on earth today descends from her.

The radical nature of this conclusion needs to be placed in perspective. There has long been an unresolved controversy between two warring tribes of palaeoanthropologists—those who believe that modern man issued from Africa in the fairly recent past, displacing earlier groups of migrants, like the Neanderthals and *Homo erectus*; and those who believe that humans have been present in Asia as well as Africa for at least a million years. If this latter view is correct, then the evolutionary transition from archaic to anatomically modern humans must have happened in parallel in different parts of the old world.

These two views carry a potent political charge. If all modern humans came from Africa less than 200 000 years ago, then we are all the same under the skin. We have barely had time, in an evolutionary sense, to diverge, but we can perhaps be held responsible for the extinction of our closest relatives, such as the Neanderthals. This theory is known as the 'Out of Africa' hypothesis. On the other hand, if the human races evolved in parallel then the differences between us are not skin deep, and our unique racial and cultural identities are firmly grounded in biology, challenging our ideals of equality. Both scenarios could have been offset by interbreeding, to an unknown degree. The dilemma is exemplified by the fate of the Neanderthals. Were they a separate subspecies driven to extinction, or did they interbreed with anatomically modern Cro-Magnons, who arrived in Europe around 40 000 years ago? Bluntly, are we guilty of genocide or gratuitous sex? Today we seem distressingly capable of both, sometimes simultaneously.

The patchy fossil record has so far proved inconclusive, not least because it is extremely difficult to tell from a few scattered fossils of widely differing ages whether one population gives rise to another in the same place, or falls extinct, or is displaced by another from a different geographical region, or indeed whether two populations interbred. Numerous fossil finds over the last century—a train of missing links—have demonstrated the possible outlines of human evolution from ape-like ancestors, to all but the most unbelieving of Creationists. Brain size, for example, more than tripled in successive hominid fossils over the last four million years. But the actual line of evolution from Australopithecines, like Lucy, around three million years ago, through *Homo erectus*, and finally to *Homo sapiens*, is fraught with unresolved issues. How can we tell if a fossil discovery represents our own ancestors, or is simply a parallel species, now fallen extinct? Was Lucy really our direct ancestor, or just an extinct, upright, knuckle-dragging ape? All we can say for sure is that there are plenty of skeletons in the closet that exhibit morphologies intermediate between the apes and humans, even if it is difficult to assign their place in our own ancestry. Charting prehistory from ancient skeletal morphology alone is at best an uncertain endeavour.

In terms of our more recent ancestry, the fossil record is also dumb. Did we interbreed with the Neanderthals? If so, we might hope one day to discover a hybrid skeleton, displaying a mosaic of features intermediate between the robust Neanderthal Man and the gracile *Homo sapiens*. Occasionally such claims are made, but rarely convince the field. Here is the kindly Ian Tattersall, commenting on one such case: 'the analysis . . . is a brave and imaginative interpretation, but it is unlikely that a majority of palaeoanthropologists will consider the case proven.'

One of the biggest problems of palaeoanthropology is its heavy reliance on morphology. This is inevitable, as little else remains. Isolating DNA would be a big help, but in most cases this is impossible. In virtually all fossil skeletons, the DNA slowly oxidizes, and very little survives beyond about 60 000 years. Even in more recent skeletons, the quantities of nuclear DNA that can be extracted are so small that sequencing is unreliable. Thus at present it seems virtually impossible to resolve our past from the fossil record alone.

Luckily, we don't necessarily need to. In principle, we can search within ourselves to read the past. All genes accumulate mutations over time, and as they do so their sequence of 'letters' slowly diverges. The longer that two groups have diverged, the more differences accumulate in the sequences of their genes. Thus, if we compare the DNA sequences of a group of people, we can calculate roughly how closely related they are, at least relative to one another. People with just a few sequence differences are more closely related than people with lots of them. By the 1970s, geneticists were becoming involved in human population studies, scrutinizing the differences between genes in different races. The results implied that there is less variation between races than had been thought—as a rule of thumb, there is more variation within races than there is between races, implying that we all share a relatively recent common ancestor. Moreover, the deepest divergences are found in sub-Saharan Africa, implying that the last common ancestor of all human races was indeed African, and lived relatively recently, certainly less than a million years ago.

Unfortunately, there are various drawbacks to this approach. Genes in the nucleus accumulate mutations very slowly, over millions of years, and indeed we still share 95 to 99 per cent of our DNA sequence with chimpanzees (depending on whether we include non-coding DNA in the sequence comparison). If gene sequences can barely tell the difference between humans and chimpanzees, then clearly we need a more sensitive measure to distinguish between human races. Another problem with gene sequences is the role of natural selection. To what extent are genes free to diverge from each other at a steady pace (neutral evolutionary drift), and when does selection constrain the rate of change, by favouring particular sequences? The answer depends not just on the gene, but also on the shifting interactions of genes with each other, and with environmental factors like climate changes, diet, infection, and migration. There is rarely an easy answer.

But the greatest problem with genes from the nucleus is sex—again. Sex recombines genes from different sources, making each of us genetically unique (apart from identical twins and clones). This in turn makes it difficult to determine our lineage. In society, the only way we can know whether we are descended from William the Conqueror, or Noah, or Ghengis Khan, is by keeping detailed records. A surname provides some indication of descent, but most genes know nothing of surnames. They could come from virtually anywhere, and any two different genes almost certainly came from two different ancestors. We are back to the problem of *The Selfish Gene*, discussed in Part 5—in a sexually reproducing species, individuals are fleeting and transitory, mere wisps of cloud; only the genes persist. So we can work out the history of genes, and gene frequencies in a population, but it is difficult to ascribe individual ancestry, and even harder to specify dates.

Down the maternal line

This is where Cann, Stoneking, and Wilson stepped in with their study of mitochondrial DNA, nearly two decades ago. They pointed out that the odd mode of mitochondrial inheritance solved many of the problems associated with nuclear genes. The differences made it possible not only to trace human lineages, but also to give a tentative estimate of dates.

The first critical difference between mitochondrial and nuclear DNA is the mutation rate. On average, the mutation rate of mitochondrial DNA is nearly twenty times faster than nuclear DNA, although the actual rate varies according to the genes sampled. This fast mutation rate equates to a fast rate of evolution (but we should beware of always equating the two, as we'll see later). The fast rate of evolution stems from the proximity of mitochondrial DNA to the free radicals generated in cellular respiration. The effect is to magnify the differences between races. While nuclear DNA can barely distinguish between chimps and humans, the mitochondrial clock ticks fast enough to reveal differences accumulating over tens of thousands of years, just the right speed for peering into human prehistory.

The second difference, said Cann, Stoneking, and Wilson, is that human mitochondrial DNA is inherited only from our mothers, by asexual reproduction. Because all our mitochondrial DNA comes from the same egg, and is replicated clonally during embryonic development, and throughout our lives, it is all (in theory) exactly the same. This means that if we take a sample of mitochondrial DNA from, say, our liver, it should be the exactly the same as a sample taken from the bone, and both should be exactly the same as a random sample taken from our mother—and hers should be exactly the same as her own mother's, and so on back into the mists of time. In other words, mitochondrial DNA works like a matriarchal surname, linking a string of individuals together down the corridor of the centuries. Unlike nuclear genes, which are shuffled and redealt every generation, mitochondrial genes allow us to track the fate of individuals and their descendents.

The third aspect of mitochondrial DNA that the Berkeley team drew on is its steady rate of evolution: the mutation rate, though fast, remains approximately constant over thousands or millions of years. This is ascribed to neutral evolution, the assumption that there is little selective pressure on mitochondrial genes, which serve a restricted and menial purpose (so the argument goes). Sporadic mutations occur at random over generations, and as averages balance out, accumulate at a steady, metronomic speed, leading to a gradual divergence between the daughters of Eve. This assumption is perhaps open to question, and later refinements of the technique have concentrated on the 'control region', a string of 1000 DNA letters that does not code for proteins, and so is claimed not be subject to natural selection (we will return to this assumption later).[1]

---

[1] Here is Bryan Sykes in *The Seven Daughters of Eve*: 'The control region mutations are not eliminated precisely because the control region has no specific function. They are neutral. It appears that this stretch of DNA has to be there in order for mitochondria to divide properly, but that its own precise sequence does not matter very much.'

So how fast does the mitochondrial clock tick? On the basis of relatively recent, approximately known colonization dates (a minimum of 30 000 years ago for New Guinea, 40 000 years ago for Australia, and 12 000 years ago for the Americas), Wilson and colleagues were able to calculate a divergence rate of about 2 per cent to 4 per cent every million years. This figure matches the rate estimated on the basis of divergence from chimpanzees, which began about six million years ago.

If this speed is correctly calibrated, then the actual measured differences between the 147 mitochondrial DNA samples give a date for their last common ancestor of about 200 000 years ago. Furthermore, in agreement with nuclear DNA studies, the deepest divergences were found among African populations, implying that our last common ancestor was indeed African. A third important conclusion of the 1987 paper related to migration patterns. Most populations from outside Africa had 'multiple origins', in other words, peoples living in the same place had different mitochondrial DNA sequences, implying that many areas were colonized repeatedly. In sum, Wilson's group concluded that Mitochondrial Eve lived fairly recently in Africa, and the rest of the world was populated by repeated waves of migration from that continent, lending support to the 'Out of Africa' hypothesis.

Not surprisingly these unprecedented findings gave birth to a dynamic new field, which dominated genealogy in the 1990s. The unresolved questions raised by skeletal morphology, by linguistic and cultural studies, by anthropology and population genetics, could at last all be answered with 'hard' scientific objectivity. Many technical refinements have been introduced, and calibrated dates modified (Mitochondrial Eve is now dated to about 170 000 years ago), but the basic tenets presented by Wilson and his colleagues underpinned the entire edifice. Wilson himself, an inspiring figure, sadly died of leukaemia at the height of his powers, at the age of 56, in 1991.

Surely Wilson would have been proud of the achievements of the field he helped found. Mitochondrial DNA has answered many questions that had seemed eternally controversial. One such question is the identity of the people living in the remote Pacific archipelagos of Polynesia. According to the famous Norwegian explorer Thor Heyerdahl, the Polynesian islands were populated from South America. To prove it he built a traditional balsa-wood raft, the *Kon-Tiki*, and set sail with five companions from Peru in 1947, arriving in the Tuamotus archipelago, 8000 km away, after 101 days. Of course, proving that a feat is possible is not the same as proving that it actually happened. Mitochondrial DNA sequences speak otherwise, corroborating earlier linguistic studies. The results suggest that the Polynesians originated from the west in at least three waves of migration. About 94 per cent of the people tested had DNA sequences similar to the peoples of Indonesia and Taiwan; 3.5 per cent seemed to have come from Vanuatu and Papua New Guinea; and 0.6 per cent from the Philippines. Interestingly, 0.3 per cent had mitochondrial DNA matching some tribes of South American Indians, so there is still a remote chance that there could have been some prehistoric contact.

Another difficult question apparently resolved was the identity of the Neanderthals. Mitochondrial DNA taken from a mummified Neanderthal corpse (found in 1856 near Düsseldorf) showed that their sequence is distinct from modern humans, and no traces of Neanderthal sequence have been found in *Homo sapiens*. This implies that the Neanderthals were a separate subspecies, which fell extinct without ever interbreeding with humans. In fact, the last common ancestor of Neanderthals and humans probably lived about 500 to 600 thousand years ago.

These findings are just two of the many fascinating insights into human prehistory afforded by mitochondrial DNA studies. But every silver lining has a cloud. A rather simplistic view of mitochondria has become the mantra, which is repeated ever more succinctly, and ever more misleadingly; the provisos are lost in the telling. We are told that mitochondrial DNA is inherited exclusively down the maternal line. There is no recombination. It is not subject to much selection because it codes for only a handful of menial genes. The mutation rate is roughly constant. The mitochondrial genes represent the true phylogeny of people and peoples because they reflect individual inheritance, not a kaleidoscope of genes.

This mantra generated unease in some quarters from the beginning, but only recently have these misgivings found substance. In particular, we now have evidence of genetic recombination between maternal and paternal mitochondria, of discrepancies in the ticking of the mitochondrial 'clock', and of strong selection on some mitochondrial genes (including the supposedly 'neutral' control region). These exceptions, while raising some doubts about the validity of our inferences into the past, sharpen our ideas of mitochondrial inheritance, and help us to grasp the real difference between the sexes.

Mitochondrial recombination

If mitochondria are passed down the maternal line exclusively, then there would seem to be little possibility for recombination. Sexual recombination refers to the random swapping of DNA between two equivalent chromosomes, to make up two new chromosomes, each of which contains a mixture of genes from both sources. Clearly DNA from two distinct sources—two parents—is needed to make recombination possible, or at least meaningful: swapping genes from two identical chromosomes makes little sense, unless one of the two chromosomes is damaged; and this, as we shall see, does raise a spectre. In general, however, during sexual reproduction, the pairs of chromosomes in the nucleus are recombined to generate new groupings of genes, mixing up genes from different parents or grandparents, but this does not happen with mitochondrial DNA, as all the mitochondrial genes are derived from the mother. Thus, according to orthodoxy, mitochondrial DNA does not recombine: we don't see a mixture of mitochondrial DNA from both the father and the mother.

Even so, for a decade some primitive eukaryotes like yeast have been known to fuse their mitochondria and recombine mitochondrial DNA. Yeast, of course, is a poor substitute for a human being, as any anthropologist will tell you, and this behaviour was no challenge to reigning orthodoxy. Other curiosities, like mussels, also showed evidence of recombination, but again these were quite easily dismissed as irrelevant to human evolution. So it came as a surprise in 1996 when Bhaskar Thyagarajan and colleagues at the University of Minnesota showed that rats too recombine mitochondrial DNA. Rats, as fellow mammals, are a little too close for comfort. It got worse. In 2001, recombination of mitochondrial DNA was shown to take place in the heart muscle of humans.

Even these studies did not rock the boat too violently, because they were limited in scope. Most mitochondria have five to ten copies of their chromosome, which act as an insurance policy against free-radical damage; it is unlikely that the same gene will be damaged on all copies so normal proteins can still be produced. But just hoarding spare copies is an inefficient method of dealing with damage, as a mixture of normal and abnormal proteins would be produced from the raggedy chromosomes. Better to repair the damage, in the standard bacterial fashion, by recombining undamaged bits of chromosomes to regenerate clean working copies. Such recombination between equivalent chromosomes in the same mitochondrion is known as 'homologous' recombination, and does not undermine the principle of uniparental inheritance— it is simply a method of repairing the damage that occurs within a single individual, as we have just noted. So even when mitochondria fuse together, and recombine DNA from different copies of their chromosome, all their DNA is still inherited only from the mother.

Nonetheless, if paternal mitochondria manage to survive in the egg, then recombination of paternal and maternal mitochondrial DNA is possible, at least in principle. We know that in humans the paternal mitochondria do get into the egg cell, so it is always possible that some survive. Does it happen? In the absence of direct evidence, various research groups tried looking for signs of mitochondrial recombination—and found them. The first evidence came in 1999 from Adam Eyre-Walker, Noel Smith, and John Maynard Smith, at the University of Sussex. Their findings were basically statistical. They argued that if mitochondrial DNA really is clonal, its sequences should continue to diverge in different populations, as they pick up new mutations. In fact this does not always happen: sometimes an 'atavistic' sequence re-emerges, which bears an uncanny resemblance to the ancestral type. There are only two ways this could happen: either by random 'back' mutations to the original sequence, which sounds inherently implausible, or else by recombination with someone who happened to have retained the original sequence. Such unexpected sequence reincarnations are known as *homoplasies*, and Eyre-Walker and colleagues found a lot of them—far more than could be realistically put down to chance. They took this to be evidence of recombination.

The paper raised an immediate storm and was attacked by establishment figures, who discovered errors in the DNA sequences that had been sampled, but not the statistical technique. After excluding these errors, they found no evidence for recombination. 'No need to panic' was the retort of Vincent Macauley and his colleagues at Oxford, and the field took a collective sigh of relief: the great edifice stood firm. But Eyre-Walker and colleagues, while accepting that there had indeed been some sampling errors, stuck to their guns. Even disregarding the errors, they said, the data were still suggestive of recombination, which 'may not lead some to panic, but surely they should, because there is a very real possibility that an assumption we have held for so long is incorrect.'

That same year, 1999 (indeed in the very same issue of the *Proceedings of the Royal Society*), Erika Hagelberg, a former student in the Oxford group and her colleagues put forward their own challenge. Their argument was based on a particular oddity, the recurrence of a rare mutation in several otherwise un-related groups living on the Pacific island of Nguna, in the Vanuatu archi-pelago. As their mitochondrial DNA was clearly inherited from different stocks, and yet the same mutation occurred repeatedly, then either it had arisen inde-pendently on several occasions, which seemed improbable, or it was evidence that the mutation arose only once, but had then been passed around to other populations—which is only possible by recombination. On closer inspection the edifice was again saved. This time the error turned out to be attributable to the sequencing machine, which had somehow become misaligned by 10 letters. After correction the mystery vanished. Hagelberg and colleagues were forced to publish a retraction, and she herself now refers to the unfortunate affair as her 'infamous mistake.'

By 2001, the evidence for recombination looked muddy, to say the least. The two major studies had both been discredited, and although the authors of both papers maintained that the rest of their data still raised doubts, that was only to be expected; they had to defend their tattered reputations. For an unbiased observer, it seemed that recombination had been disproved.

Then in 2002, a fresh challenge emerged. Marianne Schwartz and John Vissing, at Copenhagen University Hospital, reported that one of their patients, a 28-year-old man suffering from a mitochondrial disorder, had actually inherited some mitochondrial DNA from his father, and so had a mixture of both maternal and paternal DNA—the dreaded heteroplasmy. The mixture occurred as a mosaic, such that the mitochondrial DNA in his muscle cells was 90 per cent paternal and only 10 per cent maternal, whereas in his blood cells it was nearly 100 per cent maternal. This was the first time that paternal mitochondrial DNA had been shown unequivocally to be inherited in humans. Clearly some degree of 'seepage' of paternal DNA in the egg is possible, and in this case it was picked up because it caused a disease. But the study raised one overriding question: as two populations of mitochondria from the father and the mother exist in the same person, do they recombine?

The answer is: *they do.* In 2004, Konstantin Khrapko's group at Harvard reported in *Science* that 0.7 per cent of the disparate mitochondrial DNA in the patient's muscles had indeed recombined. So, given the opportunity, human mitochondrial DNA really does recombine. But that is not to say that the recombinants will be passed on. No matter if recombinant DNA is formed in the muscles, it can only influence posterity if it recombines in the fertilized egg. Only then can the recombinant form be inherited. So far, there is no evidence of this, although that is at least partly because very few groups have actually looked. On balance, the statistical evidence from population studies suggests that recombination is extremely rare. Of course, very rare recombination events might explain otherwise mystifying deviations in genetic makeup, even if such rare events are unlikely to topple the whole edifice.

But the point I want to make is that, in evolutionary terms, some degree of recombination probably does occur. Is this just a fluke, an occasional accident, or is there a deeper meaning? We'll return to this question later, but first let's consider the other exceptions to the mantra, for they, too, have a bearing on the matter.

Calibrating the clock

Mitochondrial DNA has its uses not just for reconstructing prehistory, but also in forensics, especially in establishing the identity of unknown remains. Such forensic studies are based on exactly the same assumptions—that everyone inherits a single type of mitochondrial DNA, only from their mother. Among the most celebrated forensic cases was that of the last Russian Tsar, Nicholas II, who was shot, along with his family, by a firing squad in 1918. In 1991, the Russians exhumed a Siberian grave containing nine skeletons, one of which was thought to be that of Nicholas II himself.

The trouble was that two bodies were missing; either something funny had been going on, or this was not the correct grave. Mitochondrial DNA was called to the rescue, but this didn't quite match that of the Tsar's living relatives.

Curiously, the putative Tsar's mitochondrial DNA was heteroplasmic—he had a mixture, so his true identity remained in doubt. The matter was finally laid to rest when the body of the Tsar's younger brother, Georgij Romanov, the Grand Duke of Russia, was also exhumed. He had died of tuberculosis in 1899, and his grave was known with certainty. Because both should have inherited exactly the same mitochondrial DNA from their mother, a perfect match would establish beyond doubt the identity of the Tsar; and indeed the match was perfect: the Grand Duke, too, was heteroplasmic.

While proving the utility of mitochondrial DNA analysis, the episode raised some awkward practical questions—in particular, exactly how common is heteroplasmy? Mitochondrial heteroplasmy doesn't always derive from paternal 'seepage' into the egg, but can also result from mitochondrial mutations. If the DNA in a single mitochondrion mutates, then both types can be amplified during embryonic development, leading to a mixture in the adult body. Such mixtures tend to come to light only when they cause disease, so their real incidence is not known; if they don't cause disease they can easily be overlooked. The practical bearing on forensics was important enough for several research groups to look into it; and their findings, consistent between the groups, came as a surprise. At least 10 per cent, and perhaps 20 per cent of humans, are heteroplasmic. Much of the mixture appears to come from new mutations, rather than paternal seepage.

These findings have two important implications. First, heteroplasmy is far more common than we had imagined, and this must hold consequences for the 'selfish' mitochondrial model of sexes: if we can survive quite happily with two competing populations of mitochondria (without overt disease in most cases) then clearly the conflict between mitochondria has been overstated to some extent. And second, the rate of mitochondrial mutation is far higher than expected. Attempts to calibrate the rate by comparing the sequences of distantly related family members have come to mixed conclusions, but the burden of evidence suggests that one mutation occurs every 40 to 60 generations, which is to say every 800 to 1200 years. In contrast, if we calibrate the rate of divergence on the basis of known colonization dates and fossil evidence, we calculate a rate of about 1 mutation every 6000 to 12 000 years. This discrepancy is substantial. If we use the faster clock to calculate the date of our last common ancestor, Mitochondrial Eve, we are forced to conclude she lived about 6000 years ago, more commensurate with Biblical Eve than African Eve, who supposedly lived 170 000 years ago. Clearly the recent date is not correct, but how do we explain such a big disparity?

An important fossil finding in south-west Australia may give a clue to the answer. The fossil is an anatomically modern human, and is famous as the source of the world's oldest mitochondrial DNA. It was discovered near Lake

Mungo in 1969, and later tentatively dated to about 60 000 years old. In 2001, an Australian team reported the mitochondrial DNA sequence, and it came as a shock—nothing similar has ever been found in a living person. The line has fallen extinct.[2] This raises several deep questions. In particular, we classified the Neanderthals earlier as a separate subspecies that fell extinct, on the basis of an extinct mitochondrial sequence, but we are now faced with an anatomically modern human being that had done the same thing. Applying the same rules we are obliged to say that the human too represented a separate subspecies that had fallen extinct, yet we know from the anatomical appearance that we must share nuclear genes. Presumably, there was some genetic continuity between the populations. The simplest way to reconcile the discrepancy is to conclude that a mitochondrial sequence does not invariably record the history of a population; but this forces us to question our interpretation of the past based on mitochondrial sequences alone.

What might have happened? Imagine an anatomically modern human population living in Australia. Let's say they migrated there from Africa less than 100 000 years ago. Later, a new migrant population arrives, and there is a limited degree of interbreeding. If a new-migrant mother mates with an indigenous father, and they have a healthy daughter, then her mitochondrial DNA will be 100 per cent new-migrant (assuming there is no recombination), but her nuclear genes will be 50 per cent indigenous. If everyone else fails to leave a continuous line of daughters, and our mixed child alone mothers a new population, then the indigenous mitochondrial DNA will fall extinct, while at least some indigenous nuclear genes will survive. In other words, interbreeding is quite compatible with the extinction of a lineage of mitochondrial DNA, and if we try to reconstruct history by mitochondrial DNA alone we might easily be misled. Exactly the same applies to Neanderthals, so we can't conclude from their mitochondrial DNA that they disappeared without trace. (Richard Dawkins comes to a similar conclusion, from different considerations, in *The Ancestors Tale*.) But is this scenario likely, or merely a technical possibility? It implies the survival of only a single line of daughters; would all the indigenous mitochondrial lines really fall extinct so easily?

They might. I mentioned that mitochondrial DNA works like a surname— and surnames easily fall extinct, as first shown by the Victorian polymath Francis Galton in his book *Hereditary Genius*, of 1869. It seems the average 'lifespan' of a surname is only about two hundred years. In the UK, about three

---

[2] In fact modern humans do carry a similar sequence in their nuclear DNA—a *numt* that had been transferred from the mitochondria to the nucleus (see page 132) long ago. The sequence amounts to a DNA fossil, as the mutation rate in the nucleus is some 20 times slower than in the mitochondria; it has therefore remained relatively unchanged.

hundred families claim descent from William the Conqueror, but none can prove unbroken descent down the male line. All five thousand feudal knight-hoods listed in the Domesday Book of 1086 are now extinct, and the average duration of a hereditary title in the middle ages was three generations. In Australia, the 1912 census showed that half the children descended from just one-ninth of the men and one-seventh of the women. The essential point, stressed by Australian fertility expert Jim Cummins, is that reproductive success is extremely unevenly distributed in populations. Most lines fall extinct; and just the same applies to mitochondrial DNA.

Is this just neutral drift, or does natural selection come into it? Again the Lake Mungo fossil provides a clue. In 2003, James Bowler, one of the discoverers of the fossil in 1969, and his colleagues, showed that the 60 000-year date for the fossil is incorrect. They re-dated the remains to around 40 000 years ago, based on a far more complete analysis of the stratigraphy. The new date is interesting, as it coincides with a period of climate change, when the lakes and rivers dried out and much of south-western Australia became an arid desert. In other words the Mungo line of mitochondrial DNA fell extinct at a time of changing selection pressures.

This raises the spectre of natural selection acting on mitochondrial genes. According to orthodoxy, it doesn't. If sequence changes accumulate slowly over thousands of years, and the entire trail of changes can be tracked by comparing the genomes of living people, then none of the intermediary changes could have been eliminated by natural selection—the whole succession of changes must have been random, neutral mutations. Yet this cannot explain the discrepancy between a high mutation rate and a slow rate of divergence—of evolution. Natural selection can. If the lines that evolve the fastest (which is to say diverge the most) are eliminated by natural selection, then the survivors must have smaller evolutionary variations. I mentioned earlier that we should not confound a high mutation rate with a high rate of evolution. This is a case in point. The mutation rate is fast but the evolution rate is slower, because a proportion of the mutations have negative consequences, and so are eliminated by selection. The discrepancy is squared by selection.

In the case of the Lake Mungo fossil, the extinction of the mitochondrial DNA line might be put down to natural selection, but this would go against the mantra. Could natural selection be the answer? In fact there is now good evidence that natural selection does operate on mitochondrial genes.

Mitochondrial selection

In 2004, Douglas Wallace, the guru of mitochondrial geneticists, and his group at the University of California, Irvine, published fascinating evidence that natural selection does indeed operate on mitochondrial genes. Wallace himself, in two decades at Emory University in Atlanta, pioneered the mitochondrial typing of human populations, and his work in the early 1980s underpinned the famous 1987 *Nature* paper by Cann, Stoneking, and Wilson, which we considered at the beginning of this chapter. Wallace's worldwide genetic tree defines a number of mitochondrial lineages, which he termed *haplogroups*, and which later came to be known as the daughters of Eve. To these groups he assigned alphabetical letters, known as the Emory classification. Bryan Sykes at Oxford later used the letters as the basis of personal names in his popular bestseller *The Seven Daughters of Eve*, which referred only to European lines.

Wallace (inexplicably unmentioned in Sykes's book) is not just the guru of mitochondrial population genetics, but also of mitochondrial diseases, which number in multiples of hundreds, quite out of proportion to the small number of genes. These diseases are often caused by tiny variations in the mitochondrial sequence. Not surprisingly, given his interest in the grave consequences of such variations to health, Wallace has long raised suspicions that mitochondrial genes might be subject to natural selection. Obviously, if they cause a crippling disease they are likely to be eliminated by natural selection.

Wallace and his colleagues first drew attention to statistical evidence of 'purifying selection' in the early 1990s. Over the following decade, Wallace kept these findings at the back of his mind. In many studies of mitochondrial genetics, he noted repeatedly that the geographical distribution of mitochondrial genes in human populations was not random, as predicted by the theory of neutral drift, but that particular genes thrived in certain places—often a telltale sign of selection at work. Of all the abundant lines of mitochondrial DNA in Africa, for example, but a handful ever left the dark continent; most remained strictly African. The great variety of mitochondrial DNA in the rest of the world blossomed from just a few selected groups. Similarly, in Asia, of all the mitochondrial variety, only a few types ever managed to settle in Siberia, and later migrate to the Americas. Might it be, asked Wallace, that some mitochondrial genes are adapted to particular climates, and do better there, whereas others are penalized if they leave home?

By 2002, Wallace and colleagues were beginning to look into the matter more seriously, and signalled their outlook in some thoughtful discussion papers, but it was not until 2004 that they finally found proof. The idea is breathtakingly simple, and yet holds important implications for human evolution and health. The mitochondria, they said, have two main roles: to produce energy, and to produce heat. The balance between energy generation and heat production can vary, and the actual setting might be critical to our health. Here's why.

Much of our internal heat is generated by dissipating the proton gradient across the mitochondrial membranes (see page 183). Since the proton gradient can either power ATP production *or* heat production, we are faced with alternatives: any protons dissipated to produce heat cannot be used to make ATP. (As we saw in Part 2, the proton gradient has other critical functions too, but if we assume that these remain constant, they don't affect our argument.) If 30 per cent of the proton gradient is used to produce heat, then no more than 70 per cent can be used to produce ATP. Wallace and colleagues realized that this balance could plausibly shift according to the climate. People living in tropical Africa would gain from a tight coupling of protons to ATP production, so generating less internal heat in a hot climate, whereas the Inuit, say, would gain by generating more internal heat in their frigid environment, and so would necessarily generate relatively little ATP. To compensate for their lower ATP production, they would need to eat more.

Wallace set out to find any mitochondrial genes that might influence the balance between heat production and ATP generation, and found several variants that plausibly affected heat production (by uncoupling electron flow from proton pumping). The variants that produced the most heat were favoured in the Arctic, as expected, while those that produced the least were found in Africa.

While this seems no more than common sense, the connotations conceal a twist worthy of a murder mystery. Recall from Part 4 (page 183) that the rate of free-radical formation doesn't depend on the speed of respiration, but rather on how fully the respiratory chains are packed with electrons. If electron flow is very sluggish, because there is little demand for energy, electrons build up in the chains and can escape to form free radicals. In Part 4, we saw that a fast rate of free-radical formation can be reduced if electron flow is maintained down the chains—and this can be achieved by dissipating the proton gradient to generate heat. We compared the situation with a hydroelectric dam on a river, in which the overflow channels prevent flooding. The pressing need to dissipate the proton gradient may have overridden its wastefulness, and given rise to endothermy, just as the need to prevent flooding may override the waste of water through the overflow channels. The long and short of it is that raising internal heat generation lowers free-radical formation at rest, whereas lowering internal heat production increases the risk of free-radical production when at rest.

Now think what is happening in Africans and (let's say) Inuit. Because Africans generate less internal heat than Inuit, their free-radical production ought to be higher, especially if they overeat. According to Wallace, Africans can't burn off excess food as heat as efficiently as do Inuit, so if they eat too much they will generate more free-radicals instead. This means that they ought to be more vulnerable to any diseases linked with free-radical damage, such as heart disease and diabetes, and indeed this is the case. Africans in the United

States eating an American diet are notoriously susceptible to diseases like diabetes. Conversely, Inuit ought to burn off excess food as heat, and so should suffer much less from heart disease and diabetes, and again this is found to be true. Of course there are other reasons too (such as intake of oily fish, etc.) so these conclusions are necessarily tentative. However, if there is some truth in these ideas, then another logical connotation should also be true, and there is a hint that it is: any peoples adapted to arctic climates should be more vulnerable to male infertility.

The reasoning is exactly the same. Arctic peoples divert less of their food into energy, and more into heat. This may not matter in most circumstances (they just eat more) but it matters in one instance: sperm motility. Sperm are powered by their mitochondria as they swim towards the egg, and because there are fewer than a hundred mitochondria in each cell, sperm cells are uniquely dependent on the efficiency of the remaining few—and uniquely vulnerable to energy failures. If these mitochondria fritter away their energy as heat, the sperm are more likely to be dysfunctional, and the man to suffer from *asthenozoospermia*. This means we should see patterns of male fertility that depend not on male genes, but on mitochondrial genes passed down the maternal line. In other words, male infertility should be inherited at least partly from the mother, and ought to vary according to mitochondrial haplogroup. One recent study has confirmed this to be true in Europeans: asthenozoospermia is more common in people of haplogroup T (widespread in northern Sweden) than in people of haplogroup J (more widespread in southern Europe). Whether it is also true of the Inuit, I don't know: unfortunately I can't find any data on the incidence of asthenozoospermia among the Inuit.

Altogether, these twisted relationships show that mitochondrial genes are indeed subject to natural selection.[3] The precise weighting depends on factors that include energetic efficiency, internal heat production and free-radical leakage, all of which affect our overall health and fertility, and our capacity to adapt to diverse climates and environments.

When coupled with the other findings we have discussed in this chapter, the orthodox position looks tarnished. Mitochondrial genes can be inherited from both parents, albeit rarely; they recombine, if very rarely; they mutate at vari-

---

[3] This includes the supposedly neutral control region: if recombination does not take place then the entire mitochondrial genome is a single unit, and the control-region sequences can be eliminated in a non-random manner because they are linked to regions that *do* undergo selection. And in fact it would be surprising if the control region was not subject to direct selection, as it binds factors responsible for transcribing mitochondrial proteins—a task as important as their very existence, for they may as well not exist if they are not transcribed when needed. In 2004, Wallace and colleagues showed that some control-region mutations might indeed have detrimental consequences—some are linked with Alzheimer's disease.

able rates according to circumstances, questioning the accuracy of imputed dates; and they are unquestionably subject to natural selection. If these unexpected findings don't topple the edifice of human prehistory, do they at least give a more cohesive understanding of mitochondrial inheritance? More pertinently, can these findings explain why we have two sexes?

15

Why There Are Two Sexes

In Chapter 13, we saw that the deepest biological difference between the two sexes relates to the inheritance of mitochondria. The female sex specializes to provide the mitochondria (100 000 of them in humans) in the large, immobile egg cells, while the male sex specializes to eliminate mitochondria from tiny, motile sperm cells. We looked into the reasons for this strange behaviour, and found that it often seems to boil down to conflict between genetically different populations of mitochondria. To restrict the opportunity for conflict, mitochondria are usually inherited from only one of two parents. But we also met a number of exceptions to this simple rule, including fungi, trees, bats, and even ourselves. In Chapter 14, we took a close look at ourselves to see how well the copious human data support the conflict argument. These data are controversial and arouse high passions, for they concern our own prehistory, but the coherent picture that is slowly emerging from these disputes gives fascinating insights into the deeper reasons for the difference between the two sexes. In this chapter, we'll try to draw these insights together to come up with a more satisfying answer to the puzzle of two sexes.

The essential facet of the conflict argument is that dissimilar populations of mitochondria can compete with each other for succession, and the only way to prevent such conflict is to ensure that all the mitochondria inherited in the egg are genetically identical. The only way to guarantee they are all the same is to make sure they all come from the same source—the same parent. Mixing is said to be fatal. The belief that mitochondrial mixing (heteroplasmy) is simply not tolerated underpins the mantra of human mitochondrial population genetics. According to this mantra, the male mitochondria are swiftly eliminated from the egg, and not passed on to the next generation. This means that mitochondria are passed down the maternal line by asexual replication only. Thus, mitochondrial DNA remains basically unchanged, as there is no possibility of recombination. Even so, there is a gradual divergence in the mitochondrial DNA sequence of different populations and races, as occasional neutral mutations accumulate over thousands and tens of thousands of years. These accumulated differences supposedly sit faithfully in the genome, as natural selection is said not to apply to mitochondrial genes, or at least to the 'control region' that doesn't code for proteins. Without purifying selection, the mutations are not weeded out of the genome, and so remain there forever after, mute witnesses to the flow of history.

The lessons from human evolution muddy all these tenets, and suggest there is a deeper mechanism at work. That is not to say that genomic conflict is wrong, but merely part of a larger picture. Let's rake the mud. We have seen that mitochondrial recombination does indeed occur, probably very rarely in human mitochondria, but more commonly in other species, such as yeast and mussels. It is not the taboo we had once thought. Moreover, the condition for recombination, heteroplasmy (a mixture of dissimilar mitochondria), is far more common than is assumed by the selfish conflict model. Some degree of heteroplasmy is found in 10 to 20 per cent of humans, and it is also common in many other species. Next, we have seen that there is a discrepancy in the rate of change of mitochondrial genes. The mutation rate of mitochondrial DNA in families suggests one mutation every 800 to 1200 years, whereas the long-term divergence of races suggests a mutation rate of one every 6000 to 12 000 years. The disparity can be explained if many of the variants are eliminated by natural selection. Counter to the usual mantra, there is now good evidence that natural selection does indeed act on mitochondrial genes, in ways that are subtle and pervasive.

So why are there two sexes? Think about the mitochondria. They are not independent entities but part of the larger system of the cell. Mitochondria contain proteins that are encoded by two different genomes. Genes in the nucleus encode the vast majority, some 800 proteins, whereas the handful of mitochondrial genes encodes the rest, a mere 13 proteins, all of which are critical subunits of the large protein complexes of the respiratory chains. The mitochondrial-encoded proteins are essential for respiration. It's this *necessary* interaction between two genomes, the mitochondrial and nuclear, that explains the need for two sexes. Let's see why.

The function of the mitochondria depends critically on the interaction of the proteins encoded in the nucleus with those encoded in the mitochondria. This dual control system is no frozen accident: it evolved that way, and is continuously optimized, because this is the most effective way of meeting the needs of the cell. As we saw in Part 3, the mitochondria retained a handful of genes for a positive reason: a rapid-reaction unit of genes in the mitochondria is *necessary* to maintain efficient respiration. In contrast, the genes that could be transferred successfully to the nucleus generally have been; there are many advantages to them being there, not least to quell the independence of the troublesome mitochondrial guests.

Any misalignment between proteins encoded in the nucleus and proteins encoded in the mitochondria holds potentially catastrophic consequences.

The fine control of mitochondrial function influences not just energy avail-ability, but other matters of life and death, such as apoptosis, fertility, sex, endothermy, disease, and ageing. But how well does dual genomic control work? Babies are a miraculous proof of the marvellous harmony that nature has attained, but perfection comes at a price. Many couples strive for years to have children, and infertility is common. Even for fertile couples, early (usually sub-clinical) miscarriage is the rule rather than the exception: some 70 to 80 per cent of embryos spontaneously abort in the first weeks of pregnancy, and the would-be parents may never notice. Many of these early losses occur for rea-sons that are still obscure.

The problem might often relate to the interaction of the two genomes—the need for nuclear gene products to work in cohorts with mitochondrial gene products. In mammals, the mutation rate in mitochondria is fast, on average 20 times faster than the nucleus and in places 50 times faster, owing to the proximity of mitochondrial DNA to mutagenic free radicals leaking from the respiratory chains. That's not all. In the nucleus, genes are shuffled by sex every generation. Because the genes encoding mitochondrial proteins sit on different chromosomes, they are re-dealt as a different hand each generation. The out-come is a serious mix-and-match problem. In the respiratory chains, proteins dock on to each other with nanoscopic precision. To give a single example, cytochrome c (encoded in the nucleus) must bind to a critical subunit of cytochrome oxidase (encoded in the mitochondria) to pass on its electron. If the binding is not exact, the electron is not transferred and respiration grinds to a halt. When electrons aren't passed on down the chains they form free radicals instead. These oxidize the membrane lipids and release cytochrome c to induce apoptosis. From this perspective, the unanticipated role of cytochrome c in apoptosis begins to look not like an oddity but a necessity. It puts a quick end to cells with inefficient respiration, due to a mismatch between the nuclear and mitochondrial genes.

The requirement for a close match means it is critical for mitochondrial and nuclear genes to *co-adapt* to each other in synchrony, otherwise respiration cannot work. In principle, a failure to co-adapt leads straight to an early death by apoptosis. Direct evidence of co-adaptation is mounting. If the DNA from mouse mitochondria is replaced with DNA from rat mitochondria, protein transcription proceeds normally, but respiration ceases because the rat mito-chondrial proteins can't interact properly with the mouse proteins encoded in the nucleus. In other words, control of respiration is more stringent than control of DNA transcription and translation into new proteins. Slighter differ-ences occur within species, but even small mismatches between mitochondrial and nuclear genes affect the speed and efficiency of respiration. Importantly, the evolution rate of cytochrome c mirrors that of cytochrome oxidase over evolutionary time, even though the underlying rates of change are more than 20-fold different. Presumably, any new variants that lower the efficiency of respiration are eliminated by natural selection. The imprint of selection is betrayed by the fact that many of the sequence changes that do persist are so-called *neutral substitutions*, which is to say they don't alter the sequence of the protein. The ratio of neutral substitutions to 'meaningful' substitutions is much higher than normal in mitochondrial genes, which implies that mutations that alter meaning are eliminated by natural selection. There are other hints that meaning is preserved at all costs. For example some protozoa such as *Trypanosoma*, actually edit their RNA sequences to retain the original meaning, despite changes in DNA sequence. Similarly, the fact that mitochondria harbour exceptions to the universal genetic code can be explained as an attempt to retain the original meaning despite changes in DNA sequence.

Taking all this into consideration, we can say that two sexes are needed because the dual genome system demands a close match between mitochondrial and nuclear genes. If the match is not good, respiration is impaired, and there is a much greater risk of apoptosis and developmental abnormalities. The precision of the match is continuously strained by two factors—the far higher mutation rate of mitochondrial DNA, and the randomization of new nuclear genes by sex in every generation. To ensure the match is as perfect as it can be in each generation, it is necessary to test a *single set* of mitochondrial genes against a *single set* of nuclear genes. This explains why the mitochondria need to come from just one parent. If they came from two parents, then there would be two sets of mitochondrial genes paired with one set of nuclear genes. This is like pairing two women of dissimilar build with the same man in a three-way ballroom dance. However accomplished they might be as individual dancers, the flailing threesome is likely to fall over. To dance a true metabolic waltz requires two partners—one type of mitochondria with one set of nuclear genes.

There are two important connotations to this answer. First of all, it easily encompasses existing models, while explaining the apparent aberrations noted in the studies of human evolution. To match a single mitochondrial genome with a single nuclear genome requires that (in general) the mitochondrial genome should be inherited from a single parent, hence the propensity for uniparental inheritance. If mitochondria are inherited from both parents, then the efficiency of respiration is likely to be impaired, as the two populations are obliged to dance with the same nuclear partner. This situation is exacerbated if dissimilar mitochondrial genomes compete, as in the selfish-conflict theory. Notice, however, that some degree of heteroplasmy and recombination, is feasible as it might at times provide the best genomic match. The unexpected findings from human evolution—heteroplasmy, recombination, and selection—can be explained in this way. The most important aspect is not actually the 'pureness' of the population, but rather how effectively the mitochondrial genes work against the nuclear background.

Secondly, the dual-control hypothesis gives a positive basis for natural selection. One difficulty with the selfish-conflict theory is that selection can only act to eliminate the negative consequences of genome conflict. However, we've seen that heteroplasmy exists in many circumstances without obvious competition between the two genomes—for example in angiosperms, some fungi, and bats. If the detrimental effects of genomic competition are limited, then why does natural selection *generally* favour uniparental inheritance? It would do so if uniparental inheritance were positively beneficial most of the time, rather than merely mildly detrimental part of the time. The dual-control theory gives a good reason why this would be the case: the fittest individuals generally inherit the mitochondrial DNA only from the mother, as this enables the best match of nuclear and mitochondrial genomes. And if the fittest offspring tend to inherit their mitochondrial genes from only one of two parents, we have satisfied the condition for two sexes: the female sex supplies the mitochondria, the male sex generally does not.

So where and how does selection act to ensure harmony between nuclear and mitochondrial genes? The probable answer is during the development of the female embryo, when the overwhelming majority of egg cells, or oocytes, die by apoptosis. The fittest cells apparently pass through a bottleneck, which selects for mitochondrial function. While little is known about how such a bottleneck works—and some dispute its very existence—the broad outlines conform wholly to the expectations of the dual control hypothesis. It seems that oocytes are selected on the basis of how well their mitochondria function against the nuclear background.

The mitochondrial bottleneck

The fertilized egg cell (the zygote) contains about 100 000 mitochondria, 99.99 per cent of which come from the mother. During the first two weeks of embryonic development, the zygote divides a number of times to form the embryo. Each time the mitochondria are partitioned among the daughter cells, but they don't actively divide themselves: they remain quiescent. So for the first two weeks of pregnancy, the developing embryo has to make do with the 100 000 mitochondria it inherited from the zygote. By the time the mitochondria finally begin to divide, most cells are down to a couple of hundred each. If their function is not sufficient to support development, the embryo dies. The proportion of early miscarriages caused by energetic failures is unknown, but energetic insufficiency certainly causes many failures of chromosomes to separate properly during cell division, giving rise to anomalies in the number of chromosomes such as trisomy (three, rather than two, copies of a chromosome). Virtually all of these anomalies are incompatible with full-term development; indeed only Trisomy 21 (three copies of chromosome 21) is mild enough to deliver a live birth; and even so, babies born with this anomaly have Down syndrome.

In a female embryo, the earliest recognizable egg cells (the primordial oocytes) first appear after two to three weeks of development. Exactly how many mitochondria these cells contain is controversial, and estimates range from less than 10 to more than 200. The most authoritative survey, by Australian fertility expert Robert Jansen, is at the low end of this range. Either way, this is the start of the mitochondrial bottleneck, through which selection for the best mitochondria takes place. If we persist in clinging to the idea that all the mitochondria inherited from our mother are exactly the same, then this step might seem inexplicable, but in fact there is a surprising variety of mitochondrial sequences in different oocytes taken from the same ovary. One study by Jason Barritt and his colleagues at the St Barnabus Medical Center in New Jersey, showed that more than half the immature oocytes of a normal woman contain alterations in their mitochondrial DNA. This variation is mostly inherited, and so must also have been present in the immature ovaries of the developing female embryo. What's more, this degree of variation is what remains *after* selection, so presumably the mitochondrial sequences are even more variable in the developing female embryo, where the selection takes place.

How does this selection work? The bottleneck means that there are only a few mitochondria in each cell, making it more likely that all of them will share the same mitochondrial gene sequence. Not only are there few mitochondria, but each mitochondrion has only one copy of its chromosome, rather than the usual five or six. Such restriction precludes compensation for poor function: any mitochondrial deficits are effectively paraded naked, and their inadequacies can be magnified to the point that they are detected and eliminated. The next stage is amplification—rushing out of the constraints of the bottleneck. Having established a direct match between a single clone of mitochondria and the nuclear genes, it is necessary to test how well they work together. To do so, the cells and their mitochondria must divide, and this relies on both mitochondrial and nuclear genes. The behaviour of the mitochondria is striking when examined down the electron microscope—they encircle the nucleus like a bead necklace. This remarkable configuration surely betokens some kind of dialogue between the mitochondria and the nucleus, but at present we know next to nothing about how it may work.

The replication of oocytes in the embryo over the first half of pregnancy takes their number from around 100 after 3 weeks, to 7 million after 5 months (a rise of about 2^{18}). The number of mitochondria climbs to some 10 000 per cell, or a total of about 35 billion in all the germ cells combined (a rise of 2^{29}), a massive amplification of the mitochondrial genome. Then follows some kind of selection. How this selection works is quite unknown, but by the time of birth the number of oocytes has fallen from 7 million to about 2 million, an extraordinary wastage of 5 million oocytes, or nearly three quarters of the total. The rate of loss abates after birth, but by the onset of menstruation there are only about 300 000 oocytes left; and by the age of 40, when there is a steep decline in oocyte fertility, just 25 000. After that, decline is exponential into menopause. Of the millions of oocytes in the embryo, only about 200 ovulate during a woman's entire reproductive life. It's hard not to believe that some form of competition is going on—that only the best cells win out to become mature oocytes.

There are indeed suggestions of purifying selection at work. I mentioned that half of all immature oocytes in the ovaries of a normal woman have errors in their mitochondrial sequence. Only a tiny fraction of these immature eggs mature, and only a few mature eggs are successfully fertilized to create an embryo. What selects for the best eggs is unknown, but the proportion of mitochondrial errors is known to fall to about 25 per cent in early embryos. Half of the mitochondrial error has been eliminated, implying that some kind of selection has taken place. Of course, most embryos also fail to mature (the great majority die in the first few weeks of pregnancy), and again the reasons are unknown. Nonetheless, it is known that the incidence of mitochondrial mutations in newborn babies is a tiny fraction of that in early embryos, implying that a purge of mitochondrial errors really has taken place. There is other indirect evidence of mitochondrial selection. For example, if the selection of oocytes acts as a proxy for natural selection in adults, avoiding all the costly investment to produce an adult, then species that invest their resources most heavily in a small number of offspring might be expected to have the best 'filter' for quality oocytes—they have the most to lose from getting it wrong. This does actually seem to be the case. The species that have the smallest litters also have the tightest mitochondrial bottleneck (the smallest number of mitochondria per immature oocyte), and the greatest cull of oocytes during development.

Although we don't know how such selection acts, it is plain that failing oocytes die by apoptosis, and the mechanism certainly involves the mitochondria. It is possible to preserve an oocyte otherwise destined to die simply by injecting a few more mitochondria—this is the basis of ooplasmic transfer, the technique we mentioned on page 240. The fact that such a crude manoeuvre actually does protect against apoptosis suggests that the fate of the oocyte really does depend on energy availability; and indeed there is a general correlation between ATP levels and the potential for full-term development. If energy levels are insufficient, cytochrome c is released from the mitochondria, and the oocyte commits apoptosis.

All in all, there are many tantalizing hints that selection is taking place in oocytes for the dual control system of mitochondrial and nuclear genes, although there is as yet little direct evidence. This, truly, is twenty-first-century science. But if it is shown that oocytes are the testing ground for mitochondrial performance against the nuclear genes then this would be good evidence that two sexes exist to ensure a perfect match between the nucleus and the mitochondria. Having now selected an oocyte on the basis of its mitochondrial performance, the last thing we need is this special relationship to be messed up by a big injection of sperm mitochondria adapted to a different nuclear background.[1]

We have much to learn about the relationship between the mitochondrial and nuclear genes in oocytes, but we know rather more about this relationship in other, older cells. In ageing cells, mitochondrial genes accumulate new mutations and the dual genomic control begins to break down. Respiratory function declines, free-radical leakage rises, and the mitochondria begin to promote apoptosis. These microscopic changes are writ large as we age. Our energy diminishes, we become far more vulnerable to all kinds of diseases, and our organs shrink and wither. In Part 7, we'll see that the mitochondria are central not only to the beginning of our lives, but also to their end.

---

[1] Sharp readers may notice a dilemma here, which has been articulated by Ian Ross, at UC Santa Barbara. The mitochondria are adapted to the *unfertilized* oocyte nuclear background, but this changes when the oocyte is fertilized and the father's genes are added to the mix. If the adaptation of mitochondria to nuclear genes is not to be lost, then the maternal nuclear genes should overrule the paternal genes—a process known as imprinting. Many genes are maternally imprinted but whether some of these encode mitochondrial proteins is unknown. Ross predicts they will be, and is studying mitochondrial imprinting in a fungal model.

PART 7

Clock of Life
Why Mitochondria Kill us in the End

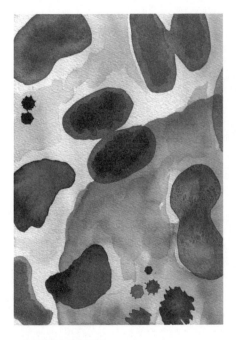

Ageing and death—mitochondria divide
or die, depending on their interactions
with the nucleus

Animals with a fast metabolic rate
tend to age quickly and succumb to
degenerative diseases such as
cancer. Birds are an exception
because they combine a fast
metabolic rate with a long lifespan,
and a low risk of disease. They
achieve this by leaking fewer free
radicals from their mitochondria. But
why does free-radical leakage affect
our vulnerability to degenerative
diseases that on the face of it have
little to do with mitochondria? A
dynamic new picture is emerging, in
which signalling between damaged
mitochondria and the nucleus plays a
pivotal role in the cell's fate, and our
own.

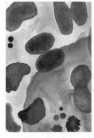

The immortal elves in Tolkein's immortal epic are as mortal as the next man. They die in droves on the battlefield. What they *don't* do is age, or at least not much. Elrond, Lord of Rivendell in *The Lord of the Rings*, was thousands of years old, dwarfing even biblical lifespans. Tolkein described his face as 'ageless, neither old nor young, though in it was written the memory of many things, both glad and sorrowful. His hair was dark as the shadows of twilight. . .'

Is this just the whimsy of an imaginative mind? Not necessarily. While ageing and the degenerative diseases it carries with it are the bane of the western world, they are not a universal currency throughout nature. Many giant trees, for example, live for thousands of years. Admittedly, trees are a long way removed from ourselves, and in any case much of the tree is just dead structural support. Better examples, far closer to home, are many birds. Parrots can live for over a hundred years, the albatross for more than a hundred and fifty. Many gulls live for seven or eight decades and show few overt signs of ageing in a way that we can recognize. A famous pair of photographs depicts the Scottish zoologist George Dunnet with a fulmar petrel that he had captured and ringed in Orkney. The first photograph shows Professor Dunnet as a handsome young man with a handsome young bird in 1952. The second was taken in 1982, and shows Dunnet with the same ringed fulmar, which he fortuitously recaptured thirty years later, again in Orkney. Dunnet is by now betraying the ravages of age, but the bird has aged not a jot, at least to the naked eye. A third photograph, which I have never managed to see, apparently pictures Dunnet with the same fulmar in 1992, just a couple of years before the death, after protracted illness, of one of them. Rest in peace, Professor Dunnet.

Yes, I hear you say, but we too may live for a hundred years or more; what is so special about a bird that does the same? The answer is that birds live far longer than they 'ought' to on the basis of their metabolic rate. If we lived as long as a lowly pigeon, relative to our own metabolic rate, we'd live happily, without much illness, for perhaps a few hundred years. So why not? Why not indeed! Given the political will to overcome the ethical dilemmas, there may be no biological reason why not. Over six million years of evolution, since we split off from the apes, we have already extended our own maximum lifespan by 5- or sixfold, from 20 or 30 years to about 120 years.[1] As depicted in the familiar evolutionary succession, from the stooped knuckle-dragging ape to the erect *homo sapiens*, we have grown in weight as well as stature, and have a lower metabolic rate. These changes were wrought by natural selection—tampering with the genes—which if we were to apply them to ourselves would be called genetic modification. But even if we lack the stomach to meddle with our genes in the interests of a vainglorious immortality, still the best way to counter the desperate degenerative diseases of old age, which debilitate an ever-growing proportion of the population, is by applying the lessons of evolution in an ethically acceptable way.

I say 'relative to metabolic rate'. Recall from Part 4 that in mammals and birds, body mass corresponds to metabolic rate: in general, the larger a species the slower its metabolic rate. For example, the cells of a rat have a metabolic rate that is seven times faster than our own. It's no coincidence that the rat also lives for a fraction of the time. The relationship between metabolic rate and lifespan can be perceived more directly in insects such as the fruit fly *Drosophila*. In this case, the metabolic rate depends on the ambient temperature, and roughly doubles for every 10°C rise in temperature; and with it, their lifespan falls from a month or more to less than a couple of weeks.

Among the warm-blooded mammals, which are relatively immune to the vicissitudes of weather, there is a broad correlation between body mass, metabolic rate, and lifespan—the larger the animal, the slower the metabolic rate, and again, the longer the life. A similar relationship holds true if we plot out the birds, but now, intriguingly, there is a gap (Figure 14). On average, if a bird and a mammal are paired so their *resting* metabolic rate is similar—we might say their *pace of life* is similar—then the bird lives three or four times longer than the mammal. In some cases the discrepancy is even greater. Thus the resting metabolic rate of a pigeon and a rat are similar, yet the pigeon lives for 35 years while the rat lives barely three or four, an order of magnitude difference. We, too, live longer than we 'should' if our lifespan is plotted against our metabolic rate—like many birds, and indeed bats, we live three or four times longer than other mammals with similar resting metabolic rates. When I say that we could extend our lifespan to perhaps several hundred years, I am comparing us with the pigeon, which lives two or three times longer than us, relative to its own metabolic rate. Put another way, a pigeon doesn't live so much longer than a rat because it has slowed down its pace of life. Rather, a pigeon lives ten times longer than a rat while maintaining exactly the same pace of living. There are, apparently, no strings attached.

---

[1] The longest recorded human life was Jeanne Calment, who died in 1997 at the age of 122. Without an heir at the age of 90, she signed a deal with a lawyer in 1965, who agreed to pay an annual retainer for her apartment, which he would inherit upon her death. The full value would be paid in 10 years. Unfortunately, the lawyer himself passed away in 1995, after 30 years of payment, and his wife had to continue paying after his death.

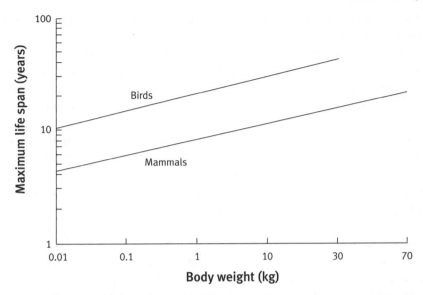

14 Graph showing lifespan against body weight in birds and mammals. Large animals have a slower metabolic rate, and live longer. This is true of both birds and mammals, and the slope of the lines on a log–log plot is very similar. However, there is a gap between the groups: birds live three to four times as long as mammals with a similar body weight and resting metabolic rate.

An important point is that ageing is usually, but not inevitably, linked to disease. The rat suffers similar diseases of old age to us. Rats become obese, get diabetes, cancer, heart disease, blindness, arthritis, stroke, dementia, you name it; but they develop these diseases within a two or three year timeframe, and not over decades. Many birds, too, suffer from equivalent diseases, but always towards the end of their lives. There is unquestionably a link between ageing and degenerative disease, but the nature of this link remains speculative and disputed. There are few things we can say for sure. One is that the link is not chronological—it does not depend on a fixed passage of time, but is relative to the lifespan of the creature concerned. It depends on age, not time; and the rate of ageing is broadly fixed for each species. While there is plenty of variation around the average, it is still not too far from the truth to say, with the bible, that our allotted time in this world is three score years and ten. This allotted time comes from within: it is controlled by our genes in some way, even though it can be modulated to a degree by diet and general health. When asked what finding would make him question his belief in evolution, J. B. S. Haldane answered: 'a Precambrian rabbit'. Likewise, I will cast all my views on ageing through the window when I meet a centenarian rat. A rat may one day evolve to live for a hundred years but only after changing a good many of its genes. It will no longer really be a rat.

There is a second point about the link between ageing and disease that is even more pertinent to our own suffering: degenerative disease is not an inevitable aspect of ageing. Some seabirds, for example, seem to sidestep the diseases of old age altogether, and do not age as 'pathologically' as ourselves. Like the elves they appear to live long and healthy lives, and somehow avoid many of the afflictions of old age. Exactly what they die *of* is not known with certainty, but it seems the incidence of crash landings rises with age; presumably, despite not succumbing to degenerative diseases, they begin to lose muscle power and coordination. There is a hint that the 'oldest old' among humans—those who live well past a hundred—are also less prone to degenerative diseases, and tend to die from muscle wastage rather than any specific illness.

There have been hundreds of theories of why we age. I discussed some of these in *Oxygen*, from a broad evolutionary point of view. Suffice to say here that many of the attributed causes of ageing fall prey to the traps of causality and circular argument. Some say, for example, that ageing is caused by a fall in the circulating levels of a hormone like growth hormone. Perhaps; but *why* do such hormone levels start to fall in the first place? Similarly, others hold that ageing stems from a decline in the function of our immune systems. Certainly this is a factor, but *why* does our immune function start to decline? One answer might be through an accumulation of wear and tear over many years, but this answer, albeit popular, will not do. Why do rats and humans accumulate wear and tear at such different rates? Could not a rat shielded from the slings and arrows of outrageous fortune live to a hundred? Absolutely not! Its rate of ageing is determined from within. We each hold within ourselves a ticking clock, and the speed at which the clock ticks is determined by our genes. In the jargon, ageing is endogenous and progressive: it comes from within, and it gets worse over time. Any explanation must account for these traits.

Most proposed clocks don't keep good time. The telomeres, for example—the 'caps' on the end of chromosomes that wear away over our lives at a steady rate—display such divergent patterns across species that they can't possibly be the primary cause of ageing. I have already dwelt on the metabolic rate as another clock. This, too, is commonly dismissed as the clock of life on the grounds that the relationship between metabolic rate and ageing can be badly distorted—as in the case of the pigeon with its long lifespan linked to a fast metabolic rate. Unlike the telomeres, however, these distortions offer deep insights into the underlying nature of ageing. The metabolic rate is a proxy, somewhat rough, for the rate of free-radical leakage from the respiratory chains within the mitochondria. Sometimes the rate of free-radical leakage is proportional to the metabolic rate, as among many mammals, but the relationship is not always consistent: there are many examples where the metabolic rate does not tally with free-radical leakage. Such transgressions potentially explain not just the long lifespan of birds, but also the exercise paradox—the fact that athletes, who consume far more oxygen than couch potatoes, do not age any faster, and indeed often age more slowly.

As an explanation for ageing, free-radical leakage from the mitochondria has been challenged repeatedly, and to convince must overcome a number of apparent paradoxes. Yet overcome them it does. The shape of the mitochondrial theory of ageing has transformed radically since its first exposition, more than thirty years ago. In its latest incarnation, however, it explains not just the broad outlines of ageing, but also many specific aspects such as muscular wastage, persistent inflammation, and degenerative diseases. In the final chapters, we'll see that the mitochondria are not only the main cause of ageing, but given the traits we have discussed in this book, it is inevitable that they should be. We'll also see what might be done about it, in our efforts to age with the grace of an elf.

16

The Mitochondrial Theory of Ageing

Denham Harman, pioneer of free radicals in biology, first proposed the mitochondrial theory of ageing in 1972. Harman's central point was simple: the mitochondria are the main source of oxygen free radicals in the body. Such free radicals are destructive, and attack the various components of the cell, including the DNA, proteins, lipid membranes, and carbohydrates. Much of this damage could be repaired or replaced in the usual way by the turnover of cell components, but hotspots of damage, most notably the mitochondria themselves, would be harder to protect simply by consuming dietary antioxidants. Thus, spake Harman, the rate of ageing and the onset of degenerative diseases should be determined by the rate of free-radical leakage from mitochondria, combined with the cell's innate ability to protect against, or repair, the damage.

Harman based his argument on the correlation between metabolic rate and lifespan in mammals. He explicitly labelled the mitochondria the 'biological clock'. In essence, he said, the faster the metabolic rate, the greater the oxygen consumption, and so the higher the free-radical production. We shall see that this relationship is often true, *but not always*. The proviso may seem trivial, yet it has confounded an entire field for a generation. Harman made a perfectly reasonable assumption, which has proved untrue. Unfortunately, his assumption has become entangled with the theory in general. To disprove it does not disprove Harman's theory, but it does overturn his single most important, and best-known, prediction—that antioxidants can prolong life.

Harman's sensible but confounding assumption was that the *proportion* of free radicals that leak from the mitochondrial respiratory chains is constant. He assumed that the leakage is basically an uncontrolled, unavoidable by-product of the mechanism of cell respiration, which juxtaposes the passage of electrons down the respiratory chains with a requirement for molecular oxygen. Inevitably, the theory goes, a proportion of these electrons escape to react directly with oxygen to form destructive free radicals. If free radicals leak at a fixed rate, let's say 1 per cent of total throughput, then the total leakage depends on the rate of oxygen consumption. The higher the metabolic rate, the faster the flux of electrons and oxygen, and the faster the free-radical leakage, even if the proportion of free radicals actually leaking never changes. So animals with a fast metabolic rate produce free radicals fast and have short lives, whereas animals with a slow metabolic rate produce free radicals slowly, and live a long time.

In Part 4, we saw that the metabolic rate of a species depends on its body mass to the power of 2/3: the larger the mass, the slower the metabolic rate in individual cells. This link is largely independent of the genes, depending instead on the power laws of biology. Now, if free-radical leakage depends *only* on the metabolic rate, then it follows that the only way to prolong the lifespan of a species *relative to metabolic rate*, is to bolster the strength of antioxidant (or anti-stress) protection. An implicit prediction of the original mitochondrial theory of ageing is therefore that creatures living a long time must have inherently better antioxidant protection. Birds, then, which live a long time, must stockpile more antioxidants. Accordingly, if we wish to live longer ourselves, we must seek to bolster our own antioxidant protection. Harman supposed that the only reason we have failed to extend our own lifespan by taking antioxidant therapy so far (this back in 1972) is because it is difficult to target antioxidants to the mitochondria. Many people still agree with this position today, despite another thirty-plus years of industrious failure.

These ideas are tenacious but erroneous, in my opinion: they cling to the mitochondrial theory of ageing like burrs. In particular, the idea that antioxidants might prolong our lives is the basis of a multibillion-dollar supplement industry, which has remarkably little solid evidence to support its claims; unlike the biblical house that was built on shifting sands, it has somehow remained standing. For over thirty years, medical researchers and gerontologists (including myself) have flung antioxidants at all kinds of failing biological systems and found that they just don't work. They may correct dietary deficiencies, and perhaps protect against certain diseases, but they don't affect maximal lifespan at all.

It is always difficult to interpret negative evidence. We are reminded, in that smug phrase, that 'absence of evidence is not evidence of absence'. The fact, if it is a fact, that antioxidants don't work, may always be related to the difficulties of targeting: the dose is wrong, or the antioxidant is wrong, or the distribution is wrong, or the timing is wrong. At what point are we entitled to walk away saying: 'No, this isn't a pharmacological problem—antioxidants really don't work'? The answer depends on the temperament, and there are some distinguished researchers who have yet to turn away. But the field as a whole did turn away in the 1990s. As two well-known free-radical gurus, John Gutteridge and Barry Halliwell, put it a few years ago: 'By the 1990s it was clear that antioxidants are not a panacea for ageing and disease, and only fringe medicine still peddles this notion.'

There are stronger reasons to challenge the standing of antioxidants too, and these come from comparative studies. I mentioned the prediction that animals with long lives should have high levels of antioxidants. For a time this prediction seemed to be true, but only after subjecting the data to a little innocent statistical jiggery-pokery. In the 1980s, Richard Cutler, at the National Institute of Ageing in Baltimore reported, somewhat misleadingly, that long-lived animals harbour more antioxidants than short-lived animals. The trouble was that he presented his data relative to the metabolic rate, and in so doing, he airbrushed out the far stronger association between metabolic rate and lifespan. In other words, a rat has a lower level of antioxidants than a human being, but *only* when antioxidant concentration is divided by metabolic rate, which is seven times faster in the rat; no wonder the poor rat seems so bereft of help. This manoeuvre concealed the true relationship between antioxidant levels and lifespan: a rat has actually far more antioxidants in its cells than a human being. A dozen independent studies have since confirmed that there is in fact a *negative* correlation between antioxidant levels and lifespan In other words, the higher the antioxidant concentration, the shorter the lifespan.

Perhaps the most intriguing aspect of this unexpected relationship is how closely antioxidant levels balance the metabolic rate. If the metabolic rate is high, then antioxidant levels are also high, presumably to prevent oxidation of the cell; yet lifespan is still short. Conversely, if the metabolic rate is low, then antioxidant levels are also low, presumably because there is a lower risk of cell oxidation; yet lifespan is still long. It seems that the body doesn't waste any time and energy in manufacturing more antioxidants than it needs—it uses them simply to maintain a balanced redox state in the cell (which means the dynamic equilibrium between oxidized and reduced molecules is kept optimal for the cell's function).[1] The cells of short-lived and long-lived animals maintain a similar, flexible, redox state by counterpoising the antioxidant concentration against the rate of free-radical generation; but lifespan is not affected by antioxidant concentration in any way. We are forced to conclude that antioxidants are virtually irrelevant to ageing.

These ideas are borne out by the birds, which live long lives in relation to their metabolic rate. According to the original version of the mitochondrial theory of ageing, birds ought to have higher antioxidant levels, but again this is not true. The relationship is inconsistent, but in general birds have lower antioxidant levels than mammals, reversing the predictions. Another test-case

---

[1] The status quo is also linked with tissue oxygen concentration. In Part 4, we noted that tissue oxygen levels are poised at 3 or 4 kilopascals across the entire animal kingdom. This means that both oxygen levels and antioxidant levels are balanced in the cell to retain a roughly constant redox state. We'll see why later in Part 7.

is calorie restriction. To date, calorie restriction is the only mechanism proved to extend the lifespan of mammals like rats and mice. Exactly how it works is debated, but the relationship with antioxidant levels in different species is ambiguous. Sometimes antioxidant concentrations go up, sometimes they go down, but there is no clearly consistent relationship. Even a piece of encouraging work from the early 1990s, suggesting that fruit flies live longer when genetically modified to express higher levels of antioxidant enzymes, turned out to be unrepeatable, at least in the hands of the original researchers (who make a distinction between strains that are long-lived and strains that are short-lived: higher antioxidant levels might extend the life of short-lived breeds of flies, in other words, they may correct a genetic deficiency). If any solid conclusion emerges from all this, it is certainly not that high levels of antioxidants prolong lifespan in healthy, well-nourished animals.

We've been confounded by the lure of antioxidants for a simple reason: the proportion of free radicals escaping from the respiratory chains is *not* constant—Harman's original assumption was wrong. While free-radical leakage often does reflect oxygen consumption, it can also be modulated up or down. In other words, far from being an uncontrolled and unavoidable by-product of cell respiration, the rate of free-radical leakage is controlled and largely avoidable. According to the pioneering work of Gustavo Barja and his colleagues at the Complutense University in Madrid, birds live long lives because they leak fewer free radicals from their respiratory chains in the first place. As a result, they don't need to have so many antioxidants, despite consuming large amounts of oxygen. Importantly, it seems that calorie restriction might work in a similar way. While there are various genetic changes, one of the most significant is a restriction in free-radical leakage from the mitochondria, despite similar oxygen consumption. In other words, in both long-lived birds and mammals the proportion of free radicals that leaks from the respiratory chains *decreases*.

This answer seems inoffensive enough but it is actually troublesome and hacks a hole in the established evolutionary theory of ageing. The problem is this. Animals that live a long time do so by restricting the free-radical leakage from their mitochondria. Because genes control the rate of ageing, it follows that in birds (and presumably in humans to a lesser degree) there has been selection to lower the rate of free-radical leakage. Fine. But if free radicals were simply damaging, why wouldn't a rat also do better by restricting its free-radical leakage? There seems to be no cost, indeed quite the contrary—there would be no need for the rat to go on manufacturing all those extra antioxidants to prevent itself from being oxidized. And surely it would have everything to gain, because a long-lived rat would have more time, and could leave behind more offspring. So rats, and by the same token humans, could live longer, *cost-free*, if they simply restricted free-radical leakage.

So why don't they? Is there a hidden cost, or do our ideas of ageing stand in need of radical revision? The cost of a long life is usually said to be a degree of impairment in sexuality. According to the disposable soma theory, first proposed by Tom Kirkwood, at the University of Newcastle, longevity is balanced against fecundity: long-lived species tend to have smaller litters, and rear them rather less frequently, than short-lived species. This is certainly true, at least in most known cases. The reason is less certain. Kirkwood suggested that the reason relates to the balance of resource use in individual cells and tissues: resources diverted towards attaining reproductive maturity, and raising litters, detract from those required to ensure longevity of the cell, such as DNA repair, antioxidant enzymes, and stress resistance—there are only so many ways to divide limited resources. Barja's data challenge this idea. Restricting free-radical leakage should have no cost on fecundity, as cellular damage is restricted without any need for better stress-resistance—the cost imputed in the disposable soma theory is negated. So, if the disposable soma theory is correct, there ought to be a hidden cost to restricting free-radical leakage; we shall see, in the final chapter, that there is indeed a hidden cost, and it holds vital connotations for our own quest to live longer.

To understand why, we need to consider another prediction of Harman's mitochondrial theory, which has also caused trouble. This held that free radicals don't necessarily damage the cell in general very much—they're mopped up by antioxidants—but they do specifically damage the mitochondria, especially their DNA. Harman actually mentioned mitochondrial DNA only in passing, but its involvement later became a fundamental tenet of the theory. Let's see why, for the gap between the predictions and hard prosaic reality reveals a great deal about what's really going on.

Mitochondrial mutations

Harman argued that, because free radicals are so reactive, those escaping from the respiratory chains should mainly affect the mitochondria themselves—they should react on the spot, where they were produced, and not damage distant locations very much. He then asked, quite perceptively, whether the gradual decay of mitochondria with increasing age 'might be mediated in part through alteration of mitochondrial DNA functions?' The chain of effect would be as follows: free radicals escape the respiratory chains and attack the adjacent mitochondrial DNA, causing mutations that undermine mitochondrial function. As mitochondria decay, the performance of the cell as a whole declines, leading to the traits of ageing.

Harman's perceptive question was addressed more explicitly a few years later by Jaime Miquel and his colleagues in Alicante, Spain. Their formulation in 1980 is still the most familiar version of the mitochondrial theory of ageing today, even though many aspects don't really fit the data, as we'll see. It goes something like this. Damage to proteins, carbohydrates, lipids, and so on can be repaired, and is not dangerous unless the rate of damage is overwhelmingly fast (as it might be, for example, after radiation poisoning). DNA is different. Although DNA damage can also be repaired, there are occasions when the damage confounds the original sequence and mutations occur. Mutations are heritable changes in DNA sequence. Except by random back-mutation to the original sequence, or recombination with another strand of un-mutated DNA, there is no way that the original sequence can be recovered. Not all mutations affect protein structure and function, but some of them certainly do. In the usual way of things, the more mutations, the greater the chance of detrimental effects.

In theory, mitochondrial mutations accumulate with age. As they do so, the efficiency of the system as a whole begins to break down. It's not possible to fashion a perfect protein from an imperfect set of instructions, so a certain degree of inefficiency is built in. Worse, if the mutations affect the respiratory chains in mitochondria, then the rate of free-radical leakage rises, spinning the whole vicious cycle faster and faster. Such positive feedback ultimately builds up into an 'error catastrophe', in which the cell loses all control over its function. When this fate has overcome a sizeable proportion of the cells in a tissue the organs fail, placing the remaining functional organs under still greater strain. The inevitable outcome is ageing and death.

So what are the chances that mutations would affect the respiratory chain proteins? It's overwhelmingly likely. We have seen that thirteen of the core respiratory proteins are encoded by mitochondrial DNA, which is anchored to the membrane right next to the respiratory chains. Any escaping free radicals are virtually bound to react with this DNA: it's only a matter of time before mutations occur. And we have seen that proteins encoded in the mitochondria interact intimately with those encoded in the nucleus. An alteration in either party can erode this intimacy, and affects the function of the respiratory chain as a whole.

If this sounds grim, it gets worse. A succession of dire findings made the whole set-up seem like a bad joke perpetrated by a satanic biochemical deity. We were told that mitochondrial DNA is not just stored in the incinerator, but it's also stripped of the normal defences: it's not wrapped in protective histone proteins; it has little ability to repair oxidative damage; and the genes are packed together so tightly, without the cushioning of 'junk' DNA, that a mutation anywhere is likely to cause havoc. This dire scenario was set off by a sense of pointlessness: most mitochondrial genes had already been transferred to the nucleus, and the handful that remained seemed to be in the wrong place. Aubrey de Gray, one of the most original and dynamic thinkers in this field, has even suggested we could cure the ravages of ageing by transferring the rest of the mitochondrial genes across to the nucleus. I disagree, for reasons that we'll come to, but it's easy to feel his point.

Why on earth did such a daft system evolve? That depends on one's view of evolution. Stephen Jay Gould used to vent his frustrations about what he called the 'adaptationist program' in biology—the assumption that all is adaptation, in other words, everything has a reason, however innocent of function it may appear, and is shaped by the processes of natural selection. Even today, biologists can be split into those who are reluctant to believe that nature does anything for nothing, and those who believe that some things are just beyond direct control. Is 'junk' DNA really junk or does it have some unknown purpose? We don't know for sure, and the answer you get will depend on whom you ask. Similarly, the 'point' of ageing is disputed. The most widely accepted view is that we are less likely to reproduce as we grow older, so natural selection is less able to weed out the genetic variants that cause damage late in life. Because mutations in mitochondrial DNA build up late in life, natural selection is unable to come up with an efficient mechanism of eliminating them. Only when the expected lifespan increases, as it does in animals isolated on an island without predators, or in birds that can fly away, or humans who use their large brains and social structures, can selection act to prolong life. If we subscribe to this view, then the sheer lunacy of storing badly protected mitochondrial genes in the incinerator is just one of those things: an accident of evolutionary history.

Is this nihilistic vision correct? I don't think so. The fault is that the line of reasoning is too stiffly chemical: it doesn't take into account the dynamism of biology. We'll see the difference this makes later. Nonetheless, it is a bold theory, and it has the great merit of making some explicit testable predictions. There are two in particular that we'll look into. First, the theory predicts that mitochondrial mutations are sufficiently corrosive to bring about the whole sorry trajectory of ageing. This, we'll see, is likely to be true. But the second prediction is probably false, at least in the full diabolical sense in which it was originally put forward—that mitochondrial mutations should accumulate with age. As they do they ought, ultimately, to bring about an 'error catastrophe'. There's little strong evidence to say that this happens. And therein lies the secret.

Mitochondrial diseases

The first report of a patient suffering from a mitochondrial disease was in 1959, a few years before the discovery of mitochondrial DNA. The patient was a 27-year-old Swedish woman, who had the highest metabolic rate ever recorded in a human being, despite a completely normal hormonal balance. It turned out the problem was a defect in mitochondrial control—her mitochondria respired at full blast even when there was no need for ATP. As a result, she ate prodigiously, yet always remained thin, and she sweated profusely even in winter. Sadly her doctors couldn't help her and she committed suicide ten years later.

A number of other patients were diagnosed with mitochondrial diseases over the following two decades, usually on the basis of their clinical records and various specific tests. For example, in many cases when the mitochondria are not functioning properly, lactic acid (a product of anaerobic respiration) accumulates in the blood, even when the patient is at rest. Biopsies of the muscles often show that some muscle fibres—typically not all—are badly damaged. They stain red on histological preparation and are known as 'ragged red fibres'. When subjected to biochemical tests, the mitochondria in these fibres are found to lack the terminal enzyme in the respiratory chain, cytochrome oxidase, rendering them incapable of respiration.

From a clinical point of view, such reports were little more than sporadic scientific curiosities. But a tidal change followed after 1981, when the dual Nobel laureate Fred Sanger and his team in Cambridge reported the complete sequence of the human mitochondrial genome. Through the 1980s and 1990s, as sequencing technology improved, it became possible to sequence the mitochondrial genes of many patients with suspected mitochondrial diseases. The results were startling, and showed not only how common mitochondrial diseases are—about 1 in 5000 people are born with mitochondrial disease—but also how bizarre. Mitochondrial diseases flout the normal rules of genetics. Their pattern of inheritance is peculiar, and often does not follow Mendel's laws.[2] The onset of symptoms may vary by decades, and occasionally diseases vanish altogether in individuals who 'should' (theoretically) have inherited them. In general, mitochondrial diseases progress with advancing age, so a disease that causes little inconvenience at the age of twenty may become utterly debilitating by the age of forty. Beyond this, however, very few generalizations can be made. Diverse tissues can be affected in individuals who carry the same mutation, while different mutations may well affect the same tissue. If you want to retain your sanity, don't try to read a textbook on mitochondrial diseases.

Despite such gross difficulties categorizing mitochondrial disease, a few general principles do help to explain the thrust of what's happening. These same principles are pertinent to ageing. Recall from Part 6 that mitochondria are nor-

---

[2] Mendel's laws govern the pattern of inheritance of 'normal' nuclear genes, in which the likelihood of a genetic trait or disease can be calculated according to the probability that an individual will inherit at random one of two copies of the same gene from each parent, giving everyone two copies of each gene. In fact, some mitochondrial diseases do follow Mendel's laws, as they are caused by nuclear genes encoding proteins destined for the mitochondria.

mally inherited from the mother, but even so, the variation in mitochondrial DNA in egg cells is surprisingly high. We saw that some degree of heteroplasmy (a mixture of genetically different mitochondria) is found in about half of the egg cells taken from the same ovary of a normal fertile woman. If these variations do not affect the functional performance of the mitochondria too seriously, then they are not eliminated during embryonic development.

But why would a defect *not* affect embryonic development? There are various possibilities. One is that the defective mitochondria are inherited in low numbers. All mitochondrial diseases are heteroplasmic, which is to say there are both normal and abnormal mitochondria. Of the 100 000 mitochondria inherited in the egg, if only 15 per cent are abnormal then the healthy majority might mask their shortcomings. Alternatively, the mutation may be less harmful, but present in a greater proportion of mitochondria, perhaps 60 per cent. If so, then the embryo can still develop normally, despite the large number of mutants. And then there is the matter of segregation. When a cell divides, its mitochondria partition themselves at random between the two daughter cells. One cell might inherit all the defective mitochondria, whereas the other gets none, or there could be any combination in between. In the developing embryo, individual cells provide the mitochondria that eventually populate different tissues with different metabolic requirements. If the cells that develop into long-lived, metabolically active tissues, such as muscle, heart, or brain, happen to inherit defective mitochondria in high numbers, then all might be lost; but if instead the defective mitochondria end up in short-lived, or less metabolically active cells, like skin cells or white blood cells, then embryonic development might well be normal. As a result of such differing thresholds, the more serious mitochondrial diseases affect long-lived, energetically active tissues, especially muscle and brain.

There are parallels with ageing here. We don't inherit all of our defective mitochondria from the egg cell: some accumulate in adult life, due to free radicals formed by normal metabolism. This generates a mixed population of mitochondria in the cells affected. What happens next depends on the type of cell. If the cell is an adult stem cell (responsible for regenerating tissue), a possible outcome would be the clonal expansion of defective mitochondria. This happens in some muscle fibres, producing the 'ragged red fibres' characteristic of mitochondrial diseases, but also found in 'normal' ageing. Conversely, if the mutation affected a long-lived cell no longer capable of division, such as a heart-muscle cell or a neurone, then the mutation could not spread beyond the bounds of that single cell. We would then expect to see different mutations in different cells, forming a 'mosaic' of disparate mitochondrial function.

Another aspect of mitochondrial disease is pertinent to normal ageing—their tendency to progress (to become more debilitating) with advancing age. The reason relates to the metabolic performance of tissues and organs. As we've seen, each organ has a threshold of function that is required for its normal performance. Symptoms only set in when an organ's performance falls below its own particular threshold. So for example we might be able to lose one kidney altogether and still function normally, but if the second kidney begins to fail too we'll die, unless we receive dialysis or have a transplant. Because all work costs energy, an organ's threshold depends on its metabolic requirements. Mitochondrial diseases are less serious if they happen to affect tissues with low metabolic requirements, like the skin, and are worse if they affect active cells like muscle cells. A similar process takes place in tissues as their cells age. A youthful muscle cell, in which 85 per cent of its mitochondria are 'normal', can handle all the energetic demands imposed on it in youth; but as its mitochondria decline with age, the energy demands placed on the remaining mitochondria rise. We draw closer to the metabolic threshold. As a result, the impairment caused by a mutant population is progressively unmasked as we get older.

But are mitochondrial mutations sufficiently corrosive to bring about the whole sorry trajectory of ageing? Some of them certainly are. An awful condition, in which apparently normal babies lose their mitochondrial DNA soon after birth, leads swiftly to liver and kidney failure. When the disease is severe, as much as 95 per cent of mitochondrial DNA can be depleted, and the afflicted babies die in weeks or months, despite having appeared completely normal at birth. More common diseases include Kearns Sayre syndrome and Pearson's syndrome, which cause substantial disability later in life, and premature death. Typical symptoms are similar to those caused by slow poisoning with a metabolic toxin like cyanide, and include loss of coordination (ataxia), seizures, movement disorders, blindness, deafness, stroke-like symptoms, and muscular degeneration. One mitochondrial mutation has even been associated with a condition similar to Syndrome X—that deadly combination of high blood pressure, diabetes, and raised cholesterol and triglycerides, said to affect 47 million Americans. Clearly, mutations in mitochondrial genes *can* have very serious effects, corrosive enough to underpin ageing. But other mitochondrial conditions are far less serious, and herein lies the problem.

The severity of any mitochondrial disease depends on the proportion of mutant mitochondria and the tissue in which they find themselves, but also on the type of mutation: on which bit of the genome it affects. If the mutation affects a gene for a particular protein, the effects may or may not be catastrophic; indeed they might even be beneficial. On the other hand, if the mutation affects a gene encoding RNA, the consequences are usually serious. Depending on the type of RNA, the mutation could alter the synthesis of all mitochondrial proteins, or all proteins containing a particular amino acid. Mutations in the control region also potentially have serious consequences, as these might alter the entire dynamic of mitochondrial replication and protein synthesis in response to changing demand.

Mutations can also occur in the nuclear genes encoding mitochondrial proteins, with similar consequences (except that these mutations follow a typical Mendelian inheritance pattern, with one gene coming from each parent; see footnote, page 281). If the nuclear mutation affects a mitochondrial transcription factor, which controls the synthesis of mitochondrial proteins, then the effects could in principle apply to all the mitochondria in the body. On the other hand, some mitochondrial transcription factors appear to be active only in particular tissues, or in response to particular hormones. A mutation in the gene for these would tend to have tissue-specific effects.

Taken together, these considerations explain the extreme heterogeneity of mitochondrial diseases. A mutation may affect a single protein, or all proteins containing a particular amino acid, or all mitochondrial proteins altogether, or the rate of protein synthesis in response to changing demands. The mutations may be tissue-specific or global, affecting all the body. They may be inherited in a classic Mendelian fashion, if the mutations are in nuclear genes, or they may be inherited from the mother only if the mutations are in the mitochondrial genes. If the latter, the effects depend on the proportion of mitochondria affected, as well as the way in which they segregate in dividing cells during embryonic development, and the metabolic threshold of the organs involved.

Given this degree of heterogeneity, the problem is the very spectrum of disease. While we tend to succumb to our own particular mixture of degenerative diseases, the underlying process of ageing is similar in all of us. How is it that we all share such a basic underlying similarity, not just with each other, but also with other animals that age at utterly different rates? If we accumulate mitochondrial mutations at random as we age, why do we not all age in very different ways and at different rates, as random and varied as the mitochondrial diseases themselves? The answer might conceivably lie in the nature of the mutations that accumulate, but this leads straight to a second problem: the magnitude and type of mutations that *do* accumulate don't seem sufficient to cause ageing. So what's going on?

The paradox of mitochondrial mutations in ageing

The pursuit of genetic mutations in ageing has proved to be a frustrating occupation. One promising theory held that nuclear mutations, accumulating over a lifetime, were the main culprit behind ageing. While nobody would dispute that mutations in nuclear genes do contribute to ageing, and especially to diseases like cancer, there is no relationship at all between lifespan and the accumulation of mutations in the nucleus, so they can't be the primary cause.

The generally accepted evolutionary theory of ageing, first postulated by J. B. S. Haldane and Peter Medawar, is a variation on the theme of mutations: if they don't accumulate over a lifetime, perhaps they accumulate over many lifetimes. Natural selection has no power to eliminate the genes that defer their detrimental effects until later in life. The classic example is Huntington's disease, which sets in well after reproductive maturity, enabling the gene to be passed on to children, so it can't be eliminated by selection. While Huntington's disease is particularly horrible, how many other genes have similar late effects? Haldane proposed that ageing was little more than a dustbin of late-acting gene mutations, hundreds or thousands of them, which could not be eliminated by natural selection, and thus accumulated over many generations. Again, there must be some truth in this idea, but I don't think it can be squared with the plasticity of ageing observed in nature. Nearly two decades of genetic studies have shown that lifespan can be extended dramatically, even in some mammals, by single point mutations in critical genes. If ageing really was written into the actual *sequence* of hundreds or thousands of genes, surely this couldn't happen. Even if one critical gene controls the activity of many others, the subordinate genes are still mutated—it is their sequence, not their activity, which is the problem. To correct the sequences, there would need to be thousands of simultaneous mutations in all the right genes, to begin to have an effect on lifespan, and this would surely take several generations at the least. For whatever reasons, the fact is that lifespan is regulated, with a surprising degree of control, throughout the animal kingdom.

Mitochondrial mutations have their own story, and this, too, is difficult to reconcile with the facts. On the face of it, though, mitochondrial genes hold more promise to explain ageing and disease. There are two reasons. First, as we saw in the last chapter, mitochondrial mutations accumulate far more quickly, from one generation to the next, than do nuclear mutations. It follows that mitochondrial mutations are more likely to build up within a single lifetime, and so in principle could tally with the rate of decline in ageing. And second, these mutations *do* have the corrosive power to undermine lives: they are not at all minor bit-players. The mitochondrial diseases underscore just how devastating such mutations can be.

So how fast do mitochondrial mutations really accumulate? It's difficult to say for sure because the rate of change over generations is restrained by natural selection. For most mitochondrial genes, the evolution rate is about 10- to 20-fold the rate found in nuclear genes, but in the control region it can be as much as fifty times faster. Because mutations are only 'fixed' in the genome if they don't cause catastrophic damage (otherwise they are eliminated by selection) the actual rate of change must be faster than this. To get an idea of how fast the change might really be, Anthony Linnane, at Monash University in Australia, and his collaborators at the University of Nagoya in Japan, considered yeast in their classic *Lancet* paper published in 1989. Yeast is revealing because, as any brewer or vintner knows, it doesn't depend on oxygen: yeast can also ferment to produce alcohol and carbon dioxide. Fermentation takes place outside the mitochondria, so yeast can tolerate serious damage to their mitochondria and still survive. Such damage was first noted in the 1940s with the discovery of 'petite' strains of yeast, whose growth is stunted. It turned out that the petite mutation involves the deletion of a large section of mitochondrial DNA, rendering the dispossessed mutant unable to respire. Critically, the petite mutation crops up spontaneously in cell cultures at a rate of about 1 in 10 to 1 in 1000 cells, depending on the yeast strain. In contrast, nuclear mutations occur at an almost infinitesimally slow rate, which is similar in both yeast and higher eukaryotes like animals—about 1 in every 100 million cells. In other words, if yeast is anything to go by, mitochondrial mutations accumulate at least 100 000 times faster than nuclear mutations. If such a fast mutation rate is true of animals too, then it could certainly account for ageing; indeed it would be hard to explain why we don't drop dead almost immediately.

The search was on: how quickly do mitochondrial mutations accumulate in the tissues of animals and people? It has to be said that this area is controversial and a consensus is only now emerging. Part of the problem is the technology used to measure the mutations: the techniques used to sequence DNA letters sometimes amplify mutated sequences at the expense of 'normal' sequences, making it very difficult to quantify the extent of damage. In consequence, results from different laboratories can be extremely variable, sometimes ranging by as much as 10 000-fold. As so often in all walks of life, those people who hope to find mitochondrial mutations tend to find them, whereas the doubters inevitably find but a few. This is almost certainly not because researchers are deliberately fabricating their data, but rather where and how they look: it's possible for both sides to be right.

Against this background, it's perhaps rash for me to attempt a categorical statement, but attempt I shall. The picture that is beginning to emerge does indeed suggest that both sides are right. It seems there is a difference in the fate of mutant mitochondria, depending on the location of the mutation—whether this lies in the control region of the mitochondrial genome, or in a coding region.

Mutations in the control region (which binds the factors responsible for copying mitochondrial DNA) can prosper, and even stage a clonal take-over of entire tissues. These mutations don't necessarily cause much functional deficit. A ground-breaking study, published in *Science* in 1999 by Giuseppe Attardi (one of the pioneers of mitochondrial DNA sequencing) and his colleagues at Caltech, showed that individual mutations in the control region can accumulate at over 50 per cent of the total mitochondrial DNA in the tissues of older people, but are largely absent in young people. We can take it that certain types of mutation *do* build up with age to high levels, but we can't say whether these mutations are harmful, as they don't affect the protein-coding genes. Certainly not all of them are harmful. Another important study published by Attardi's group in 2003 showed that one mutation in the control region is actually associated with *greater* longevity in an Italian population. In this case the mutation, a single letter change, cropped up five times more often in centenarians than in the general population, implying that it might have offered some kind of survival advantage.

In contrast, mutations in the functional protein- or RNA-coding regions very rarely accumulate at levels above about 1 per cent, which is far too low to cause a significant energy deficit. Interestingly, functional mitochondrial mutations, such as cytochrome oxidase deficiency, *are* amplified clonally within particular *cells*, so that the mutants come to predominate in those cells. This happens, for example, in some neurones, heart-muscle cells, and indeed the ragged red fibres of ageing muscles. However, the total proportion of such mutants in the tissue as a whole rarely rises above 1 per cent. There are two possible explanations for this. One is that different cells accumulate different mutations, so that any particular mutation is just the tip of an enormous iceberg of diverse mutations. The other explanation is that most mitochondrial mutations simply don't accumulate at very high levels in ageing tissues. Perhaps surprisingly, it's this latter explanation that seems closest to the truth. Several studies have shown that most mitochondria in ageing tissues have basically normal DNA, except perhaps in the control region, and moreover, are capable of virtually normal respiration. Given that the proportion of mutant mitochondria needed to undermine a cell's performance in mitochondrial diseases is as high as 60 per cent, a total mutation load of just a few per cent seems insufficient to account for ageing, at least by the tenets of the original mitochondrial theory.

So what's going on here? I find myself asking the obtuse question, 'Are we really so different to yeast?' I doubt that question will cost many readers much sleep. But it ought to! Yeasts accumulate mitochondrial mutations rapidly, yet for the most part, we don't. From an energetic point of view, we function in a similar way to yeast; the only difference is that we depend on our mitochondria, whereas yeasts don't. Perhaps this difference gives the game away—necessity. Let's say we accumulate mutations in the control region simply because they don't matter a great deal: they have little impact on function (as indeed is implicit in the studies of human inheritance discussed in Part 6), whereas most functional mutations do *not* accumulate because they *do* matter. That sounds fair enough, but it implies that selection for the best mitochondria is taking place in tissues (even in tissues composed of long-lived cells like heart and brain). So we are faced with two possibilities. Either the mitochondrial theory of ageing is completely wrong, or mitochondrial mutations do occur at a similar rate to yeasts, but the mutants are eliminated by selection within a tissue for the best mitochondria. If so, then mitochondrial function must be far more dynamic than had been envisaged in the original mitochondrial theory of ageing. Which is it?

17

Demise of the Self-Correcting Machine

After the previous chapter, you might be forgiven for supposing that the mitochondrial theory of ageing is claptrap. After all, most of its predictions seem utterly false. One prediction is that antioxidants should prolong maximal lifespan, and this does not seem to be true. Another is that mitochondrial DNA mutations should accumulate with ageing, but only the least important ones actually do. Another is that the proportion of free radicals escaping the respiratory chains is constant, so lifespan should vary with metabolic rate; but this is only true in general, and fails to explain exceptions like bats, birds, humans, and the exercise paradox (the fact that athletes, who consume more oxygen over a lifetime, don't age faster than couch potatoes). In fact, the only prediction of the original theory that seems to be true is that mitochondria are the main source of free radicals in the cell. Hardly the lineaments of a vigorous, healthy theory.

It's time to return to an idea that we parked in the previous chapter: the proportion of free radicals escaping from the respiratory chains is *not* constant and unavoidable, but is subject to natural selection. Over evolutionary time, the rate of free-radical leakage is set at the optimal level for each species. In this way, long-lived animals have a fast metabolic rate while leaking relatively few free radicals, whereas short-lived animals typically combine a fast metabolic rate with leaky mitochondria and plenty of antioxidants. We posed the question: What is the cost of well-sealed mitochondria? Why would a rat *not* benefit from cutting back its investment in antioxidants by sealing its mitochondria better? What does it have to lose?

Let's think all the way back to Part 3, and in particular to John Allen's explanation for the very existence of mitochondrial DNA (see pages 141–144). You might recall that he argued it was no fluke that a hard-core of genes survives *in every species* that relies on oxygen for respiration. The reason, he suggested, was to balance the requirements of respiration, as an imbalance in the components of the respiratory chain can lead to inefficient respiration and free-radical leakage. We saw that it is necessary to retain a local contingent of genes to pinpoint the need for reinforcements in particular mitochondria, rather than all mitochondria at once, regardless of need, which is what would happen if control were retained by the bureaucratic confederation of genes in the nucleus. Allen's essential point is that the mitochondrial genes survive because the benefits of keeping them on site outweigh the disadvantages.

How might one particular mitochondrion signal its need to produce more respiratory chain components? We're now entering into the realms of twenty-first-century science, and it's best to admit that at present little is known. As we saw in Part 3, Allen suggests that they do so by modulating the rate of free-radical production from the respiratory chains: the free radicals themselves act as the signal to start building more respiratory complexes. This suggests an immediate reason why a rat might lose out by restricting free-radical leakage— it would muffle the strength of the signal, and so require a more refined detection system. We'll see later how birds might have got around this problem, and why it wasn't worth it for rats.

What might happen if there is not enough cytochrome oxidase in a particular mitochondrion? This is the scenario that we considered in Chapter 8. Respiration becomes partially blocked, and electrons back up in the respiratory chains, making them more reactive. Oxygen levels rise, as less is consumed by respiration. The combination of high oxygen with slow electron flow means that free-radical production increases. According to Allen, this is exactly the signal required to produce more complexes, to correct the deficit. How the mitochondria detect the rise in free-radical leakage is unknown, but various plausible possibilities exist. For example, mitochondrial transcription factors (which initiate protein synthesis) might be activated by free radicals; or the stability of the RNA might depend on free-radical attack. Examples of both are known, but neither has been proved to take place in the mitochondria. Either way, the rise in free-radical leakage should lead to more core respiratory proteins being made from mitochondrial DNA. These insert themselves into the inner membrane, and once implanted they behave as beacons and assembly points for the additional proteins encoded by the nuclear genes. When the full complex is assembled, the blockage of respiration is corrected. Free-radical leakage falls again, and so the system is switched off. Overall, then, this system behaves like a thermostat, in which a fall in room temperature is itself the signal used to switch on the boiler. The rising temperature then switches off the boiler, so regulating the room temperature between two fixed limits. But of course, if the room temperature didn't fluctuate up and down, the system couldn't work at all. Similarly, if the rate of free-radical leakage from respiratory chains didn't fluctuate, there could be no self-correction to an appropriate number of respiratory complexes.

What happens if the free-radical signalling fails? If the synthesis of new respiratory proteins from the mitochondrial genes fails to stem the leak of free radicals, then the lipids of the inner membrane, such as cardiolipin, are oxidized. In Part 5, we noted that cardiolipin binds cytochrome oxidase, so if cardiolipin is oxidized, cytochrome oxidase is released from its shackles. This in turn blocks the passage of electrons down the respiratory chain altogether, so respiration grinds to a halt. Without the constant flow of electrons needed to maintain it, the membrane potential collapses, and apoptotic proteins are spilled out into the cell. If this turn of events happens in a single mitochondrion, the cell does not necessarily commit apoptosis—there appears to be a threshold. When only a few mitochondria expire at one moment, then the signal for apoptosis is not strong enough to cause the cell to die, and instead the mitochondria themselves are broken down. In contrast, if a large number of mitochondria simultaneously spill out their contents, then the cell as a whole does get the point, and goes on to commit apoptosis.

This flexible signalling system is a far cry from the spirit of the original mitochondrial theory of ageing. The original theory supposed that free radicals are purely detrimental; that the continued existence of mitochondrial DNA was a diabolical fluke of evolution; and that free-radical damage spiralled out of control, leading to the degeneration and misery of ageing. We now appreciate that free radicals are *not* purely detrimental—they carry out an essential signalling role—and that the bizarre survival of mitochondrial DNA is not a fluke, but is actually necessary for cellular and bodily health. Furthermore, mitochondria are better protected against free-radical damage than had once been assumed. Not only is mitochondrial DNA present in multiple copies (usually five to ten copies in every mitochondrion), but recent work shows that mitochondria are reasonably efficient at repairing damage to their genes, and (as we saw in Part 6) are capable of recombination to fix genetic damage.

So where does all this leave the mitochondrial theory of ageing? Perhaps surprisingly, it's not dead and buried, merely radically transformed. A new theory has emerged, phoenix-like, from the ashes, but it still places a premium on the free radicals generated by the mitochondria. This new theory is not attributable to any one mind in particular, but has gradually condensed out of the work of researchers in several related fields. Beyond being consistent with the data, the new theory has the immeasurable benefit that it gives a deep insight into the nature of the diseases of old age, and how modern medicine might set about curing them. Critically, the best way to tackle them is not to target each one individually, as medical research currently does, but to target all of them simultaneously.

The retrograde response

We have seen that the mitochondria operate a sensitive feedback system, in which the leaking free radicals themselves act as signals to calibrate and adjust performance. But the fact that free radicals play an integral part in mitochondrial function does not mean that they are not toxic too. Clearly they are, even if rather less so than is shouted about in health magazines. Lifespan *does* correlate with the rate of free-radical leakage from respiratory chains. While a good correlation doesn't necessarily imply a causal link, it's hard to claim causality in the absence of any correlation at all. If two factors are not linked in any way, then one can hardly be said to 'cause' the other; and there are remarkably few, if any, other factors which correlate with lifespan across radically different groups, such as yeast, nematodes, insects, reptiles, birds, and mammals. For the sake of argument, let's assume that free radicals *do* cause ageing. How can we square their signalling role with a more conventional idea of their toxicity— and with the evidence to date?

Yeast accumulate mitochondrial mutations at least 100 000 times faster than nuclear mutations. People, too, accumulate particular types of mitochondrial mutation with age, notably those in the 'control' region. Importantly, the control region mutations can often stage a 'take-over' of whole tissues, so that the same mutation is found in practically all the cells. In contrast, mutations in the coding regions of the mitochondrial genome can be amplified within particular cells, but only very rarely attain levels of above 1 per cent in the tissue as a whole. I suggested that this smells suspiciously of selection acting in tissues. Can we perhaps link the signalling role of free radicals with a weeding-out of detrimental mitochondrial mutations? Indeed we can, and this is the crux of the 'new' mitochondrial theory of ageing.

What might happen to the calibration of mitochondrial function if there is a spontaneous mutation in mitochondrial DNA? Let's think it out step by step. If the mutation is in the control region, it doesn't affect gene sequence, but it might affect the binding of transcription or replication factors. If the effect is not entirely neutral, the mutant mitochondrion would tend to copy its genes either more often or less often in response to an equivalent stimulus. So what is the outcome? If the mutation makes a mitochondrion 'fall asleep' on duty, so it responds sluggishly to signals for replication, the likely outcome is that the mutant mitochondria would simply disappear from the population. In response to a signal to divide, 'normal' mitochondria would divide, but the mutant mitochondria would slumber on. Their population would fall relative to normal mitochondria, and they would eventually be displaced altogether in the normal turnover of cellular components.

In contrast, if the mutation made the mitochondrion *more* alacritous in its response to an equivalent signal, we would expect to see an expansion of its DNA. At every signal to divide, the mutant mitochondria would leap into action, and so would eventually displace the 'normal' mitochondria from the population. And if the mutation occurred in a stem cell (which gives rise to replacement cells in a tissue) the mutants would be more likely to be passed on every time the stem cell divided, and so would finally take over the entire tissue. It's important to note that such mutations are most likely to stage a tissue takeover if they're not particularly detrimental to mitochondrial function. This is likely to be true, as there is nothing the matter with the respiratory complexes themselves. Energy generation can continue normally, if a degree out of synch with requirements; and as we've seen (page 287), Giuseppe Attardi's group has shown that one control-region mutation is actually beneficial.

So what happens if a mutation is in the coding region of a gene? Why do such mutations take over individual cells, but not the tissue as a whole? This time it's more likely that mitochondrial function will be altered. Let's imagine that the mutation affects cytochrome oxidase in some way. Given the need for nano-scopic precision in the interactions of different subunits, the likelihood is that respiration will be impaired, and electrons will back up in the respiratory chains. Free-radical leakage increases, and this signals the synthesis of new respiratory chain components. This time, however, building new complexes cannot correct the deficit, for these, too, would be dysfunctional (although if the deficit is modest, it may help a little). What happens next? The outcome is *not* the error catastrophe proposed in the original version of the mitochondrial theory, but more signalling. The defective mitochondrion signals its deficiency to the nucleus, by way of a feedback pathway known as the 'retrograde response', which enables the cell to compensate for its deficit.

The retrograde response was originally discovered in yeast, and was so named because it seems to reverse the normal chain of command from the nucleus to the rest of the cell. In the retrograde response, it is the mitochondria that signal to the nucleus to change its behaviour—the mitochondria, not the nucleus, set the agenda. Since its discovery in yeast, some of the same bio-chemical pathways have been found to operate in higher eukaryotes too, including humans. While the exact signals almost certainly differ in detail and meaning, the overall intention appears to be similar—to correct the metabolic deficiency. Retrograde signalling switches energy generation towards anaer-obic respiration, such as fermentation, and in the longer term stimulates the genesis of more mitochondria. It also fortifies the cell against stress, aiding survival in the more trying times ahead. Yeast, which don't depend on their mitochondria to survive, actually live longer when the retrograde response is active. With our dependence on mitochondria, it's unlikely that similar benefits would apply to people; for us, the purpose of the retrograde response is to correct the mitochondrial deficiency. But I suppose we could be said to live longer in the sense that, without it, we would certainly live 'shorter'.

Paradoxically, in the long term, a cell can only correct against energetic defi-ciency by producing more mitochondria. If the mitochondria are defective, the cell attempts to correct the problem by producing more mitochondria—hence the tendency of defective mitochondria to 'take over' cells. For many years, cells can preferentially amplify the least-damaged mitochondria. The overall mitochondrial population is in continuous flux, with a turnover time of perhaps several weeks. Mitochondria either divide, if their energy deficit is fairly mild, or they die. The mitochondria that die are broken down, and their constituents recycled by the cell. This means that the most damaged mitochondria are continuously eliminated from the population. In this way, cells can spin out their lives almost indefinitely by constantly correcting the deficit. Our neurones, for example, are usually as old as we are ourselves: they are rarely, if ever, replaced, yet their function doesn't spiral out of control in an error catastrophe, but rather declines imperceptibly. What isn't possible though, is any return to the fountain of youth. While the most devastating mitochondrial mutations can be eliminated from cells, there is no way of restoring their pristine function, short of not using the mitochondria at all (which is how egg cells, and to a degree adult stem cells, *do* reset their clocks).

The more a cell relies on defective mitochondria, the more oxidizing the intra-cellular conditions become (oxidizing means a tendency to steal electrons). When I say 'oxidizing', however, I don't mean the cell loses control of its internal environment. It retains control by adapting its behaviour, establishing a new status quo. Most proteins, lipids, carbohydrates, and DNA are not affected by the change—again, in disagreement with the predictions of the original mitochondrial theory, which anticipated evidence of accumulating oxidation. Most studies searching for such evidence have failed to find any serious difference between young and old tissues. What *is* affected is the spectrum of operative genes, and there is plenty of evidence to support this change. The shift in operative genes hinges on the activity of transcription factors—and the activity of some of the most important of these depends on their redox state (which is to say, whether they're oxidized or reduced, having lost or gained electrons). Many transcription factors are oxidized by free radicals, and reduced again by dedicated enzymes; the dynamic balance between the two states determines their activity.

The principle here is similar to lowering a canary down a mine-shaft, to test for poisonous gases. If the canary is dead on raising it from the shaft, then the miner can take an appropriate precaution, such as only venturing down if wearing a gas mask. Redox-sensitive transcription factors behave like the canary, warning the cell of impending danger, and enabling it to take evasive action. Rather than the fabric of the cell as a whole being oxidized, which is as much as to say *dead*, the 'canary' transcription factors are oxidized first. Their oxidation sets in motion the changes necessary to prevent any further oxidation. For example, NRF-1 and NRF-2 (the 'nuclear respiratory factors') are transcription factors that coordinate the expression of genes needed for generating new mitochondria. Both factors are sensitive to redox state, which dictates the strength of their binding to DNA. If the conditions in the cell become more oxidizing, then NRF-1 stimulates the genesis of new mitochondria to restore the balance, and for good measure also induces the expression of a battery of other genes, which protect against stress in the interim. NRF-2 appears to do the opposite, becoming more active when the conditions are 'reducing', and falling inactive when oxidized.

When the cell drifts to a more oxidizing internal state, then, a small posse of redox-sensitive transcription factors shifts the spectrum of active nuclear genes. The shift is away from the normal 'house-keeping' genes and towards those genes that protect the cell against stress, including some mediators that summon the help of immune and inflammatory cells. I argued in *Oxygen* that their activation helps to account for the chronic low-grade inflammation that underpins the diseases of advancing age, such as arthritis and atherosclerosis. While the exact spectrum of active genes varies from tissue to tissue, and with the degree of stress, in general the tissues establish a new 'steady-state' equilibrium, in which more resources are directed towards self-maintenance, and so fewer can be dedicated to their original tasks. This situation is liable to be stable for decades. We may notice we have less energy, or take longer to recover from minor ailments, and so on, but we're hardly in a state of terminal decline.

So, overall, what happens is this. If conditions become oxidizing within a particular mitochondrion, then the mitochondrial genes are actively transcribed to form more respiratory complexes. If this resolves the situation, then all is well and good. However, if this *fails* to resolve the situation, then conditions in the cell as a whole become more oxidizing, and this activates transcription factors like NRF-1. Their activation shifts the spectrum of nuclear genes in operation, which in turn stimulates the genesis of more mitochondria and protects the cell against stress. The new arrangement stabilizes the cell again, albeit in a new status quo that can influence vulnerability to inflammatory conditions. But there is little oxidation of the fabric of cells and tissues, and because only the least damaged mitochondria tend to proliferate there is little overt sign of mitochondrial mutations and damage. In other words, the use of free radicals to signal danger explains why we don't see the spiralling, catastrophic damage predicted by the original mitochondrial theory. And this in turn explains why the cell doesn't accumulate too many antioxidants—it needs just the right amount, so that it's sensitive to changes in the redox state of transcription factors. That is why I said earlier that biology is more dynamic than 'mere' free-radical chemistry: there is little that's accidental going on here; rather, a continuing adaptation to the metabolic undercurrents of the cell.

So how do mitochondria kill us in the end? In time, some cells run out of normal mitochondria. When the next call comes to generate more, these cells have little option but to amplify their defective mitochondria clonally, and this is why particular cells are ultimately taken over by defective clones. But why do we only see a few cells with defective mitochondria at any one time, even in tissues of elderly people? Because now another level of signalling imposes itself. When cells finally work themselves into this state they are eliminated, along with their faulty mitochondria, by apoptosis, and that's why we don't detect high levels of mitochondrial mutations across ageing tissues. But there is a high cost for such purification—the gradual loss of tissue function, and with it age-ing and death.

Disease and death

The ultimate fate of the cell depends on its ability to cope with its normal ener-getic demands, which vary with the metabolic requirements of the tissue. As in mitochondrial diseases, if the cell is normally highly active, then any significant mitochondrial deficiency will lead to a swift execution by apoptosis. What exactly constitutes the signal for apoptosis is uncertain, and again depends on the tissue, but two mitochondrial factors are probably involved—the pro-portion of damaged mitochondria, and the ATP levels in the cell as a whole. Of course, these two are interlinked. Clonal expansion of dysfunctional mitochon-dria inevitably leads to a more general failure to match ATP production to demand. In most cells, once the ATP levels fall below a particular threshold, the cell inexorably commits itself to apoptosis. Because cells with dysfunctional mitochondria eliminate themselves, it's rare to observe heavy loads of mito-chondrial mutations, even in the tissues of elderly people.

The fate of the tissue, and the function of whole organs, depends on the types of cell from which they're composed. If the cells are replaceable, by way of the division of stem cells that preserve an unsullied mitochondrial population, the loss of cells by apoptosis doesn't necessarily perturb the status quo, so long as a dynamic balance is maintained in the cell population. But if the cells forced to die are more or less irreplaceable, like neurones or heart-muscle cells, then the tissue becomes depleted of functional cells, and the survivors are placed under greater strain, pushing them closer to their limits—their own particular metabolic threshold. Any other factors that force cells closer to their limits could precipitate a specific disease. In other words, as cells draw closer to their limits, with advancing age, various random factors are more likely to push them over the abyss to apoptosis. Such factors may include environmental assaults such as smoking and infections, and physiological traumas such as heart attacks, but also any genes associated with disease.

This link between metabolic threshold and disease is critical. It explains how mitochondria can be responsible for a whole gamut of diseases, even if they appear to be completely irrelevant. This simple insight explains why rats succumb to the diseases of old age within a few years, whereas it takes decades for humans. What's more, it helps explain why birds don't age in a particularly 'pathological' way, and how we can cure many of our own diseases at a single stroke. It explains in short how we can be more like elves.

I've been enumerating the failings of the original mitochondrial theory of ageing. Here's another: it's very difficult to link the underlying process of ageing with the occurrence of age-related diseases. To be sure, there was a hypothetical relationship between free-radical production and the onset of disease, but if this is taken at face value then the theory is forced to predict that all the diseases of old age are caused by free radicals. Obviously, this is not true. Medical research has shown that most diseases of old age are an appallingly complex amalgam of genetic and environmental factors—and most of them have little to do with free radicals and mitochondria, at least not directly. Proponents of the mitochondrial theory have spent years trying to identify specific links between genes and free-radical production, but to little avail. Mutations in some genes *are* associated with free-radical production, but this is not the rule. Of more than a hundred different genetic defects known to cause degeneration of the retina, for example, only a few affect free-radical production at all.

The solution was put forward in a beautiful paper by Alan Wright and his colleagues in Edinburgh, and published in *Nature Genetics* in 2004. I personally think this paper is one of the most important to emerge for a long time, for it gives a new, unifying framework for considering the diseases of old age, which ought to replace the current paradigm—both fallacious and counterproductive, in my view.

The paradigm underlying most medical research today is gene-centric. The approach is first to pinpoint the gene, then to find out what it does and how it works, then dream up some pharmacological way of correcting the problem, and finally to apply the pharmacological solution. I think this paradigm is fallacious, as it is based on a view of ageing that now seems incorrect—the idea that ageing is little more than a dustbin of late-acting genetic mutations, which have broadly independent effects and so must be targeted individually. This is the hypothesis, you might recall, of Haldane and Medawar, which I criticized earlier on the grounds that recent genetic research shows that ageing is far more flexible. Extend the lifespan, and *all* diseases of old age are postponed by a commensurate period, if not indefinitely. More than forty different mutations extend lifespan in nematodes, fruit flies, and mice, and all of them postpone the onset of degenerative diseases in general. In other words, the diseases of old age are tied to the primary process of ageing, which is somewhat flexible. The best way of targeting the diseases of old age is therefore to tackle the underlying process of ageing itself.

Wright and colleagues considered specific mutations in genes known to increase the risk of particular neurodegenerative diseases. Rather than asking what these genes do, they wondered what happens when the *same mutation* is found in different animals with differing lifespans. Of course, the same mutations *are* often found, and not only by chance. Animal models are essential for medical research, and genetic models of diseases figure at the centre of research today. So all that Wright and colleagues needed to do was to track down data on animal models in which the same genetic mutations cause equivalent neurodegenerative diseases. Nothing else was different. They came up with ten mutations that they could flesh out with data from five species with a wide range of lifespans—mouse, rat, dog, pig, and human. The ten mutations caused different diseases, but the same mutations produced the same disease in each species. The main difference was the timing. In the case of mice, the mutations produced disease within a year or two; for people, it might take a hundred times as long to cause exactly the same disease.

It's important to appreciate that the ten mutations are all inherited genetic mutations of *nuclear* DNA. None of them have anything to do with mitochondria or free-radical production directly. Wright and colleagues considered mutations in the *HD* gene in Huntington's disease, the *SNCA* gene in familial Parkinson's disease, and the *APP* gene in familial Alzheimer's disease, plus a number of genes causing degenerative diseases of the retina, leading to blindness. In each case, the pharma industry is ploughing billions of dollars into research, as any effective treatment would recoup billions of dollars every year. More human ingenuity goes into this research than into rocket science these days. In no case has a really serious clinical breakthrough been made—the kind of breakthrough that leads to a genuine cure, or a delay in the onset of symptoms beyond months, or at best a few years. As Wright and colleagues put it, with nice understatement: 'There are few situations in which neurodegeneration rates can be altered as substantially as the differences between species shown here.' In other words, we can't begin to slow down the progression of diseases, through medical interventions, by anything like as much as happens naturally in different species.

Wright and colleagues plotted out the time of onset and the progression of disease from mild symptoms through to severe illness in the different animals. What they found was a very tight correlation between disease progression and the underlying rate of production of free radicals from mitochondria. In other words, in the species that produced free radicals quickly, the disease set in early and was quick to progress, despite no direct link with free-radical production.

Conversely, in animals that leak free radicals slowly, the onset of diseases was delayed many-fold, and they progressed more slowly. This relationship could hardly be chance, for the correlation was too tight; clearly the onset of disease is tied, in some way, to the physiological factors that regulate longevity. Nor could the relationship be ascribed to differences in the genes themselves, for in each case the defects were exactly equivalent, and the biochemical pathways were conserved. It could not be ascribed to free radicals in general, as most of the genes didn't change free-radical production directly. And it couldn't be linked to other aspects of metabolic rate, as the metabolic rate doesn't correlate with lifespan in many cases, including birds and bats—and, critically in this case, humans.

The most likely reason for the correlation, said Wright, is that in all these degenerative diseases, the cells are lost by apoptosis—and free-radical production influences the threshold for apoptosis. Each of the genetic defects creates cellular stress, culminating in the loss of cells by apoptosis. The probability of apoptosis depends on the overall degree of stress, and the ability of the cell to keep meeting its metabolic demands. If it fails to meet its demands, it commits apoptosis. And the likelihood that it will fail depends on the overall metabolic status of the cell, which is calibrated by mitochondrial free-radical leakage as we have seen. The speed at which cells activate the retrograde response and amplify defective populations of mitochondria, leading to an ATP deficit, depends on the underlying rate of free-radical leakage. Species that leak free radicals rapidly are closer to the threshold, and so more likely to lose cells by apoptosis.

Of course, all this is correlative, and it is hard to prove that a relationship is causal. But one study published in *Nature* in 2004 suggests that there is indeed a causal relationship. The study won several of its senior authors, among them Howard Jacobs and Nils-Göran Larsson, of the Karolinska Institute in Stockholm, the prestigious EU Descartes Prize for research in the life sciences. The team introduced a mutant form of a gene into mice, termed *knockin* mice, as they have a gene knocked in (which is to say that a functional gene is added to the genome, rather than knocked out, the more common approach). The knockin gene in this case encoded an enzyme known as a proof-reading enzyme. Like an editor, a proof-reading enzyme corrects any errors introduced during DNA replication. In their study, however, the researchers introduced a gene that encoded a faulty version of this enzyme, which, ironically, was prone to errors. Like a bad editor, an error-prone version of the proof-reading enzyme leaves behind more errors than usual. The gene introduced in this study encoded a proof-reading enzyme specialized to work in the mitochondria, so that the errors it left behind were in mitochondrial, rather than nuclear, DNA. Having succeeded in setting the slapdash editor to work, the investigators were duly rewarded with a several-fold rise in the usual levels of mitochondrial errors, or mutations. There were two intriguing findings. The finding that captured the headlines was that the affected mice had foreshortened lives, coupled with an early onset of several age-related conditions, including weight-loss, hair loss, osteoporosis and kyphosis (curvature of the spine), reduced fertility, and heart failure. But perhaps the most intriguing aspect of the study was that the number of mutations *did not rise* with the age of the mice. As the mice got older, the number of mitochondrial mutations in body tissues remained relatively constant, as happens in humans—there was no big increase in mutational load during ageing.

Although the reason was not ascertained, I imagine that any cells that acquired an unworkable load of mutations were simply eliminated by apoptosis, giving an impression that mitochondrial mutations did not accumulate with age. Overall, the study confirms the importance of mitochondrial mutations in ageing, but does not conform to the expectations of the original mitochondrial theory of ageing, which predicts a large accumulation of mitochondrial mutations leading into an 'error catastrophe'. But the findings do support the more subtle version of the mitochondrial theory, in which free-radical signals and apoptosis continually relive the burden of mutation.

Some critically important conclusions emerge from this thinking. First, it seems that mitochondrial mutations genuinely do contribute to the progression of ageing and disease, even if they can't always be seen—they are eliminated, along with their host cells, by apoptosis. Second, other genes associated with particular diseases add to the overall levels of cellular stress, making it more likely that the cell will die by apoptosis. From Alan Wright's work, we have seen that it makes little difference what the gene codes for, or what the particular mutation may be—the timing and mode of cell death is virtually independent of the gene itself, if we consider the differences between species; it depends on how close a cell is to the threshold for apoptosis. This means that it is pointless trying to target individual genes or mutations in clinical research— and that means that the whole caravan of medical research is bound in the wrong direction. Third, research strategies that aim to block apoptosis are also likely to fail, for apoptosis is merely a useful way of disposing of broken cells without leaving behind a bloody signature. Blocking apoptosis doesn't solve the underlying problem that the cell can no longer fulfil its task; it would be doomed to die instead by necrosis, leaving blood and gore on the pavement, and this could only make matters worse. Finally, hugely importantly, the degenerative diseases of old age, *all of them*, could be slowed down by orders of magnitude, perhaps even eliminated altogether, just by slowing down the rate of free-radical leakage from mitochondria. If some of the billions of dollars devoted to medical research were directed to the target of free-radical leakage, we could potentially cure all the diseases of old age at a stroke. Even a conservative view would put that as the greatest revolution in medicine since antibiotics. So can it be done?

18

A Cure for Old Age?

Ageing and age-related diseases can be ascribed to mitochondrial free-radical leakage. Unfortunately, or perhaps fortunately, the way in which the body deals with free-radical leakage from the mitochondria is far more complex than the rather naïve early formulation of the mitochondrial theory would have us believe. Rather than simply causing damage and destruction, free radicals play a vital role in keeping respiration fine-tuned to needs, and in signalling respiratory deficiencies to the nucleus. This is possible because the proportion of free radicals leaking from the mitochondria fluctuates. High levels of free radicals signal respiratory deficiency, which can be corrected by compensatory changes in the activity of mitochondrial genes. If the deficiency is irreversible, and the mitochondrial genes are unable to re-establish control over respiration, then the overload of free radicals oxidizes the membrane lipids, which collapses the membrane potential. Mitochondria that lose their membrane potential are effectively 'dead' and are swiftly broken down and destroyed, so free-radical overload hastens the removal of damaged mitochondria from the cell. Other, less damaged, mitochondria replicate to take their place.

Without this subtle self-correcting mechanism, the performance of the mitochondria, and the cell as a whole, would be seriously undermined. Mutations in mitochondrial DNA would escalate, and cellular function spiral out of control in an 'error catastrophe'. In contrast, the signalling role of free radicals maintains near optimal respiratory function in long-lived cells for many decades. Damaged mitochondria are removed, and undamaged mitochondria replicate to take their place. In the end, though, at least in long-lived cells, the supply of undamaged mitochondria runs out, and then a new level of signalling must take over.

If too many mitochondria become deficient in respiration simultaneously, then the general load of free radicals in the cell rises, and this signals the general respiratory failure to the nucleus. Such oxidizing conditions shift the kaleidoscope of active nuclear genes to compensate—a shift known as the retrograde response, for the mitochondria control the activity of genes in the nucleus. The cell enters a stress-resistant state and may survive like this for many years. It is limited in its powers of energy generation, but can get by so long as it is not placed under too much strain. However, any stressful episodes are likely to undermine such cells, and lead to some degree of organ failure. This in turn probably contributes to the chronic inflammation that underpins many diseases of old age.

In ageing organs, the more damaged cells are removed by means of free-radical signalling, coupled to the decline in respiratory function. When cellular ATP levels fall below a critical threshold, the cell commits apoptosis, removing itself from the firmament. The removal of damaged cells in this way contributes to the shrinkage of organs with age, but at the same time removes malfunctioning cells, so that the remainder are selected for optimal function. There is no sudden meltdown, no exponential error catastrophe, as there would inevitably be if free radicals merely played a destructive role. Likewise, the quiet elimination of cells by apoptosis, rather than the gory end of necrosis, curbs inflammation across the tissue as a whole, and so prolongs life.

So if a cell can't meet metabolic demands, it commits apoptosis. The likelihood of cell loss therefore depends in part on the metabolic requirements of the organ. Metabolically active organs, like the brain, heart, and skeletal muscle, are most likely to lose cells by apoptosis. The exact timing of cell death depends on the general stress levels. These are calibrated by the mitochondria, as we saw in Part 5; and one important factor involved in this calibration is the accumulated exposure to free radicals. As a result, long-lived animals suffer from age-related diseases late in life, while short-lived animals capitulate more quickly. The general stress levels of a cell can also be raised by particular inherited or acquired genetic mutations, or physiologically traumatic events, such as falls, heart attacks or diseases, or exposure to cigarette smoke and suchlike. The conclusion we can draw from this is hugely important: if all genetic and environmental contributions to the diseases of old age are calibrated by the mitochondria, we ought to be able to cure, or postpone, all the diseases of old age at once. Conversely, attempting to tackle them piecemeal, as we do today, is doomed to failure. All we need to do is to lower free-radical leakage over a lifetime.

Therein lies the problem. At each stage in the life of cells, the physiology of mitochondrial and cellular function *depends* on free-radical signalling. Simply attempting to quash free-radical generation with massive doses of antioxidants is likely to exacerbate the situation, if it could be made to work at all. In *Oxygen*, I put forward the idea (dubbed the 'double-agent' theory) that the body is refractory to high doses of antioxidants—we eliminate superfluous antioxidants from the body because they have the potential to play havoc with sensitive free-radical signals. I have probably down-played the possible utility of antioxidants to a degree, if only as a corrective to the usual hyperbole; it may be that they can benefit us in various ways, but frankly, I'm sceptical that they can do much more than correct dietary deficiencies. And I think that the problem with signalling means that if we wish to prolong our healthy lifespan, then we need to tear ourselves away from the lure of antioxidants, and think afresh.

So what else could we do? Relative to mammals, birds slow down the rate of free-radical leakage. By studying the differences between birds and mammals, we might gain insight into how best to cure ageing, and its attendant diseases, in everyone. So could we make ourselves more like the birds? That depends on how they do it.

According to the pioneering work of Gustavo Barja in Madrid, most free-radical leakage comes from complex I of the respiratory chains. In a series of simple but ingenious experiments using respiratory-chain inhibitors, Barja and colleagues pinpointed the site of leakage to one of more than forty subunits in complex I; and their work has been confirmed by others using different techniques. The spatial disposition of the complex means free radicals leak straight into the inner matrix, in the immediate vicinity of mitochondrial DNA. Clearly, any attempt to block leakage needs to be targeted to this complex with extraordinary precision—no wonder antioxidant therapies fail! Apart from the fact that they can potentially play havoc with signalling, it's virtually impossible to target antioxidants at a high enough concentration in such a small space. There are, after all, tens of thousands of complexes in a single mitochondrion, and typically hundreds of mitochondria in cells. And of course perhaps 50 trillion cells in the human body. Luckily, we've learnt from birds that this is not the way; they have quite low antioxidant levels. So how do they reduce free-radical leakage?

The answer is not known with any certainty, and there are a number of possibilities. It may be that birds combine bits of them all. One possibility is that the differences are written into the sequence of a handful of mitochondrial genes. The best evidence to support this possibility, ironically, derives from studies of human mitochondrial DNA, most inspiringly from Masashi Tanaka's group in Japan. In 1998 Tanaka's group reported in *The Lancet* that nearly two-thirds of Japanese centenarians shared the same variation within a mitochondrial gene—a single letter change in the code for a subunit of complex I—compared with about 45 per cent of the population as a whole. In other words, if you have the letter-change, you are 50 per cent more likely to live to a hundred. The benefits don't stop there. You are also *half* as likely to end up in hospital for any reason at all in the second half of your life: you're less likely to suffer from *any* age-related disease. Tanaka and colleagues showed that the letter change probably resulted in a tiny reduction in the rate of free-radical leakage—a tiny benefit at any one moment, but one which mounted up imperceptibly over a whole lifetime, to finally make a substantial difference. This is exactly the sort of evidence needed to confirm the theory that all age-related diseases can be targeted by a single simple mechanism. On the other hand, there are draw-backs too. The Japanese letter-change is hardly ever found outside Japan, and although its prevalence there may help explain the exceptional longevity of the Japanese, that doesn't help the rest of us a great deal. Not surprisingly, the news inspired a worldwide gene hunt, and it seems there are other mitochondrial letter changes that have a similar effect. The same problem applies to all of them, however, which is that we can only alter gene sequence by genetically modifying ourselves. Given the great rewards, this might be worth doing, but it is dangerously close to the ethically indefensible waters of choosing human traits in embryo. So at present, unless society has a major about-turn in its atti-tudes to genetic modification (GM) of humans, the best that can be said is that all of this is scientifically extremely interesting.

But GM is not the only possibility. Another way that birds might lower their rate of free-radical leakage is by uncoupling their respiratory chains. The term uncoupling refers to the dissociation of electron flow from ATP production, so that respiration dissipates energy as heat. In the same way, uncoupling a chain on a bike dissociates peddling from forward movement, which dissipates energy in a sweaty brow. The immense benefit of uncoupling the respiratory chain is that electrons can keep flowing, just as a cyclist's legs keep peddling, and this in turn reduces free-radical leakage. (I suppose that an uncoupled bike has the advantage of allowing us to burn off excess energy as heat in the same way; we now call it an exercise bike.) Because fast free-radical leakage is associated with both ageing and disease, and uncoupling reduces free-radical leakage, then uncoupling surely has the potential to extend lifespan. Like a bike chain, it is possible to partially uncouple respiration (it's called changing gear on a bike) so that some ATP synthesis can continue, but a proportion of energy is frittered as heat (just as we could still peddle while whizzing down hill, but can't engage the chain). The long and short of it is this: by enabling a constant flow of electrons down the respiratory chain, uncoupling restricts the leakage of free radicals.

We noted in Part 4 (page 183) that uncoupled mice have faster metabolic rates and do indeed live longer than their well-coupled brethren. Likewise, in Part 6, we noted that differences in the vulnerability to disease of Africans and Inuit might be related to differences in uncoupling. In a similar vein, it is quite feas-ible that birds are more uncoupled than the equivalent mammals, and that this might explain why they live longer. Uncoupling generates heat, as we just noted, so if birds really are more uncoupled, they should also generate more heat than equivalent mammals. In support of this idea, birds do indeed main-tain a higher body temperature, of about 39°C rather than 37°C, which might be the result of greater heat production from uncoupling. In fact, however, direct measurements imply that this isn't the case: the respiratory chains of birds and mammals seem to be similarly coupled, so presumably the temperature difference can be ascribed to differences in heat loss and insulation. Feathers are better than fur.

That's not to say that uncoupling couldn't help us though. Not only could it, in principle, lower free-radical leakage, and so prolong our lives, but it would also enable us to burn more calories and lose weight. We could cure obesity and all the diseases of old age in one go! Sadly, the experience so far with anti-obesity drugs is rather sorry. Dinitrophenol, for example, is a respiratory-chain uncoupler that has been tested as an anti-obesity drug. It proved toxic, at least at the high doses used. Another uncoupler is the popular recreational drug, ecstasy, which illustrates the potential dangers well: uncoupling generates heat, inducing some revellers to dance with a water supply strapped to their backs; even so, a few have died of heatstroke. Certainly more finesse is needed. Curiously, aspirin is also a mild respiratory uncoupler; I do wonder how many of its more mysterious benefits may relate to this property.

Barja's own work suggests that birds decrease free-radical leakage from complex I by lowering its reduction state. Remember that a molecule is 'reduced' when it has received electrons, and oxidized when it has lost them. Accordingly, a low reduction state means that birds tend to have relatively few electrons passing through complex I at any moment. We have noted (page 77) that there are tens of thousands of respiratory chains in each mitochondrion, and each one has its own leaky complex I. If they are in a low reduction state, then only a small proportion of them are in possession of a respiratory electron; the rest are as bare as Mother Hubbard's cupboard. If there are relatively few electrons milling around, they are less likely to escape to form free radicals. Barja argues that a similar mechanism also underpins calorie restriction, the only method so far proved to extend lifespan in mammals; there, too, the most consistent change is a fall in the reduction state, despite little change in oxygen consumption. Furthermore, this line of thinking explains the exercise paradox that I mentioned earlier—the fact that athletes consume more oxygen than the rest of us but don't age more quickly. Exercise speeds up the rate of electron flow, which in turn lowers the reduction state of complex I—the electrons are quicker to leave it again, and this makes the complex less reactive. That explains why regular activity doesn't necessarily increase the speed of free-radical leakage and might actually lower it in trained athletes.

In all these cases, the common denominator is a low reduction state. We could think of this as a half-empty cupboard, or, perhaps more helpfully, as spare capacity. But importantly, the spare capacity in birds differs from that in exercise and uncoupling (dissipating energy). In both of the latter cases, free-radical leakage is restricted because electrons are flowing down the chain, and their departure from one complex frees that complex to receive another electron: it therefore frees up some capacity. As a result, electrons are less likely to leak out as free radicals. In birds, however, spare capacity is maintained at rest, compared with mammals that have an equivalent metabolic rate and degree of uncoupling. In other words, when all else is equal, birds have more spare capacity than mammals, and for this reason they leak fewer free radicals. And because they leak fewer free radicals, they live longer lives.

If Barja is correct (and some researchers disagree with his interpretation) then the key to longer life is spare capacity. So how, indeed why, do birds retain spare capacity? To understand the answer, think of the workforce in a factory that needs to cope with fluctuating workloads. Imagine that the management have two possible strategies (no doubt they have many more, but let's focus on two): either they can retain a small workforce, and oblige them to work harder whenever extra work arrives; or they can employ a large workforce, which can cope easily with the heaviest demands, but idle away much of the year. Now consider the morale of the workforce. Let's say the small workforce becomes downright rebellious whenever they're forced to work long hours, and when in rebellious spirits, they wilfully damage equipment. On the other hand, they have short memories, and their resentment soon fades after a few beers; then they work well. Management calculates that it's worth losing a little equipment to save on labour costs. So what about the morale of the large workforce? Now, even when the extra work arrives, the large team has no difficulty accomplishing the task, and their morale remains high. But, of course, for much of the year they have too little to do, so they get bored. Their resentment is not strong (better to have a job and put up with a little boredom) but nonetheless there is a risk that the workforce may look for a less boring job, and drift away just when they're needed.

So what does all this have to do with birds and respiratory chains? Birds have adopted the strategy of the large workforce. Management preferred the high labour costs and the risk of workforce attrition, but they valued their equipment and didn't want to see it wilfully damaged. Moreover, they were ambitious, and imagined that they would win a lot of work, and so would need a large workforce. Translated into biological terms, all this means the following. Birds rely on a large number of mitochondria, and within the individual mitochondria, they have a large number of respiratory chains. They have a high workforce, and for much of the time they have a lot of spare capacity. In molecular terms, the reduction state of complex I is low: the electrons entering into the respiratory chains have plenty of space for themselves. In contrast, mammals adopt the alternative policy: they like to employ a small workforce. This means they retain as small a number of mitochondria, and respiratory chains within the mitochondria, as they can get away with. Even when the workload is low, they still pack in the electrons quite densely. The workforce becomes rebellious, and smashes up the equipment, as if they were free radicals destroying the fabric of the cell. With this degree of damage, it is only a matter of time before the factory has to close altogether.

Incidentally, it is worth noting that the rebelliousness of the workforce—the degree of damage to the equipment—would depend on the proportion of the time that the workforce is stressed and overworked to the point of rebelliousness. This depends on their workload, which in biological terms equates to metabolic rate. Animals with a fast resting metabolic rate, like rats, have a higher workload, and less spare capacity than mammals with a slow metabolic rate, like elephants. They therefore leak free radicals quickly—their workforce is rebellious much of the time—and suffer the penalty of a rapid accumulation of damage, a fast rate of ageing, and death. The same relationship applies to birds, except that, in this case, all birds have more space capacity than the equivalent mammals; small birds live longer than the equivalent mammals, but shorter than larger birds.

The idea of a rebellious workforce also helps to explain the benefits of calorie restriction, and many 'longevity' genes in nematode worms and fruit flies. In these cases, the alterations don't affect the size of the workforce, but they may reduce the workload (lower the metabolic rate, and so increase spare capacity) or they might perhaps mollify the workers, so they become less rebellious, despite continuing to handle the same volume of work (no change in spare capacity). In this sense, the effect is like religion, which Marx referred to as the opium of the masses. To continue our analogy, the management policy to curb violence in the workforce might be to offer free opium. Either way, there are costs involved—the cost of a lower capacity for work, or of an opiated work-force. In biological terms, the costs of longevity genes are usually reflected in curtailed sexuality: a shift in resource use allows the metabolic rate to remain similar, but the cost of longevity is reduced fecundity.

Birds retain spare capacity in their mitochondria without any of these draw-backs—they have a high capacity for work without a penalty in fecundity. So how come birds retain so much spare capacity? I think the answer is that powered flight requires an aerobic capacity that outstrips anything that can be achieved by even the most athletic mammals. Just to get aloft, they require more mitochondria, and more respiratory chains. If they lose these mitochon-dria, they simultaneously lose the ability to fly, or to fly as skilfully. From the management strategy point of view, the factory work can *only* be accomplished by a large workforce, so there is really no choice: management can't risk redun-dancies when the demand is low. So when birds are resting, their metabolic rate idles along and they have a large over-capacity. Technically, complex I is reduced to a lesser degree. Exactly the same reasoning applies to bats, too, which must also maintain a high aerobic capacity for powered flight.

Lest this sound theoretical, the hearts and flight muscles of birds and bats *do* have more mitochondria, and in them a greater density of respiratory chains, than mammals; but do their other organs too? After all, it is the organs, not flight muscles, which contribute most to the resting metabolic rate, as we noted in Part 4. Surprisingly little is known about the number of mitochondria in the organs of birds and bats, but it's feasible that they do have more mitochondria than do land-bound mammals, as the entire physiology of birds and bats is geared to maximal aerobic performance. To give just a single example, the number of glucose transporters in the intestine of a humming bird is far higher than in mammals, for they need to absorb glucose very quickly, to power their costly hovering flight. The extra transporters are powered by extra mitochondria. So the aerobic capacity of organs seemingly unrelated to flight is probably high too—far higher than needed to meet the low demands of resting metabolism.

Birds and bats are usually said to live long lives because flying away helps them avoid predators. No doubt there is truth in this, although many small birds have a high mortality rate in the wild, yet still live relatively long lives. The answer I've just suggested is tied in directly with the high-energy requirements of flight. High-energy expenditure demands a high density of mitochondria, not just in the flight muscles and heart but also in other organs too, to compensate. Such compensation is similar to that postulated for the origin of endothermy in the aerobic capacity hypothesis (see Part 4), but is stronger because the maximal aerobic demands of powered flight are greater than those of running, even at full stretch. Because the density of mitochondria is greater, the spare capacity at rest is greater, and this lowers the reduction state of complex I. Free-radical leakage is lowered by necessity, and this corresponds to a longer life.

So what happens in mammals other than bats: why don't they maintain lots of spare capacity, in the form of high numbers of mitochondria? One reason might be that most mammals have little to gain by having more mitochondria and aerobic power—if threatened by predators their best policy is to scarper for the nearest hole. It is in the nature of things to use it or lose it. Rats are most likely to jettison unnecessary mitochondria as a costly burden, but this leads them straight back to the problem that they have fewer complexes and a higher reduction state of complex 1. They leak more free radicals, live fast and die young. Or do they?

If rats can't gain much by hoarding more mitochondria in terms of their aerobic capacity, they ought to have another advantage: any rats that *did* hoard more mitochondria would have greater spare capacity, and so should live longer. Leaking fewer free radicals, they would not need to hoard more antioxidants or stress enzymes, and so should not be penalized under the terms of the disposable soma theory (see page 278). In fact, with such fine fettle—all those mitochondria, all that aerobic capacity—they should be physically impressive, sexually alluring to other rats, biologically 'fit', and so at an advantage in the struggle to mate. The coupling of long lifespan with strong biological fitness means that their longevity genes should spread. But none of this has happened. Rats are still rats, and they still die quickly. Is there something else to it? I think there is, and it's critically important to us, for if we wish to engineer ourselves a few genes that combine sexual allure with long life, we should know about the drawbacks.

The problem is this: having a low rate of free-radical leakage means that a more sensitive detection system is required to maintain respiratory efficiency. This after all, is why we have retained any genes in the mitochondria at all (see page 141). The costs of evolving greater refinement could explain why rats don't restrict free-radical leakage. They have the cost of elaborating a sensitive detection system, as well as that of maintaining a lot of spare capacity. The combination of the two must have been too much for rats. In the case of birds, however, the high evolutionary costs of a more sensitive detection system are counterbalanced by the strong selective advantages of improved flight. Because flight is costly, yet pays high dividends, birds *do* benefit from packing more mitochondria in all their tissues, and so have greater superfluous capacity when at rest— they benefit from retaining a large workforce, and even investing some of the profits on the latest equipment. Their spare capacity translates into a lower free-radical leakage at rest, and a longer lifespan, but *requires* a more sensitive detection system. In this case, however, the advantages of flight do outweigh the costs in terms of survival and reproduction.

If we wish to live longer, then, and to rid ourselves of the diseases of old age, we will need more mitochondria, but also perhaps a more refined free-radical detection system. That may be problematic, and will no doubt tax the ingenuity of medical researchers. But our lifespan is already several times longer than that of equivalent mammals. If my reasoning is correct, we should already have more mitochondria than mammals with an equivalent resting metabolic rate: we should have more spare capacity, coupled with a sensitive free-radical detection system. In our own case, the sophistication was perhaps worth it for a different reason to the birds—not aerobic capacity, but for its own sake, for the benefits that longevity confers on social cohesion in kin groups. The elders of the tribe passed on knowledge and experience that gave their tribe a competitive edge; and they would even have been alluring into the bargain. Have we really done this? I don't know, but it's an interesting hypothesis, and easily testable. All we need to do is to measure mitochondrial density in the organs of mammals with a comparable metabolic rate, and, with a little more difficulty, to test the sensitivity of the free-radical signalling system.

There's a tantalizing hint that we might be able to engineer longer lives in this way. Earlier on, in Chapter 17, I mentioned that a single letter change in the mitochondrial control region was five times more common in centenarians than in a cross-section of the population. The mutation seems to stimulate slightly more mitochondrial genesis in response to a signal. So if a signal arrives to say '*Mitochondria: divide!*', then people with the mutation may produce 110 new mitochondria, while those without would produce just 100. The effect might be for us to become a little more like the birds: we would have greater overcapacity when at rest. In principle, a similar effect could be achieved pharmacologically without modifying any genes at all, simply by amplifying each signal a little—so whenever a signal came for the mitochondria to divide, we could try to amplify it by, say, 10 per cent. In both cases, the extra mitochondria would share a lesser burden of work. The reduction state of the complexes would fall, and they would leak fewer free radicals. So long as we could still detect them well enough—a very delicate balancing act, but presumably the centenarians did—then we would have a good chance of living longer and better lives, less troubled by disease in our dying days.

Epilogue

More than a decade ago I spent a lot of time in the lab, trying to preserve kidneys for transplantation. The challenge was not related to rejection, the sexier end of research, but to a more pressing problem. As soon as a kidney, or any other organ, is removed from the body, the clock starts ticking—frantically. In the case of kidneys, decay renders the organ unusable within a couple of days. In the case of hearts, lungs, livers, and so on, time is even more pressing: they can't be stored for much more than a day before they are wasted. The terrible prospect of rejection sharpens the problem. It is vital, literally, to match the immune profile of the donor organ to that of the recipient, to prevent acute rejection taking place on the operating table before your eyes. That often means transporting organs over a few hundred miles to a suitable recipient. Given the constant shortage of organs, any wastage is a crime. A stride forward in preservation, giving longer to locate the most suitable recipient, to arrange transport, and to mobilize the local transplant team, would waste fewer organs. Conversely, if we could work out exactly when an organ became unusable then we could salvage organs otherwise condemned as irretrievably damaged, for example, those taken from non-heart-beating donors.

It is practically impossible to tell, just by looking at a stored organ, whether or not it will function after transplantation, even if we take a biopsy and scrutinize it down the microscope. When an organ is removed from the body, the blood is flushed out using a carefully formulated solution, and the organ stored on ice. All looks well, but appearances can be deceptive. An apparently normal organ may become irreversibly damaged after transplantation. Paradoxically, this injury is thought to be caused by the return of oxygen. The storage period primes the organ for a disastrous loss of function upon transplantation, caused by oxygen free radicals escaping the mitochondrial respiratory chains.

One day I was in an operating theatre, fixing probes to a kidney during a transplant operation, in the hope of working out what was going on inside without physically taking a sample. The machine we were using was ingenious—a near infra-red spectrometer. It shines a beam of infra-red rays, which can penetrate several centimetres across biological tissues, and measures how much comes out the other side. From this, a complicated algorithm calculates how much radiation is absorbed or reflected on route, and how much passes through. The precise wavelength of infra-red radiation chosen is critical, as different molecules absorb different wavelengths. Choose your wavelength with care, and you can focus on haem compounds—those proteins that incorporate a chemical entity known as a haem group, such as haemoglobin or cytochrome oxidase, the terminal enzyme of the respiratory chains, deep in mitochondria. Not only is it possible to work out the concentration of haemoglobin—both oxygenated and de-oxygenated forms—but you can calculate the redox state of cytochrome oxidase, which is to say you can work out what proportion of cytochrome molecules are in the oxidized versus the reduced state: what proportion is at that moment in possession of respiratory electrons. We paired this technique with a related form of spectroscopy that enabled us to work out the redox state of NADH, the compound that supplies the electrons entering into the respiratory chain. By combining the two techniques, we hoped to gain a dynamic idea of respiratory chain function in real time, without actually cutting into the kidney—obviously an immeasurable advantage during a major operation.

All this probably sounds very sophisticated, but in fact it's a nightmare of interpretation. Haemoglobin is present in massive amounts, whereas cytochrome oxidase is barely detectable. Worse still, the wavelengths of infra-red rays that the different haem compounds absorb overlap and merge with each other. It can be very hard to tell which is which. Even the machine gets confused. It measures a change in the redox state of cytochrome oxidase when what actually seems to be happening is a change in haemoglobin levels. We began to despair of ever gleaning any useful information from our contraption. Nor did the NADH levels help much. Most of the time there was a fine peak—a high concentration detected by the machine—before transplant, which vanished without trace after the organ had been transplanted, and that was that. It all sounded good on paper but the reality, as so often in research, was uninterpretable.

And then I had my own personal eureka moment, the moment I had my first inkling that mitochondria rule the world. It came about by chance, for one of the anaesthetics being used was sodium pentobarbitone. The concentration of this anaesthetic in the blood fluctuated, and on a few occasions when it did, we found we were picking it up on our machines. The levels of both oxy-haemoglobin and deoxy-haemoglobin remained unchanged, but we recorded a shift in the dynamics of the respiratory chain. Part of the NADH peak returned (it became more reduced) while the cytochrome oxidase became more oxidized. We seemed to be measuring a 'real' phenomenon, rather than the usual frustrating noise, because the levels of haemoglobin weren't changing. What was going on?

It turned out that sodium pentobarbitone is an inhibitor of complex I of the respiratory chain. When its blood levels rose, it partially blocked the passage of electrons down the respiratory chains, and this led to a back-up of electrons in the chains. The early parts, including NADH, became more reduced, while the later parts, including cytochrome oxidase, passed on their electrons to oxygen and became more oxidized. But why did this beautiful response not occur every time? This, we soon realized, depended on the quality of the organ. If the organ was fresh and functioning well, we picked up the fluctuations easily; but if it was seriously damaged it was virtually impossible to take a measurement. We saw the usual disappearance of all the peaks, never to return again. The explanation could only be that these mitochondria were as leaky as a colander—of the few electrons that entered the chain, barely any left it again at the end. Virtually all must have been dissipated as free radicals.

Without slicing out samples and subjecting them to detailed biochemical tests, we couldn't be absolutely sure about what was really happening in these mitochondria, but we could say one thing for certain—the damaged organs were losing control of their mitochondria within minutes of transplantation, and there was absolutely nothing we could do about it. We tried all kinds of antioxidants, in an attempt to improve mitochondrial function, but to no avail. Mitochondrial function in those first few minutes foretold the outcome, per- haps weeks later—if the mitochondria failed in the first few minutes, the kidney inexorably failed; if they still had some life in them, the kidney had a good chance of surviving and functioning well. The mitochondria, I realized, were masters of life and death in kidneys, and extremely resistant to being tampered with.

Since then, in considering diverse fields of research, I've come to realize that the dynamics of the respiratory chain, which I struggled to measure all those years ago, is a critical evolutionary force that has shaped not just the survival of kidneys, but the whole trajectory of life. At its heart is a simple relationship, which may have begun with the origin of life itself—the reliance of virtually all cells on a peculiar kind of energetic charge, which Peter Mitchell named the chemiosmotic, or proton-motive, force. In each chapter of this book we've examined the consequences of the chemiosmotic force, but each chapter has concentrated on the larger implications of specific aspects. In the final few pages, I'll try to tie all this together, to show how a handful of simple rules guided evolution in profound ways, from the origin of life, through the birth of complex cells and multicellular individuals, to sex, gender, ageing, and death.

The chemiosmotic force is a fundamental property of life, perhaps more ancient than DNA, RNA, and proteins. In the beginning, naturally chemi- osmotic cells might have formed from microscopic bubbles of iron-sulphur minerals, that coalesced in the mixing zone of fluids seeping up from deep in the crust, and the oceans above. Such mineral cells share some properties with living cells, and their formation needs no more than the oxidizing power of the sun—there is no call for complicated evolutionary innovations before the origin of hereditary replication through DNA. Chemiosmotic cells conduct electrons across their surface, and the current draws protons over the membrane to generate an electric charge across the membrane—a force field around the cell. This membrane charge links the spatial dimensions of the cell to the very fabric of life. All life, from the simplest bacteria to humans, still generates its energy by pumping protons across membranes, then harnessing the gradient to tasks such as motility, ATP production, heat production and the absorption of essential molecules. The few exceptions merely go to prove this general rule.

In cells today, electrons are conducted by the specialized proteins of the respiratory chains, which use the current to pump protons across the membrane. The electrons are derived from food, and pass down the respiratory chains to react with oxygen, or other molecules serving exactly the same purpose. All organisms need to control the flow of electrons down the respiratory chains. Too fast a flow fritters away energy wastefully, while too slow a flow can't match demand. The respiratory chains behave like slightly cracked drainpipes—a clear flow presents no problems, but any blockage, either at the outflow or somewhere in the middle, is likely to spring a leak through the cracks. If blocked, the chains leak electrons, and these react to form free radicals. There are just a handful of possible reasons for electron flow to block, and only a few ways to restore the flow, yet the balance between power-generation, on one hand, and free-radical formation on the other—the same problem I faced in my kidneys—has written some of the most important, if unsung, rules in biology.

First among the reasons for a blockage of electron flow is some kind of defect in the physical integrity of the respiratory chains. The chains are assembled from a large number of protein subunits, which form into large functional complexes. In eukaryotic cells, genes in the nucleus encode most of the subunits, and genes in the mitochondria encode a small number. The continued existence of mitochondrial genes in all cells containing mitochondria is a paradox, for there are many good reasons to transfer them all to the nucleus, and no obvious physical reasons why this could not have been achieved, at least in some species. The most likely reason for their persistence is a selective advantage to retaining them in the mitochondria, and this advantage seems to be related to energy generation. So, for example, an insufficient number of complexes in the second part of the respiratory chains would block electron flow, leading to a backlog of electrons in the earlier part, and free-radical leakage. In principle, the mitochondria could detect the free-radical leakage, and correct the problem by signalling to the genes to make good the deficit—to produce more complexes for the second part of the chains.

The outcome depends on the location of the genes. If the genes are in the nucleus, the cell has no means of distinguishing between different mitochondria, some of which need new complexes, and many of which don't: none are satisfied by the bureaucratic one-size-fits-all response of the nucleus. The cell loses control over energy generation, a grave penalty. Only if a small contingent of genes is retained in each mitochondrion, to code for the core protein subunits of the respiratory chains, can energy generation be controlled in a large number of mitochondria simultaneously. The additional subunits, encoded in the nucleus, fit themselves around the core mitochondrial subunits, using them as a beacon and a scaffold for construction.

The consequences of this system are profound. Bacteria pump protons across their external cell membrane, and so their size is limited by geometrical constraints: energy production slopes off with a falling surface-area-to-volume ratio. In contrast, eukaryotes internalize energy generation in mitochondria, and this frees them from the constraints facing bacteria. The difference explains why bacteria remained morphologically simple cells, while the eukaryotes were able to grow to tens of thousands of times the size, accumulated thousands of times more DNA, and developed true multicellular complexity, surely the greatest watersheds in all of life. But why did bacteria never succeed in internalizing their own energy generation? Because only endosymbiosis—a mutual, stable collaboration between partners living one inside another—is able to leave the right contingent of genes in place; and endosymbiosis is not common in bacteria. The precise concatenation of circumstances that forged the eukaryotic cell seems to have happened just once in the entire history of life on earth.

Mitochondria inverted the world of bacteria. Once cells had the ability to control energy generation across a large area of internal membranes, they could grow as large as they liked, within limits set by the distribution networks. Not only could they grow larger, but they also had good reason to do so—energetic efficiency improves with larger size in cells and multicellular organisms, just as it does in human societies, following the economies of scale. There is immediate payback for larger size—lower net production costs. The tendency of eukaryotic cells to become larger and more complex can be explained by this simple fact. The link between size and complexity is an unexpected one. Large cells almost always have a large nucleus, which ensures balanced growth through the cell cycle. But large nuclei are packed with more DNA, which provides the raw material for more genes, and so greater complexity. Unlike bacteria, which were obliged to remain small, and to jettison superfluous genes at the first opportunity, the eukaryotes became battleships—large, complicated cells with lots of DNA and genes, and as much energy as needed (and no longer any need for a cell wall). These traits made a new way of life possible, predation, in which the prey is engulfed and digested internally, a step that the bacteria never took. Without mitochondria, nature would never have been red in tooth and claw.

If the complex eukaryotic cell could only be formed by endosymbiosis, the consequences of two cells living together in mutual dependency were equally significant. Metabolic harmony may have been the rule, but there were important exceptions, and these too are attributable to the dynamics of the respiratory chain. The second reason for a block in electron flow is a lack of demand. If there is no consumption of ATP, then electron flow ceases. ATP is needed for replication of cells and DNA, and for protein and lipid synthesis—indeed, most housekeeping tasks. But demand is greatest when cells divide. Then the entire fabric of the cell must be duplicated. The dream of every living cell is to become two cells, and this applies as much to the erstwhile free-living mitochondria as to the host cells in the eukaryotic merger. If the host cell becomes genetically damaged, so that it can't divide, then the mitochondria are trapped inside their crippled host, for they are no longer able to survive independently. And if the host cell can't divide, it has little use for ATP. Electron flow slows down, and the chains become blocked and leak free radicals. This time the problem can't be resolved by building new respiratory complexes, so the mitochondria electrocute their hosts from inside with a burst of free radicals.

This simple scenario lies at the roots of two major developments in life—sex, and the origin of multicellular individuals, in which all cells in the body share a common purpose and dance to the same tune.

Sex is an enigma. Various explanations have been put forward, but none explains the primal urge of eukaryotic cells to fuse together, as do the sperm and the egg, despite the costs and dangers of doing so. Bacteria don't fuse together in this way, even though they do routinely recombine genes by lateral gene transfer, which apparently serves a similar purpose to sex. Bacteria and simple eukaryotes are often stimulated to recombine genes by various forms of physical stress, all of which involve free-radical formation. A burst of free radicals can be sufficient to induce a rudimentary form of sex, and in organisms like the green algae, *Volvox*, the free-radical signal for sex can come from the respiratory chains. In the early eukaryotic cells, mitochondria might have manipulated their hosts to fuse together and recombine their genes whenever the hosts were genetically damaged, and unable to divide by themselves. The host cell benefits, because the recombination of genes can fix or mask genetic damage, while the mitochondria themselves gain access to pastures new without killing their existing host, essential for their safe passage.

Sex may have benefited both the mitochondria and their hosts in single-celled organisms, but it no longer did in multicellular individuals. The gratuitous fusion of cells is a liability when the cells belong to an organized body, in which all constituent cells must share a common purpose. Now the same free-radical signal for sex betrays genetic damage to the host cell, which pays the penalty of death. This mechanism seems to be at the root of apoptosis, or programmed cell suicide, which is necessary for policing the integrity of multicellular individuals. Without the death penalty for cellular insurrection, multicellular colonies could never have developed the unity of purpose characteristic of the true individual—they would have been torn asunder by the selfish wars of cancer. Today, apoptosis is controlled by the mitochondria, using the same signals and machinery that they had once used to plead for sex. Much of this machinery was originally brought to the eukaryotic merger by the mitochondria. While the regulation of apoptosis is now, of course, far more complicated, at its heart the critical signal is still a burst of free radicals from a blocked respiratory chain, leading to depolarization of the mitochondrial inner membrane, and the release of cytochrome c and other 'death' proteins into the cell. Even today, it takes no more: injecting damaged mitochondria into a healthy cell is enough for that cell to kill itself.

There are ways of modulating electron flow down the respiratory chains, so these dire penalties don't happen whenever electron flow temporarily comes to a halt. The most important is to *uncouple* the chains (so the passage of electrons is not tied to the formation of ATP). Uncoupling is usually achieved by making the membrane more permeable to protons, so their passage back across the membrane is not strictly through an ATPase (the enzyme 'motor' that is responsible for generating ATP). The effect is akin to the overflow channels in a hydroelectric dam, which prevent flooding at times of low demand. The continuous circulation of protons allows a continuous passage of electrons down the respiratory chains, without regard for 'need', and this prevents the accumulation of electrons in the respiratory chains and so restricts free-radical leakage. But the dissipation of the proton gradient necessarily generates heat, and this too has been put to good use over evolution. In most mitochondria, about a quarter of the proton-motive force is dissipated as heat. When enough mitochondria are assembled, as in the tissues of mammals and birds, the heat generated is sufficient to maintain a high internal temperature, regardless of the external temperature. The origin of endothermy, or true warm-bloodedness, in birds and mammals can be ascribed to such dissipation of the proton gradient, which later made possible the colonization of the temperate and frigid regions, as well as an active nightlife. It released our ancestors from the tyranny of circumstance.

The balance between heat generation and ATP production still affects our health in surprising ways. Uncoupling of the respiratory chain is restricted in the tropics, because too much internal heat production would be detrimental in a hot climate: we could very easily overheat and die. However, this means that the 'overflow channels' are partially sealed off, so more free radicals are generated at rest, especially on a high-fat diet. This makes Africans eating a fatty western diet more vulnerable to conditions such as heart disease and diabetes, which are linked with free-radical damage. Conversely, the Inuit, who have a low incidence of such diseases, dissipate the proton gradient to generate extra internal heat in the frozen north. Accordingly, they have a relatively low free-radical leakage at rest and are less vulnerable to degenerative diseases. On the other hand, dissipating energy as heat is counterproductive in sperm, which depend on the energetic efficiency of a small number of mitochondria to power their swimming. This gives the Arctic peoples a potentially higher risk of male infertility.

In all of these circumstances, free radicals are the signal for change. The respiratory chains act like a thermostat: if free-radical leakage rises, one of several mechanisms cuts in to lower their level again, then switches itself off, just as the fluctuations in temperature switch the boiler on and off in a thermostat. In the case of the respiratory chains, free radicals are almost certainly detected in concert with other indicators of the overall 'health status' of the cell, such as ATP levels. So a rise in free-radical leakage set against falling ATP levels within one mitochondrion is the signal to build new subunits for the respiratory chains; if ATP levels are high, free radicals are the signal for greater uncoupling, or perhaps for sex in unicellular eukaryotes; and a sustained, uncorrectable rise in free-radical leakage set against falling cellular ATP levels is the signal for cell death in multicellular individuals. In each case, fluctuations in free-radical leakage are as indispensable to the feed-back loop as are the temperature fluctuations to a thermostat: free radicals are vital to life, and attempting to get rid of them, for example using antioxidants, is folly. This simple fact has forced two other major innovations on life: the origin of two sexes, and the decline and fall of organisms into ageing and death.

Free radicals are reactive and cause damage and mutations, especially to the adjacent mitochondrial DNA. In lower eukaryotes, like yeast, mitochondrial DNA acquires mutations approximately 100 000 times faster than nuclear genes. Yeasts can sustain such a high rate because they don't depend on mito-chondria to generate energy. The mutation rate is far lower in the higher eukaryotes, such as humans, because we do depend on our mitochondria. Mutations in mitochondrial DNA cause serious diseases, and tend to be eliminated by natural selection. Even so, the long-term evolution rate of mito-chondrial genes, over thousands or millions of years, is between 10 and 20 times faster than nuclear genes. What's more, the nuclear genes are reshuffled every generation to give a new hand of genes. These disparate patterns set up a serious strain. The subunits of the respiratory chain are encoded by both nuclear and mitochondrial genes, and to function properly must interact with nanoscopic precision: any changes in gene sequence might change the struc-ture or function of the subunits, and could potentially block electron flow. The only way to guarantee efficient energy generation is to match a single set of mitochondrial genes with a single set of nuclear genes in a cell, and test-drive the combination. If it crashes, the combination is eliminated; if it drives well, the cell is selected as a feasible progenitor for the next generation. But how does a cell select a single set of mitochondrial genes to test against one set of nuclear genes? Simple: it inherits its mitochondria from just one of the two parents. As a result, one parent specializes to pass on the mitochondria, in the large egg cell, whereas the other parent specializes to pass on no mitochondria—which is why sperm are so small, and why their handful of mitochondria are usually destroyed. Thus, the origin and deepest biological distinction between the two sexes, indeed the main reason for having two sexes at all, rather than none or an infinite number, relates to the passage of mitochondria from one generation to the next.

A similar problem occurs during adult life. This is the basis of ageing and the related degenerative diseases that all too often eclipse our twilight years. Mitochondria accumulate mutations through use, especially in active tissues, and these gradually undermine the metabolic capacity of the tissue. Ultimately, cells can only boost their failing energy supply by producing more mitochondria. As the supply of mint mitochondria dries up, cells are obliged to clone genetically damaged mitochondria. Cells that amplify seriously damaged mitochondria face an energy crisis and take the honourable exit—they commit apoptosis. Because damaged cells are eliminated, mitochondrial mutations don't build up in ageing tissues, but the tissue itself gradually loses mass and function, and the remaining healthy cells are under a greater pressure to meet their demands. Any additional stresses, such as nuclear gene mutations, smoking, infections, and so on, are more likely to push cells over the threshold into apoptosis.

Mitochondria calibrate the overall risk of apoptosis, which rises with age. A genetic defect that causes little stress to a young cell causes far more stress to an old cell, simply because the old cell is by now closer to its apoptotic threshold. Age, however, is not measured in years, but in free-radical leakage. Species that leak free radicals quickly, such as rats, live for a few years and succumb to age-related diseases within this brief timeframe. Species that leak free radicals slowly, like birds, may live ten times as long and succumb to degenerative diseases over this long timeframe, although they often die of other causes (such as crash landings) before these diseases set in. Critically, birds (and bats) live longer without sacrificing their 'pace of life'—their metabolic rate is similar to mammals that live but a tenth as long. The same mutations in nuclear genes cause the same age-related diseases in different species, but the rate at which they progress varies by orders of magnitude—and tallies with the underlying rate of free-radical leakage. It follows that the best way to cure, or at least postpone, the diseases of old age is to restrict free-radical leakage from the respiratory chains. This approach has the potential to cure *all* diseases of old age at once, rather than trying to tackle each independently, a tack that has so far failed to deliver a really meaningful clinical breakthrough, and is perhaps destined never to do so.

In sum, mitochondria have shaped our lives, and the world we inhabit, in ways that defy belief. All these evolutionary innovations stem from a handful of rules guiding the passage of electrons down the respiratory chains. Remarkably, we can elucidate all this after two billion years of intimate adaptations. We can do so because, despite their changes, mitochondria have retained distinctive imprints of their heritage. These clues have enabled us to trace the outlines of the story that we have followed in this book. The story is grander, more monumental, than any researcher could ever have guessed until recently. It is not the story of an unusual symbiosis, nor the tale of biological power, the industrial revolution of life. No, it is the story of life itself, not merely on Earth, but anywhere in the universe, for the morals of this story relate to the operating system that governs the evolution of all forms of complex life.

Mankind has always looked to the stars, and wondered why we are here, whether we are alone in this universe. We ask why our world is alive with plants and animals, and what were the chances against it; where we came from, who our ancestors were, what our destiny holds in store. The answer to the question of life, the universe and everything is not 42, as Douglas Adams once had it, but an almost equally cryptic shorthand: it is *mitochondria.* For mitochondria teach us how molecules sprang to life on our planet, and why bacteria dominated for so long. They show us why bacterial sludge is likely to be the climax of evolution across this lonely universe. They teach us how the first genuinely complex cells came into being, and why, since then, life on Earth has ascended a ramp of complexity to the glories we see around us, the great chain of being. They show us why energy-burning, warm-blooded creatures arose, thrusting off the shackles of the environment; why we have sex, two sexes, children, why we must fall in love. And they show us why our days in this firmament are numbered, why we must finally grow old and die. They show us, too, how we might better our twilight years, to stave off the misery of old age that curses humanity. If they don't show us the meaning of life, they do at least make some sense of its shape. And what is meaning in this world, if it doesn't make sense?

Glossary

Antioxidant any compound that protects against biological oxidation, either directly by becoming sacrificially oxidised itself in place of other molecules, or indirectly by catalysing the decomposition of biological oxidants.

Apoptosis programmed cell death, or cell suicide; a finely orchestrated and carefully controlled mechanism for removing damaged or unnecessary cells from a multicellular organism.

Archaea one of the three great domains of life, the other two being the eukaryotes and the bacteria; the archaea are similar to bacteria in their appearance down the microscope, but share a number of molecular similarities with the more complex eukaryotic cells.

Archezoa a disparate group of single-celled eukaryotic organisms that lack mitochondria; at least some were originally thought never to have had any mitochondria at all, but now all are believed to have possessed mitochondria in their past, and later lost them.

Asexual reproduction replication of a cell or organism, in which an exact clone of the parent cell or organism is produced.

ADP adenosine diphosphate, the precursor of ATP.

ATP adenosine triphosphate; the universal energy currency of life, which is formed from ADP (adenosine diphosphate) and phosphate; splitting ATP releases energy used to power many different types of biochemical work, from muscular contraction to protein synthesis.

ATPase the enzyme motor within mitochondria that harnesses the flow of protons to form ATP from ADP and phosphate. ATPase is also known as **ATP synthase.**

Cell the smallest biological unit capable of independent life, by means of self-replication and metabolism.

Cell wall the tough but permeable outer 'shell' of bacterial, archaeal and some eukaryotic cells; maintains cell shape and integrity despite changes in physical conditions.

Chemiosmosis the generation of a proton gradient across an impermeable membrane; the backflow of protons through special channels (the ATPase complexes) is used to power ATP formation.

Chemiosmotic coupling the coupling of respiration to ATP synthesis by means of a proton gradient across a membrane; the energy released by oxidation is used to pump protons across a membrane, and the passage of protons back through the drive-shaft of the ATPase is used to power ATP synthesis.

Chloroplast plant cell organelle responsible for photosynthesis; originally derived from endosymbiotic cyanobacteria.

Chromosome long molecule of DNA, often wrapped in proteins such as histones; may be circular, as in bacteria and mitochondria, or straight, as in the nucleus of eukaryotic cells.

Clonal replication alternative name for asexual reproduction.

Control region stretch of non-coding DNA in the mitochondrial genome that binds factors responsible for controlling the expression of mitochondrial genes.

Cytochrome c mitochondrial protein that shuttles electrons from complex III to complex IV of the respiratory chain; when released from the mitochondria, cytochrome c is a key initiator of apoptosis, or programmed cell death.

Cytochrome oxidase functional name for complex IV of the respiratory chain: a multi-subunit enzyme that receives electrons from cytochrome c and uses them to reduce oxygen to water, the final step of cellular respiration.

Cytoplasm the substance of the cell contained within the cell membrane, but excluding the nucleus.

Cytoskeleton network of protein fibres within the cell that provides structural support; the structure is dynamic and enables some cells to change shape and move around.

Cytosol the aqueous part of the cytoplasm, excluding organelles such as mitochondria and membrane systems.

DNA deoxyribonucleic acid, the molecule responsible for heredity; it is composed of a double helix, in which nucleotide letters are paired with each other to form a template from which an exact copy of the whole molecule can be regenerated; the sequence of nucleotide letters encodes the sequence of amino acids in proteins.

DNA sequence the sequence of nucleotide letters in DNA, which spells out the sequence of amino acids in proteins, or the binding sequences of transcription factors, or nothing at all.

Electron tiny, negatively charged wave-particle that orbits the positively charged nucleus in atoms.

Endosymbionts cells that live in a mutually beneficial relationship inside other cells.

Endosymbiosis a mutually beneficial relationship in which one type of cell lives inside another larger cell.

Enzyme a protein catalyst, responsible for speeding up biochemical reactions by many orders of magnitude, with enormous specificity.

Eukaryote an organism, either single celled or multicellular, composed of eukaryotic cells

Eukaryotic cell cells with a 'true' nucleus; all are thought to either possess, or have once possessed, mitochondria.

Evolution rate the speed at which DNA sequence changes over many generations, which equates to the rate of mutations coupled to the purging effect of natural selection, which eliminates detrimental mutations, making the evolution rate slower than the mutation rate.

Exponent a superscript number denoting the number of times a number should be multiplied by itself; the slope of a line on a log–log plot.

Fermentation chemical splitting of sugars, without net oxidation or reduction, to form alcohol (or other substances); releases sufficient energy to synthesise ATP.

Free radical an atom or a molecule with a single unpaired electron, which tends to be an unstable physical state leading to chemical reactivity.

Free-radical leakage continuous low-level production of free radicals from the respiratory chains of the mitochondria, as a result of electron carriers reacting directly with oxygen.

Gametes dedicated sex cells, such as sperm or eggs.

Gene a stretch of DNA, whose sequence of nucleotide letters encodes a single protein.

Genetic code the DNA letters, which encode the sequence of amino acids in proteins; particular combinations of letters specify particular amino acids, or other instructions such as 'start' and 'stop' reading.

Genome the complete library of genes in an organism; the term is also taken to include non-coding (i.e. non-genic) stretches of DNA.

Heteroplasmy a mixture of mitochondria (or other organelles) from two different sources, such as the father and the mother.

Histones proteins that bind to DNA in a very particular structure, found only in eukaryotic cells and a few archaea, such as methanogens.

Hydrogen hypothesis theory arguing that the eukaryotic cell originated as a chimera between two very different prokaryotic cells, in a metabolic symbiosis.

Hydrogenosome organelle found in some anaerobic eukaryotic cells, which generates energy by fermenting organic fuels to release hydrogen gas; now known to have a common ancestor with mitochondria.

Lateral gene transfer the random transfer of segments of DNA, containing genes, from one cell to another, as opposed to vertical inheritance from mother to daughter.

Lipid a type of long-chain fatty molecule found in biological membranes and as stored fuel.

Membrane the thin fatty (lipid) layer that envelops cells, and forms complex systems inside eukaryotic cells.

Metabolic rate the rate of energy consumption, measured by the rate of glucose oxidation or oxygen consumption in cells and whole organisms.

Mitochondrial organelles within cells responsible for ATP generation, and for controlling cell suicide; derived from endosymbiotic α-proteobacteria, but exactly which is still disputed.

Mitochondrial Eve the last common female ancestor of all humans living today, if judged by the slow divergence of mitochondrial DNA inherited asexually down the maternal line.

Mitochondrial DNA the chromosome found in mitochondria, typically in five to ten copies, which is usually circular, and bacterial in nature.

Mitochondrial genes the contingent of genes encoded by mitochondrial DNA; in humans there are 13 protein-coding genes, as well as genes encoding the RNA needed to manufacture proteins on site.

Mitochondrial mutation an inherited alteration in the sequence of mitochondrial DNA.

Mitochondrial disease a disease caused by mutations or deletions in mitochondrial DNA, or in nuclear genes encoding proteins targeted to the mitochondria.

Mutation an inherited alteration in the sequence of DNA, which may have a negative, positive or neutral effect on function; natural selection, acting differentially on mutations in DNA, hones the function of proteins to specific tasks.

Mutation rate the number of mutations occurring in DNA per unit time, usually a limited time period of a few generations; see also evolution rate.

NADH nicotinamide adenine dinucleotide; a molecule that carries the electrons and protons ultimately derived from glucose to complex I of the respiratory chain, for use in respiration.

Natural selection the differential survival and reproduction of individuals in a population, the members of which have heritable differences in biological fitness.

Non-coding (junk) DNA sequences of DNA that do not encode proteins or RNA.

Nucleus spherical membrane-enclosed 'control centre' of eukaryotic cells, containing chromosomes composed of DNA and protein.

Oocyte egg cell; the female sex cell, containing half the number of chromosomes as a somatic (body) cell.

Organelles tiny organs within cells dedicated to specific tasks, such as mitochondria and chloroplasts.

Oxidation loss of electrons from an atom or molecule.

Phagocytosis physical engulfment of particles by a cell, by means of changing shape and extending pseudopodia; the particles are digested in a food vacuole inside the cell.

Prokaryote a broad class of single-celled organisms that do not possess a nucleus, including both bacteria and archaea.

Protein a string of amino acids linked together in a chain, to form an almost infinite number of possible shapes and functions. Proteins form essentially all the machinery of life, including enzymes, structural fibres, transcription factors, DNA-binding proteins, hormones, receptors, and antibodies.

Proton the nucleus of a hydrogen atom, with a single positive charge.

Proton gradient a difference in proton concentration between one side of a membrane and the other.

Proton leak the drizzle of protons back through a membrane that is nearly, but not quite, impermeable to protons.

Proton pumping the physical translocation of protons from one side of a membrane to the other.

Proton-motive force the potential energy stored in a proton gradient from one side of a membrane to the other; the combination of an electrical potential difference and the pH (proton concentration) difference.

Recombination the physical crossing over and replacement of a gene from one source with the equivalent gene from another source; takes place in sexual reproduction and lateral gene transfer, and to repair a damaged chromosome by reference to a spare copy.

Redox reaction a reaction in which one molecule is oxidized at the expense of another, which is accordingly reduced.

Redox signalling the change in activity of a transcription factor as a result of its oxidation or reduction, usually by free radicals; the active transcription factor controls the expression of genes to form new proteins.

Reduction gain of electrons by an atom or molecule.

Reduction state the overall proportion of a population of molecules in the reduced state, as opposed to the oxidized state; if complex I is 70 per cent reduced, then 70 per cent of the complexes are in possession of respiratory electrons (they are reduced) and 30 per cent are oxidized.

Respiration oxidation of foodstuffs to generate energy in the form of ATP.

Respiratory chain the series of multi-subunit protein complexes embedded in the bacterial and mitochondrial membranes that pass electrons derived from glucose from one to the next, finally to react with oxygen in complex IV. The energy released by the passage of electrons is used to pump protons across the membrane.

RNA ribonucleic acid; several forms exist, including messenger RNA (an exact copy of the DNA sequence in an individual gene, which translocates to the cytoplasm); ribosomal RNA (which forms part of the ribosomes, the protein-building factories found in the cytoplasm); and transfer RNA (an adapter that couples a nucleotide code to a particular amino acid).

Sexual reproduction reproduction by means of the fusion of two sex cells, or gametes, each of which contains a random assortment of half the parental genes, to give an embryo with an equal number of genes from both parents.

Symbiosis a mutually beneficial relationship between organisms of two different species.

Transcription factor a protein that binds to a DNA sequence, to signal the transcription of a gene into an RNA copy, as the first step of protein synthesis.

Uncoupling dissociation of respiration from ATP production; instead of the proton gradient powering ATP synthesis through the ATPase, protons pass back through membrane pores, dissipating the gradient as heat; uncoupling the respiratory chain is equivalent to uncoupling a bicycle chain, thereby disconnecting peddling from forward momentum.

Uncoupling protein a protein channel in the membrane that enables the passage of protons back through the membrane, dissipating the proton gradient as heat.

Uncoupling agent any chemical that dissipates the proton gradient by shuttling protons across the membrane, thereby dissociating respiration from ATP production.

Uniparental inheritance the inheritance of mitochondria (or chloroplasts) from only one of two parents, specifically the mother.

Further Reading

Introduction

General texts

Fruton, J. *Proteins, Enzymes, Genes: The Interplay of Chemistry and Biology.* Yale University Press, New Haven, USA, 1999.

Margulis, Lynn. *Origin of Eukaryotic Cells.* Yale University Press, Yale, USA, 1970.

—— Gaia is a tough bitch. In John Brockman (ed.), *The Third Culture: Beyond the Scientific Revolution.* Simon & Schuster, New York, USA, 1995.

Sapp, Jan. *Evolution by Association: A History of Symbiosis.* Oxford University Press, Oxford, UK, 1994.

Wallin, Ivan. *Symbionticism and the Origin of Species.* Bailliere, Tindall and Cox, London, UK, 1927.

The nature of mitochondria

Attardi, G. The elucidation of the human mitochondrial genome: A historical perspective. *Bioessays* **5**: 34–39; 1986.

Baldauf, S. L. The deep roots of eukaryotes. *Science* **300**: 1703–1706; 2003.

Cooper, C. The return of the mitochondrion. *The Biochemist* **27**(3): 5–6; 2005.

Dyall, S. D., Brown, M. T., and Johnson, P. J. Ancient invasions: From endosymbionts to organelles. *Science* **304**: 253–257; 2004.

Griparic, L., and van der Bliek, A. M. The many shapes of mitochondrial membranes. *Traffic* **2**: 235–244; 2001.

Kiberstis, P. A. Mitochondria make a comeback. *Science* **283**: 1475; 1999.

Sagan, L. On the origin of mitosing cells. *Journal of Theoretical Biology* **14**: 225–274; 1967.

Schatz, G. The tragic matter. *FEBS (Federation of European Biochemical Societies) Letters* **536**: 1–2; 2003.

Scheffler, I. E. A century of mitochondrial research: achievements and perspectives. *Mitochondrion* **1**: 3–31; 2000.

Part 1

General texts

Dawkins, Richard. *The Ancestor's Tale: A Pilgrimage to the Dawn of Life.* Weidenfeld & Nicolson, London, UK, 2004.

de Duve, Christian. *Life Evolving: Molecules, Mind, and Meaning.* Oxford University Press, New York, USA, 2002.

Gould, Stephen Jay. *Wonderful Life. The Burgess Shale and the Nature of History.* Penguin, London, UK, 1989.

Knoll, Andrew H. *Life on a Young Planet: The First Three Billion Years of Evolution on Earth.* Princeton University Press, Princeton, USA, 2003.

Lane, Nick. *Oxygen: The Molecule that Made the World*. Oxford University Press, Oxford, UK, 2002.

Margulis, Lynn. *Origin of Eukaryotic Cells*. Yale University Press, Yale, USA, 1970.

Mayr, Ernst . *What Evolution Is*. Weidenfeld & Nicolson, London, UK, 2002.

Morris, Simon Conway. *Life's Solution: Inevitable Humans in a Lonely Universe*. Cambridge University Press, Cambridge, UK, 2003.

The origin of eukaryotic cells

Martin, W., Hoffmeister, M., Rotte, C., and Henze, K. An overview of endosymbiotic models for the origins of eukaryotes, their ATP-producing organelles (mitochondria and hydrogenosomes) and their heterotrophic lifestyle. *Biological Chemistry* **382:** 1521–1539; 2001.

Sagan, L. On the origin of mitosing cells. *Journal of Theoretical Biology* **14:** 255–274; 1967.

Vellai, T., and Vida, G. The origin of eukaryotes: The difference between prokaryotic and eukaryotic cells. *Proceedings of the Royal Society of London B: Biological Sciences* **266:** 1571–1577; 1999.

Catastrophic loss of the cell wall

Cavalier-Smith, T. The phagotrophic origin of eukaryotes and phylogenetic classification of Protozoa. *International Journal of Systematic and Evolutionary Microbiology* **52:** 297–354; 2002.

Maynard-Smith, John, and Szathmáry, Eörs. *The Origins of Life*, Chapter 6: The Origin of Eukaryotic Cells. Oxford University Press, Oxford, UK, 1999.

Bacterial cytoskeleton

van den Ent, F., Amos, L. A., and Lowe, J. Prokaryotic origin of the actin cytoskeleton. *Nature* **413:** 39–44; 2001.

Jones, L. J., Carballido-Lopez, R., and Errington, J. Control of cell shape in bacteria: Helical, actin-like filaments in *Bacillus subtilis*. *Cell* **104:** 913–922; 2001.

Discovery of the archaea

Keeling, P. J., and Doolittle, W. F. Archaea: Narrowing the gap between prokaryotes and eukaryotes. *Proceedings of the National Academy of Sciences of the USA* **92:** 5761–5764; 1995.

Woese, C. R., and Fox, G. E. Phylogenetic structure of the prokaryotic domain: The primary kingdoms. *Proceedings of the National Academy of Sciences of the USA* **74:** 5088–5090; 1977.

The archezoa

Cavalier-Smith, T. A 6-kingdom classification and a unified phylogeny. In H. E. A. Schenk and W. Schwemmler (eds.), *Endocytobiology II*, pp. 1027–1034. Walter de Gruyter, Berlin, Germany, 1983.

—— Eukaryotes with no mitochondria. *Nature* **326:** 332–333; 1987.

—— Archaebacteria and Archezoa. *Nature* **339:** 100–101; 1989.

Rickettsia as the ancestor of mitochondria

Andersson, J. O., and Andersson, S. G. A century of typhus, lice and *Rickettsia*. *Research in Microbiology* **151:** 143–150; 2000.

Andersson, S. G., Zomorodipour, A., Andersson J. O. , Sicheritz-Ponten, T., Alsmark U. C., Podowski, R. M., Naslund, A. K., Eriksson, A. S., Winkler, H. H., Kurland, C. G. The genome sequence of *Rickettsia prowazekii* and the origin of mitochondria. *Nature* **396:** 133–140; 1998.

Andersson, S. G. E., Karlberg, O., Canback, B., and Kurland, C. G. On the origin of mitochondria: A genomics perspective. *Philosophical Transactions of the Royal Society of London B: Biological Sciences* **358:** 165–179; 2003.

Collapse of the archezoa

Clark, C. G., and Roger, A. J. Direct evidence for secondary loss of mitochondria in *Entamoeba hitolytica*. *Proceedings of the National Academy of Sciences of the USA* **92:** 6518–6521; 1995.

Keeling, P. J. A kingdom's progress: Archezoa and the origin of eukaryotes. *Bioessays* **20:** 87–95; 1998.

Methanogens as the host cell

Martin, W., and Embley, T. M. Early evolution comes full circle. *Nature* **431:** 134–136; 2004.

Pereira, S. L., Grayling, R. A., Lurz, R., and Reeve, J. N. Archaeal nucleosomes. *Proceedings of the National Academy of Sciences of the USA* **94:** 12633–12637; 1997.

Rivera, M., Jain, R., Moore, J. E., and Lake, J. A. Genomic evidence for two functionally distinct gene classes. *Proceedings of the National Academy of Sciences of the USA* **95:** 6239–6244; 1998.

Rivera, M. C., and Lake, J. A. The ring of life provides evidence for a genome fusion origin of eukaryotes. *Nature* **431:** 152; 2004.

Hydrogen hypothesis

Akhmanova, A., Voncken, F., van Alen, T., van Hoek, A., Boxma, B., Vogels, G., Veenhuis, M., and Hackstein, J. H. A hydrogenosome with a genome. *Nature* **396:** 527–528; 1998.

Boxma, B., de Graat, R. M., and van der Staay, G. W., et al. An anaerobic mitochondrion that produces hydrogen. *Nature* **434:** 74–79; 2005.

Embley, T. M., and Martin, W. A hydrogen-producing mitochondrion. *Nature* **396:** 517–519; 1998.

Gray, M. W. Evolutionary biology: The hydrogenosome's murky past. *Nature* **434:** 29–31; 2005.

Martin, W., and Müller, M. The hydrogen hypothesis for the first eukaryote. *Nature* **392:** 37–41; 1998.

—— Russell, M. J. On the origins of cells: A hypothesis for the evolutionary transitions from abiotic geochemistry to chemoautotrophic prokaryotes, and from prokaryotes to nucleated cells. *Philosophical Transactions of the Royal Society of London B* **358:** 59–85; 2003.

Müller, M., and Martin, W. The genome of *Rickettsia prowazekii* and some thoughts on the origin of mitochondria and hydrogenosomes. *Bioessays* **21:** 377–381; 1999.

Anaerobic mitochondria

Horner, D. S., Heil, B., Happe, T., and Embley, T. M. Iron hydrogenases—ancient enzymes in modern eukaryotes. *Trends in Biochemical Sciences* **27**: 148–153; 2002.

Sutak, R., Dolezal, P., Fiumera, H. L., Hardy, I., Dancis, A., Delgadillo-Correa, M., Johnson, P. J., Mujller, M., and Tachezy, J. Mitochondrial-type assembly of FeS centers in the hydrogenosomes of the amitochondriate eukaryote *Trichomonas vaginalis*. *Proceedings of the National Academy of Sciences of the USA* **101**: 10368–10373; 2004.

Theissen, U., Hoffmeister, M., Grieshaber, M., and Martin, W. Single eubacterial origin of eukaryotic sulfide: Quinone oxidoreductase, a mitochondrial enzyme conserved from the early evolution of eukaryotes during anoxic and sulfidic times. *Molecular Biology and Evolution* **20**(9): 1564–1574; 2003.

Tielens, A. G., Rotte, C., van Hellemond, J. J., and Martin, W. Mitochondria as we don't know them. *Trends in Biochemical Sciences* **27**: 564–572; 2002.

Van der Giezen, M., Slotboom, D. J., Horner, D. S., Dyal, P. L., Harding, M., Xue, G. P., Embley, T. M., and Kunji, E. R. Conserved properties of hydrogenosomal and mitochondrial ADP/ATP carriers: A common origin for both organelles. *EMBO (European Molecular Biology Organization) Journal* **21**: 572–579; 2002.

Ocean chemistry

Anbar, A. D., and Knoll, A. H. Proterozoic ocean chemistry and evolution: A bioinorganic bridge? *Science* **297**: 1137–1142; 2002.

Canfield, D. E. A new model of Proterozoic ocean chemistry. *Nature* **396**: 450–452; 1998.

—— Habicht K. S., and Thamdrup B. The Archean sulfur cycle and the early history of atmospheric oxygen. *Science* **288**: 658–661; 2000.

Part 2

General texts

de Duve, Christian. *Life Evolving: Molecules, Mind, and Meaning*. Oxford University Press, New York, USA, 2002.

Harold, Franklin M. *The Way of the Cell. Molecules, Organisms, and the Order of Life*. Oxford University Press, New York, USA, 2001.

—— *The Vital Force: A Study of Bioenergetics*. W. H. Freeman and Co., New York, USA, 1986.

Lane, Nick. *Oxygen: The Molecule that Made the World*. Oxford University Press, Oxford, UK, 2002.

Nicholls, David, and Ferguson, Stuart J. *Bioenergetics 3*. Academic Press, Oxford, UK, 2002.

Prebble, John, and Weber, Bruce. *Wandering in the Gardens of the Mind—Peter Mitchell and the Making of Glynn*. Oxford University Press, Oxford, UK, 2003.

Wolpert, Lewis and Richards, Alison. *Passionate Minds: The Inner World of Scientists*. Oxford University Press, Oxford, UK, 1997.

Energy production and the sun

Schatz, G. The tragic matter. *FEBS (Federation of European Biochemical Societies) Letters* **536**: 1–2; 2003.

Lavoisier and the discovery of respiration

Jaffe, Bernard. *Crucibles*. Newton Publishing Co., New York, USA, 1932.

Lavoisier, A. *Elements of Chemistry*. Dover Publications Inc., New York, USA, 1965.

Morris, R. *The Last Sorcerers: The Path from Alchemy to the Periodic Table*. Joseph Henry Press, Washington DC, USA, 2003.

Discovery of the respiratory chain

Gest, H. Landmark discoveries in the trail from chemistry to cellular biochemistry, with particular reference to mileposts in research on bioenergetics. *Biochemistry and Molecular Biology Education* **30:** 9–13; 2002.

Keilin, D. *The History of Cell Respiration and Cytochrome*. Cambridge University Press, Cambridge, UK, 1966.

—— Cytochrome and respiratory enzymes. *Proceedings of the Royal Society of London B: Biological Sciences* **104:** 206–252; 1929.

Lahiri, S. Historical perspectives of cellular oxygen sensing and responses to hypoxia. *Journal of Applied Physiology* **88:** 1467–1473; 2000.

Warburg, O. *The Oxygen-Transferring Ferment of Respiration*. In *Nobel Lectures, Physiology or Medicine 1922–1941*, Nobel Lecture, 1931. Elsevier Publishing Company, Amsterdam, Holland, 1965 (and available online at the Nobel e-Museum).

Fermentation

Buchner, E. *Cell-Free Fermentation*. In *Nobel Lectures, Chemistry 1901–1921*, Nobel Lecture, 1907. Elsevier Publishing Company, Amsterdam, Holland, 1966 (and available online at the Nobel e-Museum).

Discovery of ATP

Engelhardt, W. A. Life and Science. Autobiography. *Annual Review of Biochemistry* **51:** 1–19; 1982.

Fruton, J. *Proteins, Enzymes, Genes: The Interplay of Chemistry and Biology*. Yale University Press, New Haven, USA, 1999.

Gest, H. Landmark discoveries in the trail from chemistry to cellular biochemistry, with particular reference to mileposts in research on bioenergetics. *Biochemistry and Molecular Biology Education* **30:** 9–13; 2002.

Rate of ATP production

Rich, P. The cost of living. *Nature* **421:** 583; 2003.

The elusive squiggle

Gest, H. Landmark discoveries in the trail from chemistry to cellular biochemistry, with particular reference to mileposts in research on bioenergetics. *Biochemistry and Molecular Biology Education* **30:** 9–13; 2002.

Harold, F. M. The 1978 Nobel Prize in Chemistry. *Science* **202:** 1174–1176; 1978.

Peter Mitchell and chemiosmotics

Chappell, J. B. Nobel Prize: Chemistry. *Trends in Biochemical Sciences* **4:** N3–N4; 1979.

Harold, F. M. The 1978 Nobel Prize in Chemistry. *Science* **202:** 1174–1176; 1978.

Matzke, M. A, and Matzke, A. J. M. Kuhnian revolutions in biology: Peter Mitchell and the chemiosmotic theory. *Bioessays* **19:** 91–93; 1997.

Mitchell, P. *David Keilin's Respiratory Chain Concept and its Chemiosmotic Consequences.* In *Nobel Lectures in Chemistry 1971–1980*, Nobel Lecture, 1978, Sture Forsén (ed.), World Scientific Publishing Company, Singapore, 1993 (and available online at the Nobel e-Museum).

—— Coupling of phosphorylation to electron and hydrogen transfer by a chemi-osmotic type of mechanism. *Nature* **191:** 144–148; 1961.

Orgel, L. E. Are you serious, Dr Mitchell? *Nature* **402:** 17; 1999.

Prebble, J. Peter Mitchell and the ox phos wars. *Trends in Biochemical Sciences* **27:** 209–212; 2002.

Schatz, G. *Efraim Racker.* In *Biographical Memoirs*, vol. 70. National Academies Press, Washington DC, USA, 1996.

The Jagendorf-Uribe experiment

Jagendorf, A. T., and Uribe, E. ATP formation caused by acid-base transition of spinach chloroplasts. *Proceedings of the National Academy of Sciences USA* **55:** 170–177; 1966.

—— Chance, luck and photosynthesis research: An inside story. *Photosynthesis Research* **57:** 215–229; 1998.

Structure of the ATPase

Walker, J. E. *ATP Synthesis by Rotary Catalysis.* In *Nobel Lectures in Chemistry 1996–2000*, Nobel Lecture, 1997, Ingmar Grenthe (ed.), World Scientific Publishing Company, Singapore, 2003 (and available online at the Nobel e-Museum).

Wider function of the proton current

Harold, Franklin M. *The Way of the Cell. Molecules, Organisms, and the Order of Life.* Oxford University Press, New York, USA, 2001.

—— Gleanings of a chemiosmotic eye. *Bioessays* **23:** 848–855; 2001.

Origin of life

Martin, W., and Russell, M. J. On the origins of cells: A hypothesis for the evolutionary transitions from abiotic geochemistry to chemoautotrophic prokaryotes, and from prokaryotes to nucleated cells. *Philosophical Transactions of the Royal Society of London B: Biological Sciences* **358:** 59–85; 2003.

Russell, M. J., and Hall, A. J. The emergence of life from iron monosulphide bubbles at a submarine hydrothermal redox and pH front. *Journal of the Geological Society of London* **154:** 377–402; 1997.

—— —— Cairns-Smith, A. G., and Braterman, P. S. Submarine hot springs and the origin of life. *Nature* **336:** 117; 1988.

Wächtershäuser, G. Groundworks for an evolutionary biochemistry: The iron-sulphur world. *Progress in Biophysics and Molecular Biology* **58:** 85–201; 1992.

Part 3

General texts

Dennett, Daniel. *Darwin's Dangerous Idea*. Penguin, London, UK, 1995.

Maynard Smith, John, and Szathmáry, Eörs. *The Origins of Life*. Oxford University Press, Oxford, UK, 1999.

Monod, Jacques. *Chance and Necessity*. Penguin, London, UK, 1997 (first published in English 1971).

Prescott, L. M., Harley, J. P., and Klein, D. A. *Microbiology* (5th edition). McGraw-Hill Education, Maidenhead, UK, 2001.

Ridley, Mark. *Mendel's Demon*. Weidenfeld & Nicolson, London, UK, 2000.

Speed of bacterial proliferation

Jensen, P. R., Loman, L., Petra, B., van der Weijden, C., and Westerhoff, H. V. Energy buffering of DNA structure fails when *Escherichia coli* runs out of substrate. *Journal of Bacteriology* 177: 3420–3426; 1995.

Koedoed, S., Otten, M. F., Koebmann, B. J., Bruggeman, F. J., Bakker, B. M., Snoep, J. L., Krab, K., van Spanning, R. J. M., van Verseveld, H. W., Jensen, P. R., Koster, J. G., and Westerhoff, H. V. A turbo engine with automatic transmission? How to marry chemicomotion to the subtleties and robustness of life. *Biochimica et Biophysica Acta* 1555: 75–82; 2002.

O'Farrell, P. H. Cell cycle control: Many ways to skin a cat. *Trends in Cell Biology* 2: 159–163; 1992.

Genome size of soil bacteria

Konstantinidis, K. T., and Tiedje, J. M. Trends between gene content and genome size in prokaryotic species with larger genomes. *Proceedings of the National Academy of Sciences USA* 101: 3160–3165; 2004.

Genome of *Rickettsia*

Andersson, J. O., and Andersson, S. G. A century of typhus, lice and *Rickettsia*. *Research in Microbiology* 151: 143–150; 2000.

Andersson, S. G., Zomorodipour, A., Andersson, J. O., Sicheritz-Ponten, T., Alsmark, U. C., Podowski, R. M., Naslund, A. K., Eriksson, A. S., Winkler, H. H., and Kurland, C. G. The genome sequence of *Rickettsia prowazekii* and the origin of mitochondria. *Nature* 396: 133–140; 1998.

Gross, L. How Charles Nicolle of the Pasteur Institute discovered that epidemic typhus is transmitted by lice: Reminiscences from my years at the Pasteur Institute in Paris. *Proceedings of the National Academy of Sciences USA* 93: 10539–10540; 1996.

Gene loss and lateral gene transfer

Frank, A. C., Amiri, H., and Andersson, S. G. E. Genome deterioration: Loss of repeated sequences and accumulation of junk DNA. *Genetica* 115: 1–12; 2002.

Vellai, T., Takács, K., and Vida, G. A new aspect to the origin and evolution of eukaryotes. *Journal of Molecular Evolution* 46: 499–507; 1998.

——— Vida, G. The origin of eukaryotes: The difference between prokaryotic and eukaryotic cells. *Proceedings of the Royal Society of London B: Biological Sciences* 266: 1571–1577; 1999.

Difficulties in defining bacterial 'species'

Doolittle, W. F., Boucher, Y., Nesbo, C. L, Douady, C. J., Andersson, J. O., and Roger, A. J. How big is the iceberg of which organellar genes in nuclear genomes are but the tip? *Philosophical Transactions of the Royal Society of London B: Biological Sciences* **358:** 39–58; 2003.

Martin, W. Woe is the Tree of Life. In J. Sapp (ed.), *Microbial Phylogeny and Evolution: Concepts and Controversies.* Oxford University Press, New York, USA, 2005.

Maynard Smith, J., Feil, E. J., and Smith, N. H. Population structure and evolutionary dynamics of pathogenic bacteria. *Bioessays* **22:** 1115–1122; 2000.

Spratt, B. G., Hanage, W. P., and Feil, E. J. The relative contributions of recombination and point mutation to the diversification of bacterial clones. *Current Opinion in Microbiology* **4:** 602–606; 2001.

Respiratory efficiency and genome size

Konstantinidis, K., and Tiedje, J. M. Trends between gene content and genome size in prokaryotic species with larger genomes. *Proceedings of the National Academy of Sciences USA* **101:** 3160–3165; 2004.

Vellai, T., Takács, K., and Vida, G. A new aspect to the origin and evolution of eukaryotes. *Journal of Molecular Evolution* **46:** 499–507; 1998.

Giant bacteria

Schulz, H. N., Brinkhoff, T., Ferdelman, T. G., Hernández Mariné, M., Teske, A., and Jørgensen, B. B. Dense populations of a giant sulfur bacterium in Namibian shelf sediments. *Science* **284:** 493–495; 1999.

Bacteria without cell walls

Ruepp, A., Graml, W., Santos-Martinez, M. L., Koretke, K. K., Volker, C., Mewes, H. W., Frishman, D., Stocker, S., Lupas, A. N., and Baumeister, W. The genome sequence of the thermoacidophilic scavenger *Thermoplasma acidophilum. Nature* **407:** 508–513; 2000.

Taylor-Robinson, D. *Mycoplasma genitalium*—an update. *International Journal of STD and AIDS* **13:** 145–151; 2002.

Gene transfer to the nucleus

Bensasson, D., Feldman, M. W., and Petrov, D. A. Rates of DNA duplication and mitochondrial DNA insertion in the human genome. *Journal of Molecular Evolution* **57:** 343–354; 2003.

Huang, C. Y., Ayliffe, M. A., and Timmis, J. N. Direct measurement of the transfer rate of chloroplast DNA into the nucleus. *Nature* **422:** 72–76; 2003.

Martin, W. Gene transfer from organelles to the nucleus: Frequent and in big chunks. *Proceedings of the National Academy of Sciences USA* **100:** 8612–8614; 2003.

Turner, C., Killoran, C., Thomas, N. S., Rosenberg, M., Chuzhanova, N. A., Johnston, J., Kemel, Y., Cooper, D. N., and Biesecker, L. G. Human genetic disease caused by de novo mitochondrial-nuclear DNA transfer. *Human Genetics* **112:** 303–309; 2003.

Origin of the nucleus

Martin, W. A. briefly argued case that mitochondria and plastids are descendents of endosymbionts but that the nuclear compartment is not. *Proceedings of the Royal Society of London B: Biological Sciences* **266:** 1387–1395; 1999.

Berry, S. Endosymbiosis and the design of eukaryotic electron transport. *Biochimica et Biophysica Acta* **1606:** 57–72; 2003.

Fungi as the first eukaryotes
Martin, W., Rotte, C., Hoffmeister, M., Theissen, U., Gelius-Dietrich, G., Ahr, S., and Henze, K. Early cell evolution, eukaryotes, anoxia, sulfide, oxygen, fungi first (?), and a tree of genomes revisited. *IUBMB (International Union of Biochemistry and Molecular Biology) Life* **55:** 193–204; 2003.

Why mitochondrial genes still exist
Allen, J. F. Control of gene expression by redox potential and the requirement for chloroplast and mitochondrial genes. *Journal of Theoretical Biology* **165:** 609–631; 1993.

—— The function of genomes in bioenergetic organelles. *Philosophical Transactions of the Royal Society of London B: Biological Sciences* **358:** 19–38; 2003.

—— Raven, J. A. Free-radical-induced mutation vs redox regulation: Costs and benefits of genes in organelles. *Journal of Molecular Evolution* **42:** 482–492; 1996.

Chomyn, A. Mitochondrial genetic control of assembly and function of complex I in mammalian cells. *Journal of Bioenergetics and Biomembranes* **33:** 251–257; 2001.

Race, H. L., Herrmann, R. G., and Martin, W. Why have organelles retained genomes? *Trends in Genetics* **15:** 364–370; 1999.

Phylogeny of the ATP exporters
Andersson, S. G. E., Karlberg, O., Canbäck, B., and Kurland, C. G. On the origin of mitochondria: A genomics perspective. *Philosophical Transactions of the Royal Society of London B: Biological Sciences* **358:** 165–179; 2003.

Löytynoja, A., and Milinkovitch, M. C. Molecular phylogenetic analyses of the mitochondrial ADP-ATP carriers: The plantae/fungi/metazoa trichotomy revisited. *Proceedings of the National Academy of Sciences USA* **98:** 10202–10207; 2001.

Why bacteria are still bacteria
Lane, N. Mitochondria: Key to complexity. In W. Martin (ed.), *Origins of Mitochondria and Hydrogenosomes.* Springer, Heidelberg, Germany, 2006.

Part 4

General texts
Ball, Philip. *The Self-Made Tapestry.* Oxford University Press, Oxford, UK, 1999.

Gould, Stephen Jay. *Full House.* Random House, New York, USA, 1997.

Haldane, J. B. S. *On Being the Right Size,* ed. John Maynard-Smith. Oxford University Press, Oxford, UK, 1985.

Mandelbrot, Benoit. *The Fractal Geometry of Nature.* W. H. Freeman, New York, 1977.

Ridley, Mark. *Mendel's Demon.* Weidenfeld & Nicolson, London, UK, 2000.

The power laws of biology
Bennett, A. F. Structural and functional determinates of metabolic rate. *American Zoologist* **28:** 699–708; 1988.

Heusner, A. Size and power in mammals. *Journal of Experimental Biology* **160**: 25–54; 1991.

Kleiber, M. *The Fire of Life*. Wiley, New York, USA, 1961.

Fractal geometry and scaling

Banavar, J., Damuth, J., Maritan, A., and Rinaldo, A. Supply-demand balance and metabolic scaling. *Proceedings of the National Academy of Sciences USA* **99**: 10506–10509; 2002.

West, G. B., Brown J. H., and Enquist B. J. A general model for the origin of allometric scaling in biology. *Science* **276**: 122–126; 1997.

—— —— —— The fourth dimension of life: Fractal geometry and allometric scaling of organisms. *Science* **284**: 1677–1679; 1999.

—— Woodruff, W. H., and Brown, J. H. Allometric scaling of metabolic rate from molecules and mitochondria to cells and mammals. *Proceedings of the National Academy of Sciences USA* **99**: 2473–2478; 2002.

Questioning the universal constant

Dodds, P. S., Rothman, D. H., and Weitz, J. S. Re-examination of the '3/4-law' of metabolism. *Journal of Theoretical Biology* **209**: 9–27; 2001.

White, C. R., and Seymour, R. S. Mammalian basal metabolic rate is proportional to body mass$^{2/3}$. *Proceedings of the National Academy of Sciences USA* **100**: 4046–4049; 2003.

Resting and maximal metabolic rates

Bishop, C. M. The maximum oxygen consumption and aerobic scope of birds and mammals: Getting to the heart of the matter. *Proceedings of the Royal Society of London B: Biological Sciences* **266**: 2275–2281; 1999.

Tissue oxygen concentration in marine invertebrates and mammals

Massabuau, J. C. Primitive and protective, our cellular oxygenation status? *Mechanisms of Ageing and Development* **124**: 857–863; 2003.

Components of metabolic rate in mammals

Porter, R. K. Allometry of mammalian cellular oxygen consumption. *Cellular and Molecular Life Sciences* **58**: 815–822; 2001.

Rolfe, D. F. S., and Brown, G. C. Cellular energy utilization and molecular origin of standard metabolic rate in mammals. *Physiological Reviews* **77**: 731–758; 1997.

Scaling of metabolic rate

Darveau, C. A., Suarez, R. K., Andrews, R. D., and Hochachka, P. W. Allometric cascade as a unifying principle of body mass effects on metabolism. *Nature* **417**: 166–170; 2002.

Hochachka, P. W., Darveau, C. A., Andrews, R. D., and Suarez, R. K. Allometric cascade: A model for resolving body mass effects on metabolism. *Comparative Biochemistry and Physiology A: Molecular and Integrative Physiology* **134**: 675–691; 2003.

Storey, K. B. Peter Hochachka and oxygen. In R. C. Roach et al. (eds.), *Hypoxia: Through the Lifecycle*. Kluwer Academic/Plenum Publishers, New York, USA, 2003.

Weibel, E. R. The pitfalls of power laws. *Nature* **417**: 131–132; 2002.

Evolution of endothermy

Bennett, A. F., and Ruben, J. A. Endothermy and activity in vertebrates. *Science* 206: 649–653; 1979.

—— Hicks, J. W., and Cullum, A. J. An experimental test for the thermoregulatory hypothesis for the evolution of endothermy. *Evolution* 54: 1768–1773; 2000.

Hayes, J. P., and Garland, T., Jr. The evolution of endothermy: Testing the aerobic capacity hypothesis. *Evolution* 49: 836–847; 1995.

Ruben, J. The evolution of endothermy in mammals and birds: From physiology to fossils. *Annual Review of Physiology* 57: 69–85; 1995.

Mitochondria in muscles and organs of lizards and mammals

Else, P. L., and Hulbert, A. J. An allometric comparison of the mitochondria of mammalian and reptilian tissues: The implications for the evolution of endothermy. *Journal of Comparative Physiology B: Biochemical, Systemic, and Environmental Physiology* 156: 3–11; 1985.

Hulbert, A. J., and Else, P. L. Evolution of mammalian endothermic metabolism: Mitochondrial activity and cell composition. *American Journal of Physiology* 256: R63–R69; 1989.

Proton leak

Brand, M. D., Couture, P., Else, P. L., Withers, K. W., and Hulbert, A. J. Evolution of energy metabolism: Proton permeability of the inner membrane of liver mitochondria is greater in a mammal than in a reptile. *Biochemical Journal* 275: 81–86; 1991.

Brookes, P. S., Buckingham, J. A., Tenreiro, A. M., Hulbert, A. J., and Brand, M. D. The proton permeability of the inner membrane of liver mitochondria from ectothermic and endothermic vertebrates and from obese rats: Correlations with standard metabolic rate and phospholipid fatty acid composition. *Comparative Biochemistry and Physiology* 119B: 325–334; 1998.

Speakman, J. R., Talbot, D. A., Selman, C., Snart, S., McLaren, J. S., Redman, P., Krol, E., Jackson, D. M., Johnson, M. S., and Brand, M. D. Uncoupled and surviving: Individual mice with high metabolism have greater mitochondrial uncoupling and live longer. *Aging Cell* 3: 87–95; 2004.

Elastic strain in kangaroos

Bennett, M. B., and Taylor, G. C. Scaling of elastic strain energy in kangaroos and the benefits of being big. *Nature* 378: 56–59; 1995.

Cell volume, nuclear volume and DNA content

Cavalier-Smith, T. Economy, speed and size matter: Evolutionary forces driving nuclear genome miniaturization and expansion. *Annals of Botany* 95: 147–175; 2005.

Part 5

General texts

Buss, Leo. *The Evolution of the Individual.* Princeton University Press, New Jersey, USA, 1987.

Dawkins, Richard. *The Selfish Gene.* Oxford University Press, Oxford, UK, 1976.

Dawkins, Richard *The Extended Phenotype*. Oxford University Press, Oxford, UK, 1984.
—— *The Ancestor's Tale: A Pilgrimage to the Dawn of Life*. Weidenfeld & Nicolson, London, UK, 2004.
Harold, Franklin. *The Vital Force: A Study of Bioenergetics*. W. H. Freeman and Co., New York, USA, 1986.
Klarsfeld, André, and Revah, Frédéric. *The Biology of Death: Origins of Mortality*. Cornell University Press, Ithaca, USA, 2004.
Margulis, Lynn. Gaia is a tough bitch. In John Brockman (ed.), *The Third Culture: Beyond the Scientific Revolution*. Simon & Schuster, New York, USA, 1995.
Maynard-Smith, John, and Szathmáry, Eörs. *The Major Transitions of Evolution*. W. H. Freeman, San Francisco, USA, 1995.

Levels of selection

Blackstone, N. W. A units-of-evolution perspective on the endosymbiont theory of the origin of the mitochondrion. *Evolution* **49:** 785–796; 1995.
Maynard-Smith, J. The units of selection. *Novartis Foundation Symposium* **213:** 203–211; 1998.
Mayr, E. The objects of selection. *Proceedings of the National Academy of Sciences USA* **94:** 2091–2094; 1997.

Apoptosis and the evolution of multicellularity

Huettenbrenner, S., Maier, S., Leisser, C., Polgar, D., Strasser, S., Grusch, M., and Krupitza, G. The evolution of cell death programs as prerequisites of multicellularity. *Mutation Research* **543:** 235–249; 2003.
Michod, R. E., and Roze, D. Cooperation and conflict in the evolution of multicellularity. *Heredity* **86:** 1–7; 2001.

Discovery of apoptosis

Featherstone, C. Andrew Wyllie: From left field to centre stage. *The Lancet* **351:** 192; 1998.
Kerr, J. F. History of the events leading to the formulation of the apoptosis concept. *Toxicology* **181–182:** 471–474; 2002.
—— Wyllie A. H., and Currie A. R. Apoptosis: A basic biological phenomenon with wide-ranging implications in tissue kinetics. *British Journal of Cancer* **26:** 239–257; 1972.

Caspase enzymes

Barinaga, M. Cell suicide: By ICE, not fire. *Science* **263:** 754–756; 1994.
Horvitz, H. R. Nobel lecture: Worms, life and death. *Bioscience Reports* **23:** 239–303; 2003.
—— Sulston, J. E. Joy of the worm. *Genetics* **126:** 287–292; 1990.
Wiens, M., Krasko, A., Perovic, S., and Muller, W. E. G. Caspase-mediated apoptosis in sponges: cloning and function of the phylogenetic oldest apoptotic proteases from Metazoa. *Biochimica et Biophysica Acta* **1593:** 179–189; 2003.

Mitochondrial involvement in apoptosis

Brown, G. C. Mitochondria and cell death. *The Biochemist* **27**(3): 15–18; 2005.
Zamzami, N., Marchetti, P., Castedo, M., Zanin, C., Vayssière, J. L., Petit P. X., and Kroemer G. Reduction in mitochondrial potential constitutes an early irreversible

step of programmed lymphocyte death in vivo. *Journal of Experimental Medicine* 181: 1661–1672; 1995.

—— —— —— Decaudin, D., Macho, A., Hirsch, T., Susin, S. A., Petit, P. X., Mignotte, B., and Kroemer, G. Sequential reduction of mitochondrial transmembrane potential and generation of reactive oxygen species in early programmed cell death. *Journal of Experimental Medicine* 182: 367–377; 1995.

—— Susin, S. A., Marchetti, P., Hirsch, T., Gómez-Monterret, I., Castedo, M., and Kroemer, G. Mitochondrial control of nuclear apoptosis. *Journal of Experimental Medicine* 183: 1533–1544; 1996.

Cytochrome c release

Balk, J., and Leaver, C. J. The PET-1–CMS mitochondrial mutation in sunflower is associated with premature programmed cell death and cytochrome c release. *The Plant Cell* 13: 1803–1818; 2001.

Kluck, R. M., Bossy-Wetzel, E., Green, D. R., and Newmeyer, D. D. The release of cytochrome c from mitochondria: A primary site for Bcl-2 regulation of apoptosis. *Science* 275: 1132–1136; 1997.

Liu, X., Kim, C. N., Yang, J., Jemmerson, R., and Wang, X. Induction of apoptotic program in cell-free extracts: Requirement for dATP and cytochrome c. *Cell* 86: 147–157; 1996.

Ott, M., Robertson, J. D., Gogvadze, V., Zhitotovsky, B., and Orrenius, S. Cytochrome c release from mitochondria proceeds by a two-step process. *Proceedings of the National Academy of Sciences USA* 99: 1259–1263; 2002.

Yang, J., Liu, X., Bhalla, K., Kim, C. N., Ibrado, A. M., Cai, J., Peng, T. I., Jones, D. P., and Wang, X. Prevention of apoptosis by Bcl-2: Release of cytochrome c from mitochondria blocked. *Science* 275: 1129–1132; 1997.

Other mitochondrial apoptotic proteins

Candé, C., Cecconi, F., Dessen, P., and Kroemer, G. Apoptosis-inducing factor (AIF) pathway: Key to the conserved caspase-independent pathways of cell death? *Journal of Cell Science* 115: 4727–4734; 2002.

van Gurp, M., Festjens, N., van Loo, G., Saelens, X., and Vandenabeele, P. Mitochondrial intermembrane proteins in cell death. *Biochemical and Biophysical Research Communications* 304: 487–497; 2003.

Bcl-2 family

Adams, J. M., and Cory, S. Life-or-death decisions by the Bcl-2 protein family. *Trends in Biochemical Sciences* 26: 61–66; 2001.

Orrenius, S. Mitochondrial regulation of apoptotic cell death. *Toxicology Letters* 149: 19–23; 2004.

Zamzami, N., and Kroemer, G. Apoptosis: Mitochondrial membrane permeabilization— the (w)hole story? *Current Biology* 13: R71–R73; 2003.

Link between intrinsic and extrinsic pathways of apoptosis

Sprick, M. R., and Walczak, H. The interplay between the Bcl-2 family and death receptor-mediated apoptosis. *Biochemica et Biophysica Acta* 1644: 125–132; 2004.

Bacterial origin of apoptotic genes

Ameisen, J. C. On the origin, evolution, and nature of programmed cell death: A timeline of four billion years. *Cell Death and Differentiation* 9: 367–393; 2002.

Koonin, E. V., and Aravind, L. Origin and evolution of eukaryotic apoptosis: The bacterial connection. *Cell Death and Differentiation* 9: 394–404; 2002.

Host-symbiont relationships in the evolution of apoptosis

Blackstone, N. W., and Green, D. R. The evolution of a mechanism of cell suicide. *Bioessays* 21: 84–88; 1999.

—— Kirkwood, T. B. L. Mitochondria and programmed cell death: 'Slave revolt' or community homeostasis? In P. Hammerstein (ed.), *Genetic and Cultural Evolution of Cooperation.* MIT Press, Cambridge MA, USA 2003.

Frade, J. M., and Michaelidis, T. M. Origin of eukaryotic programmed cell death: A consequence of aerobic metabolism? *Bioessays* 19: 827–832; 1997.

Müller, A., Günther, D., Düx, F., Naumann, M., Meyer T. F., and Rudel, T. Neisserial porin (PorB) causes rapid calcium influx in target cells and induces apoptosis by the activation of cysteine proteases. *EMBO (European Molecular Biology Organization) Journal* 18: 339–352; 1999.

Naumann, M., Rudel, T., and Meyer, T. Host cell interactions and signalling with *Neisseria gonorrhoeae. Current Opinion in Microbiology* 2: 62–70; 1999.

Free radicals and recombination

Brennan, R. J., and Schiestl, R. H. Chloroform and carbon tetrachloride induce intrachromosomal recombination and oxidative free radicals in *Saccharomyces cerevisiae. Mutation Research* 397: 271–278; 1998.

Filkowski, J., Yeoman, A., Kovalchuk, O., and Kovalchuk, I. Systemic plant signal triggers genome instability. *Plant Journal* 38: 1–11; 2004.

Nedelcu, A. M., Marcu, O., and Michod, R. E. Sex as a response to oxidative stress: A twofold increase in cellular reactive oxygen species activates sex genes. *Proceedings of the Royal Society of London B: Biological Sciences* 271: 1591–1592; 2004.

Sex and the origin of death

Blackstone, N. W., and Green, D. R. The evolution of a mechanism of cell suicide. *Bioessays* 21: 84–88; 1999.

—— —— Redox control and the evolution of multicellularity. *Bioessays* 22: 947–953; 2000.

Part 6

General texts

Ridley, Mark. *Mendel's Demon: Gene Justice and the Complexity of Life.* Phoenix, London, UK, 2001.

Sykes, Bryan. *The Seven Daughters of Eve.* Corgi, London, UK, 2001.

Evolution of the sexes

Charlesworth, B. The evolution of chromosomal sex determination. *Novartis Foundation Symposium* 244: 207–224; 2002.

Whitfield, J. Everything you always wanted to know about sexes. *PLoS (Public Library of Science) Biology* 2: 0718–0721; 2004.

Uniparental inheritance
Birky, C. W., Jr. Uniparental inheritance of mitochondrial and chloroplast genes: Mechanisms and evolution. *Proceedings of the National Academy of Sciences USA* 92: 11331–11338; 1995.

Hoekstra, R. E. Evolutionary origin and consequences of uniparental mitochondrial inheritance. *Human Reproduction* 15 (suppl. 2): 102–111; 2000.

Selfish conflict
Cosmides, L. M., and Tooby, J. Cytoplasmic inheritance and intragenomic conflict. *Journal of Theoretical Biology* 89: 83–129; 1981.

Hurst, L., and Hamilton, W. D. Cytoplasmic fusion and the nature of sexes. *Proceedings of the Royal Society of London B: Biological Sciences* 247: 189–194; 1992.

Partridge, L., and Hurst, L. D. Sex and conflict. *Science* 281: 2003–2008; 1998.

Male sterility in plants
Budar, F., Touzet, P., and de Paepe, R. The nucleo-mitochondrial conflict in cytoplasmic male sterilities revisited. *Genetica* 117: 3–16; 2003.

Sabar, M., Gagliardi, D., Balk, J., and Leaver, C. J. ORFB is a subunit of F1F(O)-ATP synthase: Insight into the basis of cytoplasmic male sterility in sunflower. *EMBO (European Molecular Biology Organization) Reports* 4: 381–386; 2003.

Drosophila giant sperm
Pitnick, S., and Karr, T. L. Paternal products and by-products in Drosophila development. *Proceedings of the Royal Society of London B: Biological Sciences* 265: 821–826; 1998.

Heteroplasmy in angiosperms
Zhang, Q., Liu, Y., and Sodmergen. Examination of the cytoplasmic DNA in male reproductive cells to determine the potential for cytoplasmic inheritance in 295 angiosperm species. *Plant Cell Physiology* 44: 941–951; 2003.

Ooplasmic transfer
Barritt, J. A., Brenner, C. A., Malter, H. E., and Cohen, J. Mitochondria in human offspring derived from ooplasmic transplantation. *Human Reproduction* 16: 513–516; 2001.

St John, J. C. Ooplasm donation in humans: The need to investigate the transmission of mitochondrial DNA following cytoplasmic transfer. *Human Reproduction* 17: 1954–1958; 2002.

Mitochondrial DNA and human evolution
Ankel-Simons, F., and Cummins, J. M. Misconceptions about mitochondria and mammalian fertilisation: Implications for theories on human evolution. *Proceedings of the National Academy of Sciences USA* 93: 13859–13863; 1996.

Cann, R. L., Stoneking, M., and Wilson, A. C. Mitochondrial DNA and human evolution. *Nature* 325: 31–36; 1987.

Krings, M., Stone, A., Schmitz, R. W., Krainitzki, H., Stoneking, M., and Pääbo, S. Neanderthal DNA sequences and the origin of modern humans. *Cell* 90: 19–30; 1997.

Mitochondrial recombination

Eyre-Walker, A., Smith, N. H., and Smith, J. M. How clonal are human mitochondria? *Proceedings of the Royal Society of London B* 266: 477–483; 1999.

Hagelberg, E. Recombination or mutation rate heterogeneity? Implications for Mitochondrial Eve. *Trends in Genetics* 19: 84–90; 2003.

Kraytsberg, Y., Schwartz, M., Brown, T. A., Ebralidse, K., Kunz, W. S., Clayton, D. A., Vissing, J., and Khrapko, K. Recombination of human mitochondrial DNA. *Science* 304: 981; 2004.

Calibrating the mitochondrial clock

Gibbons, A. Calibrating the mitochondrial clock. *Science* 279: 28–29; 1998.

Cummins, J. Mitochondria DNA and the Y chromosome: Parallels and paradoxes. *Reproduction, Fertility and Development* 13: 533–542; 2001.

Lake Mungo fossil

Adcock, G. J., Dennis, E. S., Easteal, S., Huttley, G. A., Jermiin, L. S., Peacock, W. J., and Thorne, A. Mitochondrial DNA sequences in ancient Australians: Implications for modern human origins. *Proceedings of the National Academy of Sciences USA* 98: 537–542; 2001.

Bowler, J. M., Johnston, H., Olley, J. M., Prescott, J. R., Roberts, R. G., Shawcross, W., and Spooner, N. A. New ages for human occupation and climatic change at Lake Mungo, Australia. *Nature* 421: 837–840; 2003.

Mitochondrial selection

Coskun, P. E., Ruiz-Pesini, E., and Wallace, D. C. Control region mtDNA variants: Longevity, climatic adaptation, and a forensic conundrum. *Proceedings of the National Academy of Sciences USA* 100: 2174–2176; 2003.

—— Beal, M. F., and Wallace, D. C. Alzheimer's brains harbor somatic mtDNA control-region mutations that suppress mitochondrial transcription and replication. *Proceedings of the National Academy of Sciences USA* 101: 10726–10731; 2004.

Ruiz-Pesini, E., Mishmar, D., Brandon, M., Procaccio, V., and Wallace, D. C. Effects of purifying and adaptive selection on regional variation in human mtDNA. *Science* 303: 223–226; 2004.

—— Lapeña, A. C., Díez-Sánchez, C., Pérez-Martos, A., Montoya, J., Alvarez, E., Díaz, M., Urriés, A., Montoro, L., López-Pérez, M. J., and Enríquez J. A. Human mtDNA haplogroups associated with high or reduced spermatozoa motility. *American Journal of Human Genetics* 67: 682–696; 2000.

The dual genomic control system (co-adaptation)

Ballard, J. W. O., and Whitlock, M. C. The incomplete natural history of mitochondria. *Molecular Ecology* 13: 729–744; 2004.

Blier, P. U., Dufresne, F., and Burton, R. S. Natural selection and the evolution of mtDNA-encoded peptides: Evidence for intergenomic co-adaptation. *Trends in Genetics* 17: 400–406; 2001.

Ross, I. K. Mitochondria, sex and mortality. *Annals of the New York Academy of Sciences* **1019:** 581–584; 2004.

The mitochondrial bottleneck

Barritt, J. A., Brenner, C. A., Cohen, J., and Matt, D. W. Mitochondrial DNA rearrangements in human oocytes and embryos. *Molecular Human Reproduction* **5:** 927–933; 1999.

Cummins, J. M. The role of mitochondria in the establishment of oocyte functional competence. *European Journal of Obstetrics and Gynecology and Reproductive Biology* 115S: S23–S29; 2004.

Jansen, R. P. S. Germline passage of mitochondria: Quantitative considerations and possible embryological sequelae. *Human Reproduction* 15 (suppl. 2): 112–128; 2000.

Krakauer, D. C., and Mira, A. Mitochondria and germ-cell death. *Nature* **400:** 125–126; 1999.

Perez, G. I., Trbovich, A. M., Gosden, R. G., and Tilly, J. L. Mitochondria and the death of oocytes. *Nature* **403:** 500–501; 2000.

Part 7

General texts

Halliwell, B., and Gutteridge, J. *Free Radicals in Biology and Medicine.* Oxford University Press, Oxford, UK, 1999.

Holliday, Robin. *Understanding Ageing.* Cambridge University Press, Cambridge, UK, 1995.

Lane, Nick. *Oxygen: The Molecule that Made the World.* Oxford University Press, Oxford, UK, 2002.

Lifespan and metabolic rate

Barja, G. Mitochondrial free-radical production and aging in mammals and birds. *Annals of the New York Academy Sciences* **854:** 224–238; 1998.

Brunet-Rossinni, A. K., and Austad, S. N. Ageing studies on bats: A review. *Biogerontology* **5:** 211–222; 2004.

Skulachev, V. P. Mitochondria, reactive oxygen species and longevity: Some lessons from the Barja group. *Ageing Cell* **3:** 17–19; 2004.

Speakman, J. R., Selman, C., McLaren, J. S., and Harper, E. J. Living fast, dying when? The link between ageing and energetics. *Journal of Nutrition* **132** (suppl. 2): 1583S–1597S; 2002.

Mitochondrial theory of ageing

Harman, D. The biologic clock: The mitochondria? *Journal of the American Geriatrics Society* **20:** 145–147; 1972.

Miquel, J., Economos, A. C., Fleming, J., and Johnson, J. E., Jr. Mitochondrial role in cell ageing. *Experimental Gerontology* **15:** 575–591; 1980.

Failure of antioxidants

Barja, G. Free radicals and aging. *Trends in Neurosciences* **27:** 595–600; 2004.

Cutler, R. G. Antioxidants and longevity of mammalian species. *Basic Life Sciences* **35:** 15–73; 1985.

Orr, W. C., Mockett, R. J., Benes J. J., and Sohal, R. S. Effects of overexpression of copper-zinc and manganese superoxide dismutases, catalase, and thioredoxin reductase genes on longevity in *Drosophila melanogaster. Journal of Biological Chemistry* **278**: 26418–26422; 2003.

Mitochondrial diseases

Chinnery, P. F., DiMauro, S., Shanske, S., et al. Risk of developing a mitochondrial DNA deletion disorder. *Lancet* **364**: 591–596; 2004.

Fernández-Moreno, M., Bornstein, B., Petit, N., and Garesse, R. The pathophysiology of mitochondrial biogenesis: Towards four decades of mitochondrial DNA research. *Molecular Genetics and Metabolism* **71**: 481–495; 2000.

Marx, J. Metabolic defects tied to mitochondria gene. *Science* **306**: 592–593; 2004.

Schapira, A. Mitochondrial DNA and disease. *The Biochemist* **27**(3): 24–27; 2005.

Wallace, D. C. Mitochondrial diseases in man and mouse. *Science* **283**: 1482–1488; 1999.

Mitochondrial mutations in ageing

Coskun, P. E., Ruiz-Pesini, E., and Wallace, D. C. Control region mtDNA variants: Longevity, climatic adaptation, and a forensic conundrum. *Proceedings of the National Academy of Sciences USA* **100**: 2174–2176; 2003.

Lightowlers, R. N., Jacobs, H. T., and Kajander, O. A. Mitochondrial DNA—all things bad? *Trends in Genetics* **15**: 91–93; 1999.

Linnane, A. W., Marzuki, S., Ozawa, T., and Tanaka, M. Mitochondria DNA mutations as an important contributor to ageing and degenerative diseases. *Lancet* **1 (8639)**: 642–645; 1989.

Michikawa, Y., Mazzucchelli, F., Bresolin, N., Scarlato, G., and Attardi, G. Aging-dependent large accumulation of point mutations in the human mtDNA control region for replication. *Science* **286**: 774–779; 1999.

Zhang, J., Asin-Cayuela, J., Fish, J., Michikawa, Y., Bonafè, M., Olivieri, F., Passarino, G., De Benedictis, G., Franceschi, C., and Attardi, G. Strikingly higher frequency in centenarians and twins of mtDNA mutation causing remodeling of replication origin in leukocytes. *Proceedings of the National Academy of Sciences USA* **100**: 1116–1121; 2003.

Redox signalling in mitochondria

Allen, J. F. Control of gene expression by redox potential and the requirement for chloroplast and mitochondrial genes. *Journal of Theoretical Biology* **165**: 609–631; 1993.

—— The function of genomes in bioenergetic organelles. *Philosophical Transactions of the Royal Society of London B: Biological Sciences* **358**: 19–38; 2003.

Landar, A. L., Zmijewski, J. W., Oh, J. Y., and Darley Usmar, V. M. Message from the cell's powerhouse. *The Biochemist* **27**(3): 9–14; 2005.

The retrograde response

Butow, R. A., and Avadhani, N. G. Mitochondrial signaling: The Retrograde response. *Molecular Cell* **14**: 1–15; 2004.

De Benedictis, G., Carrieri, G., Garastro, S., Rose, G., Varcasia, O., Bonafè, M., Franceschi, C., and Jazwinski, S. M. Does a retrograde response in human aging and longevity exist? *Experimental Gerontology* **35**: 795–801; 2000.

Apoptosis and neurodegenerative diseases

Coskun, P. E., Ruiz-Pesini, E., and Wallace, D. C. Control region mtDNA variants: Longevity, climatic adaptation, and a forensic conundrum. *Proceedings of the National Academy of Sciences USA* 100: 2174–2176; 2003.

Wright, A. F., Jacobson, S. G., Cideciyan, A. V., Roman, A. J., Shu, X., Vlachantoni, D, McInnes, R. R., and Riemersma, R. A. Lifespan and mitochondrial control of neurodegeneration. *Nature Genetics* 36: 1153–1158; 2004.

Proof-reading in mice

Balaban, R. S., Nemoto, S., and Finkel, T. Mitochondria, oxidants, and aging. *Cell* 120: 483–495; 2005.

Trifunovic, A., Wredenberg, A., Falkenberg, M., Spelbrink, J. N., Rovio, A. T., Bruder, C. E., Bohlooly-Y, M., Gidlof, S., Oldfors, A., Wibom, R., Tornell, J., Jacobs, H. T., and Larsson, N. G. Premature ageing in mice expressing defective mitochondrial polymerase. *Nature* 429: 417–423; 2004.

Source of leakage at complex I

Herrero, A., and Barja, G. Localization of the site of oxygen radical generation inside complex I of heart and nonsynaptic brain mammalian mitochondria. *Journal of Bioenergetics and Biomembranes* 32: 609–615; 2000.

Kushnareva, Y., Murphy, A. N., and Andreyev, A. Complex I-mediated reactive oxygen species generation: Modulation by cytochrome c and $NAD(P)^+$ oxidation state. *Biochemical Journal* 368: 545–553; 2002.

Japanese centenarians

Tanaka, M., Gong, J. S., Zhang, J., Yoneda, M., and Yagi, K. Mitochondrial genotype associated with longevity. *Lancet* 351: 185–186; 1998.

—— —— —— Yamada, Y., Borgeld, H. J., and Yagi, K. Mitochondrial genotype associated with longevity and its inhibitory effect on mutagenesis. *Mechanisms of Ageing and Development* 116: 65–76; 2000.

Uncoupling, ageing and obesity

Ruiz-Pesini, E., Mishmar, D., Brandon, M., Procaccio, V., and Wallace, D. C. Effects of purifying and adaptive selection on regional variation in human mtDNA. *Science* 303: 223–226; 2004.

Speakman, J. R., Talbot, D. A., Selman, C., Snart, S., McLaren, J. S., Redman, P., Krol, E., Jackson, D. M., Johnson, M. S., and Brand, M. D. Uncoupled and surviving: Individual mice with high metabolism have greater mitochondrial uncoupling and live longer. *Aging Cell* 3: 87–95; 2004.

Exercise paradox

Herrero, A., and Barja, G. ADP-regulation of mitochondrial free-radical production is different with complex I- or complex II-linked substrates: Implications for the exercise paradox and brain hypermetabolism. *Journal of Bioenergetics and Biomembranes* 29: 241–249; 1997.

Calorie restriction and free-radical leakage

Gredilla, R., Barja, G., and López-Torres, M. Effect of short-term caloric restriction on

H$_2$O$_2$ production and oxidative DNA damage in rat liver mitochondria and location of the free radical source. *Journal of Bioenergetics and Biomembranes* **33:** 279–287; 2001.

Sex versus survival

Kirkwood, T. B., and Rose, M. R. Evolution of senescence: Late survival sacrificed for reproduction. *Philosophical Transactions of the Royal Society of London B: Biological Sciences* **332:** 15–24; 1991.

Aerobic capacity of birds

Maina, J. N. What it takes to fly: The structural and functional respiratory refinements in birds and bats. *Journal of Experimental Biology* **203:** 3045–3064; 2000.

Index

Italic numbers denote references to illustrations. References to footnotes are followed by 'n.'.